A Level
Salters' Advanced
Chemistry
for OCR
Year 1 and AS
Fourth Edition

B

Project director
Chris Otter

Authors
Frank Harriss
Dave Waistnidge
Lesley Johnston
Mark Gale
Dave Newton

THE SALTERS' INSTITUTE

OXFORD
UNIVERSITY PRESS

OXFORD
UNIVERSITY PRESS

Great Clarendon Street, Oxford, OX2 6DP, United Kingdom

Oxford University Press is a department of the University of Oxford. It furthers the University's objective of excellence in research, scholarship, and education by publishing worldwide. Oxford is a registered trade mark of Oxford University Press in the UK and in certain other countries

British Library Cataloguing in Publication Data
Data available

978-0-19-833289-3

10 9 8 7 6 5 4 3

Paper used in the production of this book is a natural, recyclable product made from wood grown in sustainable forests. The manufacturing process conforms to the environmental regulations of the country of origin.

Printed in Great Britain by Bell and Bain Ltd, Glasgow

This resource is endorsed by OCR for use with specification H033 AS Level GCE Chemistry B (Salters) and year 1 of H433 A Level GCE Chemistry B (Salters). In order to gain endorsement this resource has undergone an independent quality check. OCR has not paid for the production of this resource, nor does OCR receive any royalties from its sale. For more information about the endorsement process please visit the OCR website www.OCR.org.uk/

Index compiled by INDEXING SPECIALISTS (UK) Ltd., Indexing House, 306A Portland Road, Hove, East Sussex BN3 5LP United Kingdom.

FSC MIX Paper from responsible sources FSC® C007785

First edition

George Burton, Margaret Ferguson, John Holman, Gwen Pilling, David Waddington, Malcolm Churchill, Derek Denby, Frank Harriss, Miranda Stephenson, Brian Ratcliff, Ashley Wheway

Second edition

John Lazonby, Gwen Pilling, David Waddington, Derek Denby, John Dexter, Margaret Ferguson, Frank Harriss, Gerald Keeling, Dave Newton, Brian Ratcliff, Mike Shipton, Terri Vine

Third edition

Chris Otter, Adelene Cogill, Frank Harriss, Dave Newton, Gill Saville, Kay Stephenson, David Waistnidge, Ashley Wheway,

We are grateful for sponsorship from the Salters' Institute, which has contributed to support the Salters Advanced Chemistry project and has enabled the development of these materials.

AS/A Level course structure

This book has been written to support students studying for OCR AS Chemistry B (Salters) and for students in their first year of studying for OCR A Level Chemistry B (Salters). It covers the AS content from the specification, the content of which will also be examined at A Level. The content covered is shown in the contents list, which also shows you the page numbers for the main topics within each chapter. There is also an index at the back to help you find what you are looking for. If you are studying for OCR AS Chemistry B (Salters), you will only need to know the content in the blue box.

AS exam / **A level exam**

Year 1 content

1 Elements of life
2 Developing fuels
3 Elements from the sea
4 The ozone story
5 What's in a medicine

Year 2 content

6 The chemical industry
7 Polymers and life
8 Oceans
9 Developing metals
10 Colour by design

A Level exams will cover content from Year 1 and Year 2 and will be at a higher demand. You will also carry out practical activities throughout your course.

Contents by chemical storylines

Chapter 4 The ozone story — 192

Chapter 5 What's in a medicine — 246

Techniques and procedures — 282

Contents by chemical ideas

How to use this book

This book contains many different features. Each feature is designed to support and develop the skills you will need for your examinations, as well as foster and stimulate your interest in chemistry.

Each Topic has storylines content which contains engaging and contemporary contexts that are relevant to the concepts you will cover. Storylines content is highlighted by a coloured background.

Chemical ideas:

The chemical ideas sections contain the concepts you need to know for your examinations. Each chemical ideas section starts with a chemical ideas reference. This tells you what concepts you will be covering and links to the Contents page by chemical ideas. You can use the Contents page by chemical ideas to see how the different chapters interlink - an important skill for your examinations.

Study Tips

Study tips contain prompts to help you with your understanding and revision.

Extension features

These features contain material that is beyond the specification. They are designed to stretch and provide you with a broader knowledge and understanding and lead the way into the types of thinking and areas you might study in further education. As such, neither the detail nor the depth of questioning will be required for the examinations. But this book is about more than getting through the examinations.

1 Extension features also contain questions that link the off-specification material back to your course.

Synoptic link

These highlight the key areas where topics relate to each other. As you go through your course, knowing how to link different areas of chemistry together becomes increasingly important. Many exam questions, particularly at A Level, will require you to bring together your knowledge from different areas. Synoptic links that link to the Techniques and procedures chapter have a practical symbol

Summary Questions

1 These are short questions at the end of each topic.

2 They test your understanding of the topic and allow you to apply the knowledge and skills you have acquired.

3 The questions are ramped in order of difficulty. Lower-demand questions have a paler background, with the higher-demand questions having a darker background. Try to attempt every question you can, to help you achieve your best in the exams.

Introduction at the beginning of each chapter summarises what you will cover.

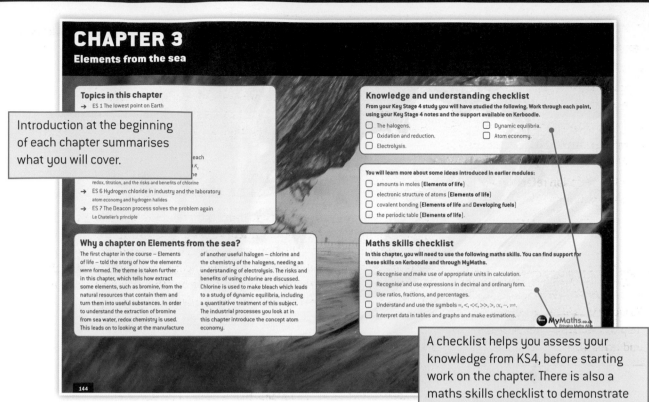

CHAPTER 3
Elements from the sea

Topics in this chapter
→ ES 1 The lowest point on Earth

redox, titration, and the risks and benefits of chlorine
→ ES 6 Hydrogen chloride in industry and the laboratory
atom economy and hydrogen halides
→ ES 7 The Deacon process solves the problem again
Le Chatelier's principle

Why a chapter on Elements from the sea?
The first chapter in the course – Elements of life – told the story of how the elements were formed. The theme is taken further in this chapter, which tells how extract some elements, such as bromine, from the natural resources that contain them and turn them into useful substances. In order to understand the extraction of bromine from sea water, redox chemistry is used. This leads on to looking at the manufacture of another useful halogen – chlorine and the chemistry of the halogens, needing an understanding of electrolysis. The risks and benefits of using chlorine are discussed. Chlorine is used to make bleach which leads to a study of dynamic equilibria, including a quantitative treatment of this subject. The industrial processes you look at in this chapter introduce the concept atom economy.

Knowledge and understanding checklist
From your Key Stage 4 study you will have studied the following. Work through each point, using your Key Stage 4 notes and the support available on Kerboodle.
☐ The halogens.
☐ Oxidation and reduction.
☐ Electrolysis.
☐ Dynamic equilibria.
☐ Atom economy.

You will learn more about some ideas introduced in earlier modules:
☐ amounts in moles (**Elements of life**)
☐ electronic structure of atoms (**Elements of life**)
☐ covalent bonding (**Elements of life** and **Developing fuels**)
☐ the periodic table (**Elements of life**).

Maths skills checklist
In this chapter, you will need to use the following maths skills. You can find support for these skills on Kerboodle and through MyMaths.
☐ Recognise and make use of appropriate units in calculation.
☐ Recognise and use expressions in decimal and ordinary form.
☐ Use ratios, fractions, and percentages.
☐ Understand and use the symbols =, <, <<, >>, >, ∝, ~, ⇌.
☐ Interpret data in tables and graphs and make estimations.

MyMaths.co.uk

144

A checklist helps you assess your knowledge from KS4, before starting work on the chapter. There is also a maths skills checklist to demonstrate the skills you will learn in that chapter.

Visual summaries emphasise how the key concepts of that chapter interlink, and how they interlink with content covered in other chapters.

Application task brings together some of the key concepts of the chapter in a new context.

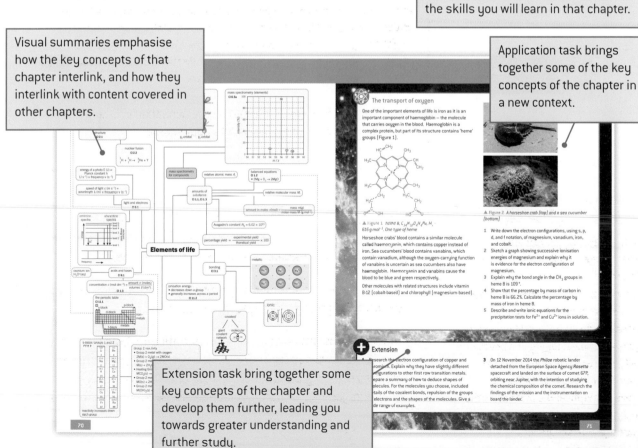

The transport of oxygen
One of the important elements of life is iron as it is an important component of haemoglobin – the molecule that carries oxygen in the blood. Haemoglobin is a complex protein, but part of its structure contains 'heme' groups [Figure 1].

▲ Figure 1 heme B, $C_{34}H_{32}O_4N_4Fe$, $616\,g\,mol^{-1}$. One type of heme

Horseshoe crabs' blood contains a similar molecule called haemocyanin, which contains copper instead of iron. Sea cucumbers' blood contains vanabins, which contain vanadium, although the oxygen-carrying function of vanabins is uncertain as sea cucumbers also have haemoglobin. Haemocyanin and vanabins cause the blood to be blue and green respectively.

Other molecules with related structures include vitamin B-12 (cobalt-based) and chlorophyll (magnesium-based).

▲ Figure 2 A horseshoe crab (top) and a sea cucumber (bottom)

1 Write down the electron configurations, using s, p, d, and f notation, of magnesium, vanadium, iron, and cobalt.
2 Sketch a graph showing successive ionisation energies of magnesium and explain why it is evidence for the electron configuration of magnesium.
3 Explain why the bond angle in the CH_3 groups in heme B is 109°.
4 Show that the percentage by mass of carbon in heme B is 66.2%. Calculate the percentage by mass of iron in heme B.
5 Describe and write ionic equations for the precipitation tests for Fe^{2+} and Cu^{2+} ions in solution.

Elements of life

⊕ Extension
Research the electron configuration of copper and bromine. Explain why they have slightly different configurations to other first row transition metals.
Prepare a summary of how to deduce shapes of molecules. For the molecules you choose, included details of the covalent bonds, repulsion of the groups of electrons and the shapes of the molecules. Give a wide range of examples.

3 On 12 November 2014 the *Philae* robotic lander detached from the European Space Agency *Rosetta* spacecraft and landed on the surface of comet 67P, orbiting near Jupiter, with the intention of studying the chemical composition of the comet. Research the findings of the mission and the instrumentation on board the lander.

70

71

Extension task bring together some key concepts of the chapter and develop them further, leading you towards greater understanding and further study.

Practice questions

1 The amount of energy needed to raise 1 g of a substance 1 °C, is called its:

> **Practice questions at the end of each chapter including questions that cover practical and maths skills.**

3 $10\,cm^3$ of $2.0\,mol\,dm^{-3}$ NaOH is mixed with $10\,cm^3$ $2.0\,mol\,dm^{-1}$ H_2SO_4. The temperature rises by y °C.

The enthalpy change of neutralisation of the reaction of NaOH with H_2SO_4 (in kJ) is given by:

A 4.18 y   C 20×4.18 y

B $4.18\frac{y}{2}$ D $15 \times 4.18\frac{y}{2}$ *(1 mark)*

4 Chloroethene has the formula of CH_2=CHCl. Poly(chloroethene) can be represented as:

A $-(CH_2Cl)-$ C $-(CHCHCl)-$

B $-(CHClCHCl)-$ D $-(CH_2CHCl)-$

5 Which of the following *cannot* be formed from ethene in a one-step reaction?

A CH_2BrCH_2Br C CH_3CH_2OH

B CH_3CHBr_2 D CH_3CH_2Br *(1 mark)*

6 Which of the following might be found in the exhaust when hydrogen burns in an internal combustion engine?

A NO_x C hydrocarbons

B CO   D SO_2 *(1 mark)*

7 A fuel is burnt in a small lamp and heats up a copper calorimeter containing water. Look at these three statements.

1 The specific heat capacity of the water is not accurately known.

2 Fuel may evaporate from the wick.

3 The fuel may not burn completely.

Which statements are limitations to accuracy in measuring the energy transferred?

A 1, 2, and 3 correct   C 2 and 3 correct

B 1 and 2 correct   D 1 correct *(1 mark)*

8 Which of the following are isomers of propan-1-ol?

1 $CH_3CH(OH)CH_3$

2 $CH_3CH_2OCH_3$

3 CH_3CH_2CHO

A 1, 2, and 3 correct   C 2 and 3 correct

B 1 and 2 correct   D 1 correct *(1 mark)*

9 Disadvantages of fossil fuels compared with biofuels include:

1 they give off CO_2 when burned

2 they have not (recently) absorbed CO_2 from the atmosphere

3 supplies are running out.

A 1, 2, and 3 correct   C 2 and 3 correct

B 1 and 2 correct   D 1 correct *(1 mark)*

10 Poly(propene) is a polymer which is now being used to make 'polymer banknotes'. One useful property is that it is unreactive to reagents such as acids.

a Draw **full** structural formulae for propene and poly(propene) *(2 marks)*

b Why is poly(propene) less reactive than propene? *(1 mark)*

c Propene reacts with bromine.

(i) Describe a test for propene based on this reaction. *(1 mark)*

(ii) Give the mechanism for the reaction of propene with bromine, using curly arrows and showing lone pairs and partial charges. Give the formula of the product. *(4 marks)*

(iii) Say what you understand by the term *electrophile* and identify an electrophile from your mechanism. *(2 marks)*

(iv) Some chloride ions are added to the reaction in (ii).

Give the formula of a product containing chlorine that would be formed. *(1 mark)*

d Butene is the next alkene in the homologous series after propene.

(i) What do you understand by the term *homologous series*? *(1 mark)*

(ii) What term is used to describe compounds like alkenes that have C=C bonds? *(1 mark)*

(iii) One structural isomer of butene has two stereoisomers.

Draw the structures for these two stereoisomers and name them. *(2 marks)*

11 Sherbet sweets get their fizz from the reaction between sodium bicarbonate and citric acid.

$C_6H_8O_7(aq) + 3NaHCO_3(s) \rightarrow$
$Na_3C_6H_5O_7 + 3CO_2(g) + 3H_2O$
Equation 12.1

a (i) Some students decide to investigate this **endothermic** reaction and measure the enthalpy change of reaction.

They have available 10.0 g portions of sodium bicarbonate and 25.0 cm³ portions of a solution of citric acid. The citric acid portions represent an excess over the sodium bicarbonate portions in the reaction in **Equation 12.1**.

Describe how they would carry out their experiment and how they would work out ΔH per mole of sodium bicarbonate from their results. *(6 marks)*

(ii) A student says that the temperature change they measure would be inaccurate because of heat losses. Comment on this statement. *(2 marks)*

b Calculate the volume of carbon dioxide (measured at room temperature and pressure) that the students would collect if they reacted 10.0 g of sodium bicarbonate with excess citric acid. Assume none of the carbon dioxide dissolves.

Give your answer to an appropriate number significant figures. *(3 marks)*

12 Petrol cars produce less NO_x and particulates than diesel cars but more CO and hydrocarbons.

a (i) Suggest why diesel cars produce fewer hydrocarbons. *(1 mark)*

(ii) Write an equation to show how CO is formed from the combustion of hexane in a petrol engine. *(1 mark)*

(iii) Give a reason why particulates are a pollutant. *(1 mark)*

b NO and CO can be removed from the exhaust of a petrol engine by reacting them together over a catalytic converter. This uses a heterogeneous catalyst.

(i) Give the equation for this reaction. *(1 mark)*

(ii) Explain the term *heterogeneous* in the context of catalysis and describe the first stage in the mechanism of this type of catalysis. *(2 marks)*

c Heterogeneous catalysts are also used for cracking hydrocarbons. Write the equation for a reaction in which nonane is cracked to produce ethene and one other product. *(1 mark)*

d Ethanol is one example of a biofuel.

(i) Write the equation for the complete combustion of ethanol.

Show state symbols under standard conditions. *(1 mark)*

(ii) Use the bond enthalpy values in the table to calculate a value for the enthalpy change of combustion of ethanol. *(3 marks)*

Bond	Average bond enthalpy / kJ mol⁻¹
C—H	413
C—C	347
C—O	358
O—H	464
C=O	805
O=O	498

(iii) The enthalpy change of combustion of ethanol in a Data Book is different from your answer to (ii). Suggest **two** reasons for this. *(2 marks)*

(iv) Biofuels are said to be sustainable. Explain the word *sustainable* in this context. *(1 mark)*

Techniques and procedures

> **Dedicated Techniques and procedures chapter, detailing all of the practical skills you need to know for your examinations. This chapter is referenced to throughout the book by practical synoptic links.**

Understanding and being familiar with practical techniques and procedures is an important part of being an effective chemist. This section outlines the techniques and procedures you need to know about as part of your course.

Measurement

Weighing a solid

An accurate weighing will use a balance that records to two decimal places.

- Zero the balance (sometimes called tare).
- Add a weighing bottle or similar container onto the balance and add approximately the required mass of solid.
- Accurately weigh the mass of solid plus weighing bottle and record the information.
- Tip the solid into the glassware where you will be using it.
- Accurately reweigh the empty weighing bottle.
- Subtract the recorded mass for the empty weighing bottle from the mass recorded for the solid and the weighing bottle.

Measuring volumes of liquids

Either measuring cylinders can be used to give rough measurements of liquids. In order to measure volumes of liquids more exactly two methods can be used – a pipette or a burette.

▲ **Figure 1** *Measuring the volume of liquid in a pipette*

A pipette is used for accurately dispensing a *fixed* volume of a liquid (typically 1.0 cm³ to 50 cm³ or 25 cm³).

1 Ensure the pipette is completely clean by rinsing out with water and then a small volume of the solution to be pipetted.

2 Dip the pipette into the solution to be pipetted and, using a pipette filler, draw enough liquid into the pipette until is it exactly the right volume – when the bottom of the meniscus is level with the line on the neck of the pipette when viewed at eye level.

3 Run the liquid out of the pipette into the piece of glassware the solution is being transferred to. When all of the liquid has run out

4 Allow the liquid to run out of the pipette until it stops. Touch the end of the pipette on the side of the conical flask and remove. There will still be a drop in the pipette – this is how it should be. The precise volume you require will have been dispensed.

Burette

1 Clean the burette by rinsing out with water and then a small volume of liquid of the solution to be used.

2 Make sure the burette tap is closed. Pour the solution into the burette using a small funnel. Fill the burette above the zero line.

3 Use a clamp to hold your burette in place and allow some of the solution to run into a beaker until there are no air bubbles in the jet of the burette. Record the burette reading to the nearest 0.05 cm³.

4 Carry out the titration to the end point.

5 Record the reading on the burette to the nearest 0.05 cm³. Subtract the reading taken at the beginning of the titration from this reading taken at the end. This is known as the titre.

Measuring volumes of gases

The volume of a gas produced in a chemical reaction can be measured using either a gas syringe or an inverted burette. The latter is called 'collecting gas over water'.

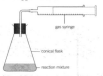

▲ **Figure 3** *Collecting gas using a gas syringe*

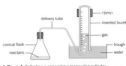

▲ **Figure 4** *Collecting a gas using a measuring cylinder*

In order for as much as possible of the gas to be collected the system needs to be gas tight.

The volume of gas collected in an inverted burette is the initial volume minus the final volume of gas.

▲ **Figure 2** *Measuring the volume of liquid using a burette*

Study tip

If the gas is soluble in water you need to use the gas syringe method, as some of the gas would dissolve before reaching the burette in the second method.

Synthesis

Heating under reflux

Heating under reflux is used for reactions involving volatile liquids. It ensures that reactants and/or products do not escape whilst the reaction is in progress.

Kerboodle

This book is supported by next generation Kerboodle, offering unrivalled digital support for independent study, context, differentiation, assessment, and the new practical endorsement.

If your school subscribes to Kerboodle, you will also find a wealth of additional resources to help you with your studies and with revision.

- Study guides
- Activity sheets for hands on application of your knowledge and practicals to try
- Maths skills boosters and calculation worksheets
- Animations and revision podcasts

Apply your knowledge by trying the activities signposted throughout the book.

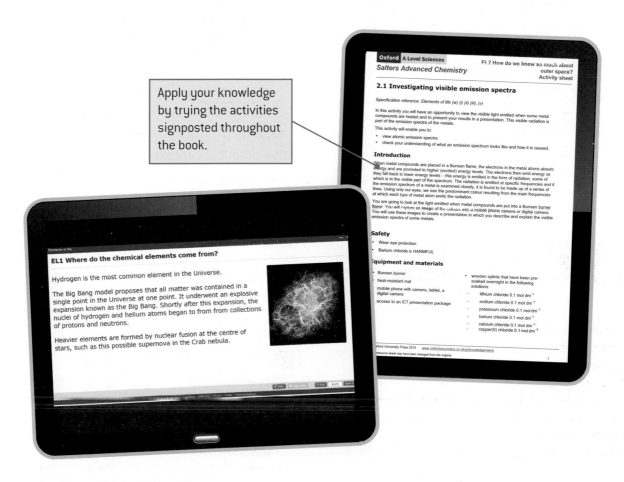

For teachers, Kerboodle also has plenty of further support, including answers to the questions in the book and a digital markbook. There are also full teacher notes for the activities and worksheets, which include suggestions on how to support students and engage them in their own learning. All of the resources are pulled together into teacher guides that suggest a route through each chapter.

Development of practical skills in chemistry

Chemistry is a practical subject and experimental work provides you with important practical skills, as well as enhancing your understanding of chemical theory. You will be developing practical skills by carrying out experiments including investigative work in the laboratory throughout both the AS and the A level Chemistry course. You will be assessed on your practical skills is two different ways:

- written examinations (AS and A level)
- practical endorsement (A level only)

Practical coverage throughout this book

Practical skills are a fundamental part of a complete education in science, and you should keep a record of your practical work from the start of your A level course that you can later use as part of your practical endorsement. You can find more details of the practical endorsement from your teacher or from the specification.

In this book and its supporting materials practical skills are covered in a number of ways. By studying the Techniques and procedures chapter and the Exam-style questions in this student book, and by using the Practical activities and Skills sheets in Kerboodle, you will have many opportunities to learn about the scientific method and carry-out practical activities.

1.1 Practical skills assessed in written examinations

In the written examination papers for AS and A level, at least 15% of the marks will be from questions that assess practical skills. The questions will cover four important skill areas, all based on the practical skills that you will develop by carrying out experimental work during your course.

- Planning – your ability design experiments including solving problems in a practical context.

- Implementing – your ability to use important practical techniques and processes and to present observations and data.

- Analysing – your ability to process, anaylse, and interpret experimental results.

- Evaluating – your ability to evaluate experimental results and to draw conclusions from them including suggesting improvements to procedures and apparatus.

1.1.1 Planning

- Designing experiments
- Identifying variables to be controlled
- Evaluating the experimental method

Skills checklist

- ☐ Selecting apparatus and equipment
- ☐ Selecting appropriate techniques
- ☐ Selecting appropriate quantities of chemicals and scale of working
- ☐ Solving chemical problems in a practical context
- ☐ Applying chemistry concepts to practical problems

1.1.2 Implementing

- Using a range of practical apparatus
- Carrying out a range of techniques
- Using appropriate units for measurements
- Recording data and observations in an appropriate format

Skills checklist

- ☐ Understanding practical techniques and processes
- ☐ Identifying hazards and safe procedures
- ☐ Using appropriate units
- ☐ Recording qualitative observations accurately
- ☐ Recording a range of quantitative measurements
- ☐ Using the appropriate apparatus

1.1.3 Analysis

- Processing, analysing, and interpreting results
- Analysing data using appropriate mathematical skills
- Using significant figures appropriately
- Plotting and interpreting graphs

Skills checklist

- ☐ Interpreting qualitative observations
- ☐ Analysing quantitative experimental data
- ☐ For graphs,
  - selecting and labelling axes with appropriate scales, quantities, and units
  - drawing tangents and measuring gradients and intercepts

1.1.4 Evaluation

- Evaluating results to draw conclusions
- Identify anomalies
- Explain limitations in experimental procedures
- Identifying uncertainties and calculating percentage errors
- Suggesting improvements to procedures and apparatus

Skills checklist

- ☐ Reaching conclusions from qualitative observations
- ☐ Identifying uncertainties and calculating percentage errors
- ☐ Identifying procedural limitations and measurement errors
- ☐ Refining procedures and apparatus to suggest improvements

1.2 Practical skills assessed in practical endorsement

You will also be assessed on how well you carry out a wide range of practical work and how to record the results of this work. These hands-on skills are divided into 12 categories and form the practical endorsement. This is assessed for A level Chemistry qualification only.

The endorsement requires a range of practical skills from both years of your course. If you are taking only AS Chemistry, you will not be assessed through the practical endorsement but the written AS examinations will include questions that relate to the skills that naturally form part of the AS common content to the A level course.

1.2.1 Practical skills

By carrying out experimental work through the course, you will develop your ability to:

- design and use practical techniques to investigate and solve problems
- use a wide range of experimental and practical apparatus, equipment, and materials, including chemicals and solutions
- carry out practical procedures skillfully and safely, recording and presenting results in a scientific way
- research using online and offline tools.

Along with the experimental work, these skills are covered in practical skills questions throughout the book.

1.2.2 Use of apparatus and techniques

To meet the requirements for the practical endorsement, you will be assessed in at least 12 practical experiments to enable you to experience a wide range of apparatus and techniques. These practical experiments are incorporated throughout the book in practical application boxes.

This will help to give you the necessary skills to be a competent and effective practical chemist.

Practical Activity Group (PAG) overview

The PAG labels identify activities that could count towards the practical endorsement. The table below shows where these PAG references are covered throughout this course. You will find further details on each PAG in the Techniques and procedures chapter.

Specification reference		Chapter reference
PAG1	*Moles determination EL(b)(i)*	EL, DF
PAG2	*Acid–base titration EL(c)(i)*	EL, OZ
PAG3	*Enthalpy determination DF(f), O(b)*	DF, O
PAG4	*Qualitative analysis of ions EL(s), ES(k), CI(j)*	EL, ES, CI
PAG5	*Synthesis of an organic liquid WM(f)*	WM
PAG6	*Synthesis of an organic solid WM(e)*	WM
PAG7	*Qualitative analysis of organic functional groups DF(c), WM(c), WM(d), PL(h), PL(j), PL(n), CD(f)*	DF, WM, PL, CD
PAG8	*Electrochemical cells DM(d)*	DM
PAG9	*Rates of reaction – continuous monitoring method CI(c), DM(n)*	CI, DM
PAG10	*Rates of reaction – initial rates method CI(c)*	CI
PAG11	*pH measurement O(k)*	O
PAG12	*Research skills*	EL, DF, OZ, CI, DM, O, CD

Maths skills and How Science Works across Module 1

Maths skills are very useful for scientists and as you study your course you will learn maths techniques and equations that support the development of your science knowledge. Each module opener in this book has an overview of the maths skills that relate to the theory in the chapter. There are also questions using maths skills throughout the book that will help you practice.

How Science Works skills help you to put science in a wider context, and to develop your critical and creative thinking skills and help you solve problems in a variety of contexts. How Science Works is embedded throughout this book, particularly in application boxes and practice questions.

You can find further support for maths and how science works on Kerboodle.

CHAPTER 1

Elements of life

Topics in this chapter

- EL 1 Where do the chemical elements come from?
  atomic structure, mass spectrometry, and nuclear fusion
- EL 2 How do we know so much about outer space?
  wave and particle models of light
- EL 3 Electrons, where would we be without them?
  shells, sub-shells, and orbitals
- EL 4 Organising the elements of life
  periodicity
- EL 5 The molecules of life
  covalent bonding and shapes of molecules
- EL 6 What are we made of?
  A_r, M_r, percentage yield, and balancing equations
- EL 7 Blood, sweat, and seas
  bonding, structure, properties, and precipitates
- EL 8 Spectacular metals
  Group 1 and Group 2 chemistry
- EL 9 How salty?
  reacting masses, acids and bases, neutralisation, and concentration of solutions

Why a chapter on Elements of life

This chapter tells the story of the elements of life – what they are, how they originated, and how they can be detected and measured. It shows how studying the composition of stars can throw light on the formation of the elements that make up our own bodies and considers how these elements combine to form the molecules of life. The chapter also takes the opportunity to look at some aspects of how science works, in particular, developing models and seeing how the scientific community validates work.

The chapter begins with a journey through the Universe. Starting with deep space, the story unfolds through the galaxies, the stars, and our own Sun and solar system. This section looks at the origin of the elements, introducing ideas about the structure of atoms, and briefly considers how elements combine to form compounds and the formation of molecules in the apparently inhospitable dense gas clouds of space, such molecules possibly being the origin of the molecules of life, which make up our bodies.

Later in the chapter you are brought back down to Earth. You learn how to measure amounts of elements (in terms of atoms) and how to calculate chemical formulae. The story then leads into learning about patterns in the properties of elements and the periodic table.

Knowledge and understanding checklist

From your Key Stage 4 study you should have studied the following. Work through each point, using your Key Stage 4 notes and the support available on Kerboodle.

☐ the periodic table

☐ protons, neutrons, and electrons

☐ chemical bonding

☐ writing chemical equations

☐ the wave model of light

☐ the electromagnetic spectrum

☐ relative atomic masses, relative molecular masses, and relative formula masses

☐ chemical formulae

☐ ionic and covalent bonding

☐ acids

☐ precipitation.

Maths skills checklist

In this chapter, you will need to use the following maths skills. You can find support for these skills on Kerboodle and through MyMaths.

☐ Identify and use appropriate units in calculations.

☐ Recognise and use expressions in decimal and ordinary form.

☐ Use ratios, fractions, and percentages.

☐ Use calculators to find and use power functions.

☐ Use appropriate numbers of significant figures.

☐ Find arithmetic means.

☐ Identify uncertainty in measurements.

☐ Change the subject of an equation.

☐ Substitute numerical values into algebraic equations using appropriate units for physical quantities.

☐ Solve algebraic equations.

☐ Translate information between graphical, numerical, and algebraic form.

☐ Use angles and shapes in 2D and 3D structures.

MyMaths.co.uk
Bringing Maths Alive

EL 1 Where do the chemical elements come from?

Specification reference: EL(a), EL(g), EL(h), EL(x)

Various models have been proposed to explain the origin of the Universe, but the Big Bang theory is the one agreed on by the majority of cosmologists.

The Big Bang theory

At some moment, all matter in the Universe was contained in a single point. This matter underwent an explosive expansion known as the Big Bang.

After about three minutes, the nuclei of hydrogen and helium formed from hot collections of tiny particles such as protons and neutrons.

After about 10 000 years, the Universe cooled sufficiently so that electrons moved slowly enough to be captured by oppositely charged protons in nuclei to form atoms. The Universe was made of mainly hydrogen and helium.

As the Universe continued to cool, dust and gas was pulled together by their gravity forming gas clouds. Particles had low kinetic energy and moved around relatively slowly, so gravitational forces were able to keep them together.

Parts of the clouds contracted in on themselves, compressing the gases and forming clumps of denser gas. The densest part of a clump was at the centre, where temperatures were hot enough so that atoms could not retain their electrons. Matter became a **plasma** of ionised atoms and unbound electrons.

Nuclear reactions, such as nuclear fusion, occur releasing vast amounts of energy and causing the dense gas cloud to glow – dense gas cloud becomes a star.

Nuclear reactions generate a hot wind that drives away some of the dust and gas leaving behind the stars. Planets condense out of the remaining dust cloud around these stars.

Nuclear fusion is common at the centre of stars. In nuclear fusion, lighter nuclei are fused together to form heavier nuclei, such as hydrogen atoms joining together to form helium. Other heavier elements are also produced in stars by fusion. Nuclei approach each other at high speed, with a large kinetic energy to overcome repulsion by positive charges on the two nuclei. It requires extreme conditions of temperature and gravitational pressure for the reacions to occur. However, the vast amount of energy released, with no pollution, could make it a useful source of energy if such conditions could be controlled. This is currently being researched by scientists.

▲ Figure 1 *Formation of stars in the Eagle Nebula – the thin 'fingers' at the top of the pillar of gas are embryonic stars*

The life cycle of stars

Hydrogen is still the most common element in the Universe. All stars turn hydrogen into helium by nuclear fusion. The theory of the evolution of the stars shows how heavy elements can be formed from lighter ones, and helps to explain the way elements are distributed throughout the Universe.

Heavyweight stars

The temperatures and pressures at the centre of heavyweight stars mean that, along with hydrogen fusion, other fusion reactions can take place producing heavier elements than helium. Layers of elements form within the star, with the heaviest elements near the centre (Figure 2).

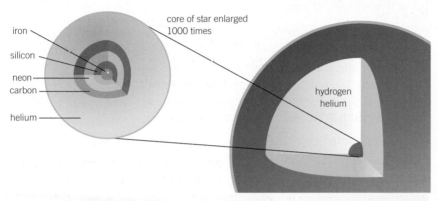

iron
silicon
neon
carbon
helium

core of star enlarged
1000 times

hydrogen
helium

▲ Figure 2   *A model of the core of a typical heavy-weight star after a few million years – long enough for extensive fusion to have taken place*

Eventually, the element at the centre of the core is iron. When iron nuclei fuse together they do not release energy but *absorb* it. When the core of a heavyweight star contains mostly iron it becomes unstable and explodes – a supernova (Figure 3). A supernova causes the elements in the star to be dispersed into the Universe as clouds of dust and gas and the life cycle begins again.

Lightweight stars

The Sun is a lightweight star – it is not as hot as most other stars and will last longer than heavyweight stars. It will keep on shining until all the hydrogen has been used up and the core stops producing energy – there will be no supernova. Once the hydrogen is used up, the Sun will expand into a red giant, swallowing up the planets Mercury and Venus. The oceans on Earth will start to boil and eventually it too will be engulfed by the Sun. The good news for Earth is that the Sun still has an estimated 5000 million years' supply of hydrogen left.

▲ Figure 3   *A possible supernova in the Crab nebula*

As red giants get bigger they become unstable and the outer gases drift off into space, leaving behind a small core called a white dwarf, about $\frac{1}{100}$ of the size of the original star.

In order to understand the nuclear reactions in stars, you need to look at atomic structure and fusion reactions.

Activity EL 1.1

This activity allows you to explore the nature of science and changing models.

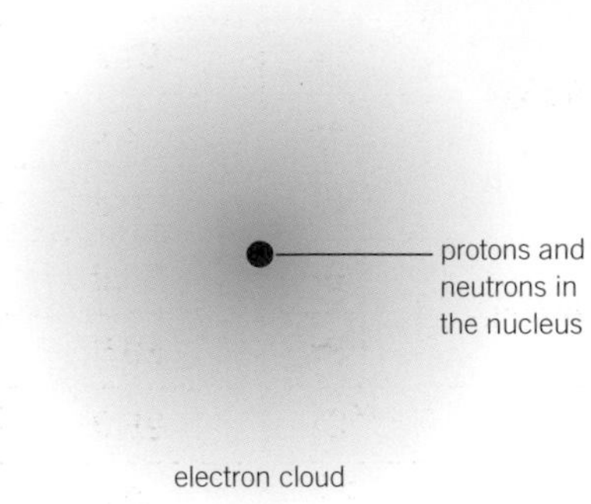

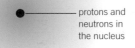

Figure 4 shows: protons and neutrons in the nucleus, electron cloud

▲ **Figure 4** *A simple model of the atom – not to scale*

Chemical Ideas: Atomic structure 2.1

A simple model of the atom

Scientists cannot see inside atoms, but experimental evidence provides a working model of atomic structure. Models are useful as they help to explain observations. Scientists spend lots of time building, testing, and revising scientific models. They should not be considered as the truth.

In a simplified model of atomic structure, atoms can be considered to be made of three types of sub-atomic particle – protons, neutrons, and electrons. Many chemical and nuclear processes can be explained using this simplified model.

Protons and neutrons form the nucleus of atoms. Electrons move around the nucleus. The nucleus is tiny compared with a volume occupied by the electrons. If an atom was the size of Wembley Stadium, the nucleus would be the size of a pea!

Sub-atomic particles

Table 1 summarises some properties of protons, neutrons, and electrons.

▼ **Table 1** *Some properties of sub-atomic particles*

Particle	Mass on relative atomic mass scale	Charge (relative to neutron)	Location in atom
proton	1	+1	in nucleus
neutron	1	0	in nucleus
electron	0.000 55	−1	around nucleus

Protons and electrons have equal but opposite electrical charges. Neutrons have no charge. Protons and neutrons have almost equal masses, and are much more massive than electrons. The nucleus accounts for almost all the mass of the atom but hardly any of its volume. Most of the atom is empty space.

It is the electrons in the outer parts of atoms that interact together in chemical reactions.

Nuclear symbols

The nucleus can be described by just two numbers – the **atomic number** Z and the **mass number** A. The atomic number is the number of protons in the nucleus. It is numerically equal to the charge on the *nucleus*. The atomic number is the same for every atom of an element, for example, $Z = 6$ for *all* carbon atoms. The mass number is the number of protons *and* neutrons in the nucleus.

mass number A = atomic number Z + number of neutrons N

Nuclear symbols identify the mass number and the atomic number as well as the symbol for the element (Figure 5).

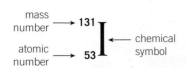

mass number → 131, atomic number → 53, chemical symbol, I

▲ **Figure 5** *The nuclear symbol for iodine-131*

Study tip

The atomic number can be omitted because the chemical symbol tells us which element it is. So $^{12}_{6}C$ can be simplified to ^{12}C or carbon-12.

What are isotopes?

Isotopes are atoms of the *same* element with different *mass* numbers. All atoms of an element have the same number of protons. Differences in mass are caused by different numbers of neutrons.

Most elements exist naturally as a mixture of isotopes (Table 2). The **relative atomic mass** A_r is an average of the relative isotopic masses, taking into account their abundances. A technique called **mass spectrometry** is used to find the atomic mass of elements and compounds.

▼ Table 2 *Isotopes of some elements*

Element	Isotope	Abundance /%
chlorine	^{35}Cl	75.0
	^{37}Cl	25.0
iron	^{54}Fe	5.8
	^{56}Fe	91.7
	^{57}Fe	2.2
	^{58}Fe	0.3
bromine	^{79}Br	50.0
	^{81}Br	50.0
calcium	^{40}Ca	96.9
	^{42}Ca	0.7
	^{43}Ca	0.1
	^{44}Ca	2.1
	^{48}Ca	0.2

Chemical ideas: Radiation and matter 6.5a

Mass spectrometry with elements

Mass spectrometry measures the atomic or molecular mass of different particles (i.e., atoms or molecules) in a sample and the relative abundance of different **isotopes** in an element.

In a mass spectrometer, sample atoms or molecules are ionised to positively charged **cations**. These **ions** are separated according to their mass m to charge z ratios, m/z. The separated ions are detected, together with their relative abundance.

Using mass spectra to calculate relative atomic mass

Figure 6 shows the mass spectrum for naturally occurring iron. If you assume that all the ions are singly charged, $z = 1$, m/z is the same as the mass of the ion detected. The relative abundance of each ion can be calculated from the height of each peak.

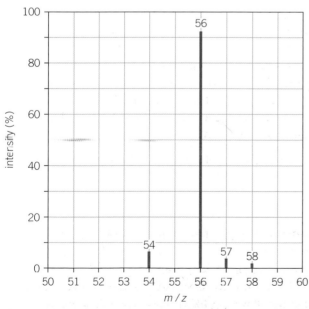

Relative isotopic mass	Relative abundance /%
54	5.8
56	91.7
57	2.2
58	0.3

▲ Figure 6 *The mass spectrum for naturally occurring iron (left) and a table summarising the information from the mass spectrum (right)*

Samples of an element may vary slightly in average A_r depending on their source. This is due to slightly different abundances of individual isotopes between samples.

Synoptic link

You will find out more about mass spectrometry in Chapter 5, What's in a medicine?

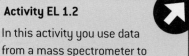

Activity EL 1.2

In this activity you use data from a mass spectrometer to calculate an A_r.

 Worked example: Calculating the relative atomic mass of iron

What is the relative atomic mass of iron?

Step 1: Using the table in Figure 6, calculate the average mass of 100 atoms.

(relative isotopic mass × relative abundance)

$(54 \times 5.8) + (56 \times 91.7) + (57 \times 2.2) + (58 \times 0.3)$
$= 313.2 + 5135.2 + 125.4 + 17.4$
$= 5591.2$

Step 2: Calculate the mass of one atom.

$$5591.2 \div 100 = 55.91$$

Average mass of one atom of iron is 55.9 (1 d.p.).

Chemical ideas: Atomic structure 2.2

Nuclear fusion

In a **nuclear fusion** reaction, two light atomic nuclei fuse together forming a single heavier nucleus of a new element, releasing enormous quantities of energy. For two nuclei to fuse, they must come very close together. At the normal temperatures found on the Earth, the positive nuclei repel so strongly that fusion cannot happen. At very high temperatures, such as in a star, the nuclei are moving much more quickly and collide with so much energy that this repulsion is overcome.

Atomic numbers and mass numbers must balance in a nuclear equation. The equation below shows fusion of different isotopes of hydrogen.

$$^1_1H + {}^2_1H \rightarrow {}^3_2He + \gamma$$

The γ symbol indicates the release of energy when the fusion reaction occurs. This is a common occurrence in fusion reactions.

Summary questions

1 Copy and complete the table. (*5 marks*)

Isotope	Symbol	Atomic number	Mass number	Number of neutrons
carbon-12	$^{12}_{6}C$	6	12.0	6
carbon-13			13.0	7
oxygen-16	$^{16}_{8}O$	8	16.0	
strontium-90	$^{90}_{38}Sr$	38		52
iodine-131	$^{131}_{53}I$			
	$^{121}_{53}I$			

2 State the number of protons, neutrons, and electrons present
   in each of the following atoms.
   a $^{79}_{35}Br$ (*1 mark*)   c $^{35}_{17}Cl$ (*1 mark*)

   b $^{81}_{35}Br$ (*1 mark*)   d $^{37}_{17}Cl$ (*1 mark*)

3 Use the data in Table 2 to calculate the relative atomic masses of:
   a bromine   (*2 marks*)
   b calcium.   (*2 marks*)

4 The relative atomic mass of the element iridium is 192.2. Iridium
   occurs naturally as a mixture of iridium-191 and iridium-193.
   Calculate the percentage of each isotope in naturally occurring
   iridium using the following steps.
   a If the percentage of iridium-193 in naturally occurring iridium
      is x%, there must be x atoms of iridium-193 in every 100 atoms
      of the element. How many atoms of iridium-191 must there
      be in every 100 atoms of iridium?   (*1 mark*)
   b Write an expression for the total relative mass of the
      atoms of iridium-193 in the 100-atom sample. (*1 mark*)
   c Write an expression for the total relative mass of the
      atoms of iridium-191 in the 100-atom sample. (*1 mark*)
   d Write an expression for the total relative mass of all
      100 iridium atoms.   (*1 mark*)
   e Write an expression for the relative atomic mass of
      one iridium atom.   (*1 mark*)
   f Use your answer to part e and the value of the relative
      atomic mass of iridium to calculate the percentage of
      each isotope in naturally occurring iridium. (*3 marks*)

5 Write a nuclear equation for each of the following.
   a A $^{7}_{3}Li$ nucleus absorbs a colliding proton and then
      disintegrates into two identical fragments. (*2 marks*)
   b The production of carbon-14 by collision of a neutron with
      an atom of nitrogen-14.   (*2 marks*)

6 The relative atomic mass of antimony is 121.8. Antimony
   exists as two isotopes – antimony-121 and antimony-123.
   Calculate the relative abundances of the two isotopes. (*4 marks*)

The work of chemists has made a vital contribution to the understanding of the origin, structure, and composition of our Universe. To do this, they have used a method called **spectroscopy**.

Spectroscopy

Many spectroscopic techniques exist but all are based on the same scientific principle – under certain conditions, a substance can absorb or emit electromagnetic radiation in a characteristic way. There are different types of electromagnetic radiation (Figure 1). By analysing electromagnetic radiation you can identify a substance or find information about it, such as its structure and the way atoms are held together.

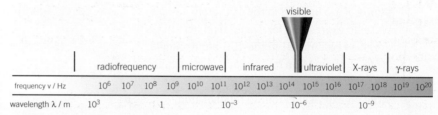

▲ Figure 1 *The electromagnetic spectrum*

Absorption spectra

Glowing stars emit all of the light frequencies between the ultraviolet and infrared parts of the electromagnetic spectrum. The Sun emits mainly visible light, whereas stars hotter than the Sun emit mainly ultraviolet radiation.

Outside a star's surface – the photosphere – is a region called the chromosphere that contains ions, atoms, and small molecules. These particles *absorb* some of the emitted radiation so the light analysed from stars is missing certain frequencies. The absorption lines appear black as these are the missing frequencies of light – they have been absorbed by the particles in the chromosphere (Figure 2).

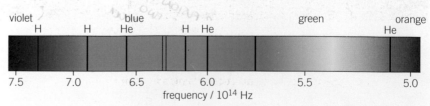

▲ Figure 2 *Absorption spectrum of a B-type star (e.g., β Centauri). This spectrum allows scientists to detect the presence of hydrogen and helium. Black lines occur where frequencies are missing from the otherwise continuous spectrum. (The frequency increases from right to left, which is a common convention for absorption spectra)*

Emission spectra

When the atoms, molecules, and ions in the chromosphere absorb energy, they are raised from their ground state (their lowest energy state) to higher energy states called excited states. The particles can lose their extra energy by *emitting* electromagnetic radiation. The resulting emission spectra can also be detected. Emission spectra appear as coloured lines on a black background – corresponding to the frequencies emitted (Figure 3).

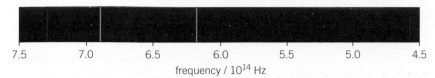

7.5	7.0	6.5	6.0	5.5	5.0	4.5

frequency / 10^{14} Hz

▲ Figure 3 *The hydrogen emission spectrum in the visible region. Coloured lines occur where frequencies are emitted*

Continuous and atomic spectra

White light contains all the visible wavelengths and its spectrum is normally continuous, like a rainbow. However light seen from stars is not continuous. It consists of lines, corresponding to the absorption or emission of specific frequencies of light – atomic spectra.

The atomic spectrum of hydrogen atoms

The Sun's chromosphere consists mainly of hydrogen and helium atoms and so hydrogen atoms dominate the chromosphere's *emission* spectrum, but helium emission lines can also be seen. Comparing the emission spectrum of hydrogen (Figure 3) with the absorption spectrum shown in Figure 2, the emitted frequencies match up exactly with those of the absorbed frequencies for hydrogen.

Hydrogen atoms also have a characteristic emission spectrum in the ultraviolet region of the electromagnetic spectrum (Figure 4). The hydrogen emission spectrum in visible light is known as the Balmer series. The hydrogen emission spectrum in ultraviolet light is known as the Lyman series.

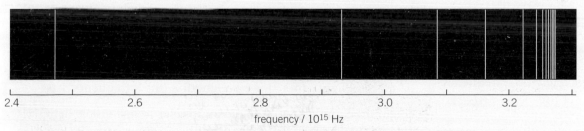

2.4	2.6	2.8	3.0	3.2

frequency / 10^{15} Hz

▲ Figure 4 *The Lyman series in the hydrogen atomic spectrum*

These spectra are the result of the interaction of light and matter.

Chemical ideas: Radiation and matter 6.1a

Light and electrons

Spectroscopy is the study of how light and matter interact. In order to understand spectroscopy you need to know about the behaviour of light.

Bohr's theory and wave-particle duality

Chemists use two models to describe the behaviour of light – the wave theory and the particle theory. Neither fully explains all the properties of light, but some are explained by the wave model and others by the particle model.

The wave theory of light

Light is one form of electromagnetic radiation. Like all electromagnetic radiation, it behaves like a wave with a characteristic **wavelength** λ and **frequency** v.

A wave of light travels the distance between two points in a certain time – it doesn't matter what kind of light it is, the time is always the same. The **speed** the wave moves, the speed of light c, is the same for all kinds of light and electromagnetic radiation. It has a value of $3.00 \times 10^8 \, \text{m s}^{-1}$ when the light is travelling in a vacuum.

Different colours of light have different wavelengths. Waves can also have different frequencies. Both waves in Figure 5 travel at the same speed, so they both travel the same *distance* per second but wave B has twice the frequency of wave A.

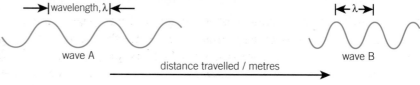

▲ **Figure 5** *Wavelength measures the distance (in metres) travelled by the wave during one cycle. Wave A has twice the wavelength of wave B*

Frequency and wavelength are very simply related. In Figure 5, wave B has twice the frequency but half the wavelength of wave A. Multiplying the wavelength and frequency together give the speed of light c. A simple equation links together c, λ, and v.

speed of light $c \, (\text{m s}^{-1})$ = wavelength λ (m) × frequency $v \, (\text{s}^{-1})$

Though the term light is often meant to refer to just visible light, it is only one small part of the **electromagnetic spectrum** (Figure 1). Other regions include radio waves, ultraviolet, infrared, and gamma rays.

The particle theory of light

In some situations, the behaviour of light is easier to explain by thinking of it not as waves but as particles. This regards light as a stream of tiny packets of energy called **photons**. The energy of the photons is

---

Synoptic link

You will look further at this in Topic OZ 2, Screening the Sun.

Study tip

As the wavelength increases, the frequency decreases.

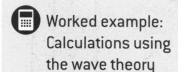 **Worked example: Calculations using the wave theory**

Calculate the wavelength of an ultraviolet wave of frequency $1.4 \times 10^{15} \, \text{s}^{-1}$.

Step 1: Rearrange the equation

$$\text{speed of light} = \underset{\lambda \, (\text{m})}{\text{wavelength}} \times \underset{v \, (\text{s}^{-1})}{\text{frequency}}$$
$$c \, (\text{m s}^{-1})$$

to make wavelength the subject.

$$\lambda = \frac{c}{v}$$

Step 2: Calculate the wavelength.

$$\frac{3.00 \times 10^8}{1.4 \times 10^{15}} = 2.14 \times 10^{-7} \, \text{m}$$

related to the position of the light in the electromagnetic spectrum, for example, photons with energy 3×10^{-19} J correspond to red light.

The wave and the photon models of light are linked.

energy of a photon E (J) = Planck constant h (Js^{-1}) × frequency v (s^{-1})

Js not Js^{-1} !

The value of the Planck constant is 6.63×10^{-34} Js^{-1}.

Bohr's theory

Using the hydrogen emission spectrum (Figure 3) the scientist Niels Bohr came up with a theory to explain why the hydrogen atom only emits a limited number of specific frequencies.

When an atom is **excited**, electrons jump into higher energy levels. Later, they drop back into lower levels, emitting the extra energy as electromagnetic radiation and giving off an **emission spectrum**.

When white light is passed through a relatively cool sample of a gaseous element, black lines appear in the otherwise continuous absorption spectrum. The black lines in the **absorption spectrum** correspond to light that has been *absorbed* by the atoms in the sample. Electrons have been raised to higher levels without then dropping back again. They correspond exactly with the coloured lines in the emission spectrum of that element.

The sequence of lines in an atomic spectrum is characteristic of the atoms of an element. They can be used to identify the element, even when the element is present in a compound or is part of a mixture. The *intensities* of the lines provide a measure of the element's abundance.

Bohr's theory not only explains how you get absorption spectra and emission spectra, it also gave scientists a model for the electronic structure of atoms. It was controversial when first proposed as it relied on the new theory of the *quantisation* of energy. However, Bohr's theory predicted experimental observations accurately and lent support to the new *quantum theory*.

The main points of Bohr's theory were:

- the electron in the hydrogen atom exists only in certain definite energy levels or electron shells
- a photon of light is emitted or absorbed when the electron changes from one energy level to another
- the energy of the photon is equal to the difference between the two energy levels ΔE
- since $E = hv$ it follows that the frequency of the emitted or absorbed light is related to ΔE by $\Delta E = hv$.

Energy levels and quanta

An electron can only possess definite quantities of energy, or **quanta**. The electron's energy cannot change continuously – it is not able to change to any value, only those values that are allowed.

▼ **Table 1** *The flame colours of some common metal ions*

Ion	Colour
Li$^+$	bright red
Na$^+$	yellow
K$^+$	Lilac
Ca^{2+}	brick red
Ba^{2+}	apple green
Cu^{2+}	blue green

▲ **Figure 6** *Flame tests for sodium chloride, potassium chloride, and barium sulphate*

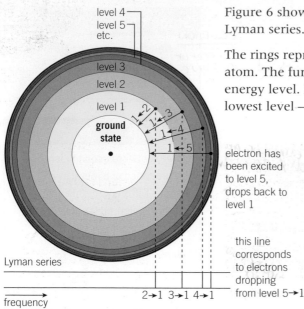

Figure 6 shows how Bohr's ideas explained the emission lines of the Lyman series.

The rings represent the energy levels of the electron in the hydrogen atom. The further away from the nucleus an electron is, the higher the energy level. Levels are labelled with numbers, starting at one for the lowest level – the ground state.

The frequencies of the lines of the Lyman series correspond to changes in electronic energy from various upper levels to one common lower level, level 1. Each line corresponds to a particular energy level change, such as level 4 to level 1.

An alternative way of representing emission spectra is by an energy level diagram. Each arrow represents an electron dropping from a *higher* energy level to a *lower* energy level. It can be seen from Figure 7 that the larger the energy gap ΔE between the two levels, the higher the frequency of electromagnetic radiation emitted.

▲ **Figure 6** *How the Lyman series in the emission spectrum is related to energy levels in the hydrogen atom*

Summary questions

1 Lithium carbonate is used to give the bright red colour to a firework.

The visible region of the emission spectrum of lithium contains several coloured lines with a particularly intense line at the red end of the spectrum. The emission spectrum is shown below.

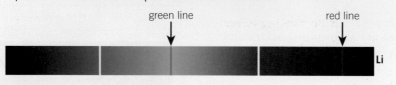

Which line, green or red, has a greater frequency? (*1 mark*)

2 Flame tests can be used to identify some metal cations.

Describe the flame colours for the following cations.

a Li^+ b Na^+ c K^+ d Ca^{2+} e Ba^{2+} f Cu^{2+} (*6 marks*)

3 Describe the similarities and differences between the above emission spectrum of lithium and the absorption spectrum of lithium. (*2 marks*)

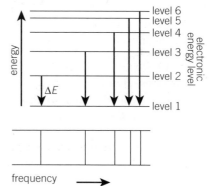

▲ **Figure 7** *An energy level diagram and corresponding emission spectrum for the Lyman series. A similar diagram can be used to represent an absorption spectrum – the difference between the diagrams would be that the arrows would point upwards as the electron is raised from a lower energy level to a higher energy level*

4 Copy the diagram that represents electronic energy levels in a lithium atom.
   a Draw an arrow to represent an electron energy level change that might give rise to the red line seen in the emission spectrum. (*1 mark*)
   b Draw an arrow to represent an electron energy level change that might give rise to the green line seen in the emission spectrum. (*1 mark*)

5 Calculate the frequency of electromagnetic radiation with a wavelength of 5.5×10^{-7} m (*1 mark*)

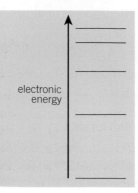

EL 3 Electrons, where would we be without them?

Specification reference: EL(e), EL(f)i, EL(f)iii

Electrons are involved in our lives and the world around us in many ways. Figure 1 to 4 show some examples.

▲ Figure 1 Laser light shows

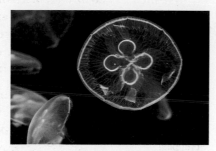
▲ Figure 2 Bioluminescence in jellyfish

▲ Figure 3 Modern technology

▲ Figure 4 Chemical reactions like cooking

Learning outcomes

Demonstrate and apply knowledge and understanding of:

→ conventions for representing the distribution of electrons in atomic orbitals – the shapes of s- and p-orbitals

→ the electronic configuration, using sub-shells and atomic orbitals, of:

- atoms from hydrogen to krypton

- the outer sub-shell structures of s- and p-block elements of other periods.

In order to understand why electrons have such wide-ranging effects, you need to find out a bit more about atomic structure.

Determining atomic structure

Both models of electron structure – negative particles or negative wave form – can be used to understand the structure of atoms. Electrons can be thought of as particles when filling the shells of atoms, but the electron orbitals can be thought of as standing (or stable) waves with a maximum overall negative charge equivalent to two electrons.

Chemical ideas: Atomic structure 2.3

Electronic structure: shells, sub-shells, and orbitals

Shells of electrons

Bohr's theory to describe the emission spectrum of the hydrogen atom needed expanding to describe the structure of atoms with more than one electron. For these atoms, the energy levels 2, 3, 4, and so on have a more complex structure than the single levels that exist in hydrogen. It is more appropriate to talk about the first, second, third electron shells rather than energy level 1, 2, 3. The shells are labelled by giving each

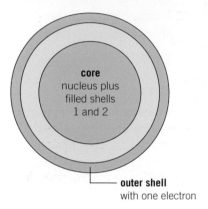

core
nucleus plus
filled shells
1 and 2

outer shell
with one electron

▲ Figure 5 *The core and outer shell model of a sodium atom [2.8.1]*

a principal quantum number n. For the first shell, $n = 1$, for the second shell $n = 2$, and so on. The higher the value of n, the further the shell is from the nucleus and the higher the energy associated with the shell.

Although each shell can hold more than one electron, there is a limit. The maximum numbers of electrons which can be held in the first four shells are:

- first shell   ($n = 1$)   2 electrons
- second shell   ($n = 2$)   8 electrons
- third shell   ($n = 3$)   18 electrons
- fourth shell   ($n = 4$)   32 electrons

A shell that contains its maximum number of electrons is called a filled shell. Electrons are arranged so that the lowest energy shells are filled first. Electron shell configurations can be written, for example, the electron shell configuration of sodium is 2, 8, 1. This means:

- two electrons in first shell
- eight electrons in second shell
- one electron in third shell.

Chemists explain many of the properties of atoms without needing to use a detailed theory of atomic structure. Much chemistry is decided only by the outer shell electrons, and one very useful model treats the atom as being composed of a core of the nucleus and the inner electrons shells, surrounded by an outer shell (Figure 5).

Sub-shells of electrons

Much of the knowledge of electron shells has come from studying the emission spectrum of hydrogen. The hydrogen atom has only one electron and its spectrum is relatively simple to interpret. The spectra of elements other than hydrogen are much more complex – electron shells are not the whole story. The shells are themselves split up into sub-shells.

The sub-shells are labelled s, p, d, and f. The $n = 1$ shell has only an s-sub-shell. The $n = 2$ shell has two sub-shells – s and p. The $n = 3$ shell has three sub-shells – s, p, and d. The $n = 4$ shell has four sub-shells – s, p, d, and f.

The different types of sub-shells can hold different numbers of electrons (Table 1).

- the $n = 1$ shell can hold two electrons in the s-sub-shell
- the $n = 2$ shell can hold two electrons in the s-sub-shell and six electrons in the p-sub-shell – a total of eight electrons
- the $n = 3$ shell can hold two electrons in the s-sub-shell, six electrons in the p-sub-shell, and 10 electrons in the d-sub-shell – a total of 18 electrons
- the $n = 4$ shell can hold two electrons in the s-sub-shell, six electrons in the p-sub-shell, 10 electrons in the d-sub-shell, and 14 electrons in the f-sub-shell – a total of 32 electrons.

▼ Table 1 *Maximum number of electrons in the s-, p-, d-, and f-sub-shells*

Sub-shell	Maximum number of electrons
s	2
p	6
d	10
f	14

In atoms other than hydrogen, the sub-shells within a shell have different energies. The energy of a sub-shell is not fixed, but falls as the charge on the nucleus increases from one element to the next in the periodic table. Figure 6 shows the relative energies of the sub-shell for each of the shells $n = 1$ to $n = 4$ in a typical many-electron atom. The overlap in energy between the $n = 3$ and $n = 4$ shells has important consequences that you will meet later.

Atomic orbitals

The s- p- d- and f-sub-shells are themselves divided further into atomic orbitals. An electron in a given orbital can be found in a particular region of space around the nucleus.

- An s-sub-shell always contains *one* s-orbital
- A p-sub-shell always contains *three* p-orbitals
- A d-sub-shell always contains *five* d-orbitals
- An f-sub-shell always contains *seven* f-orbitals

In an isolated atom, orbitals in the same sub-shell have the same energy (Figure 7).

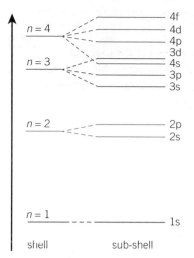

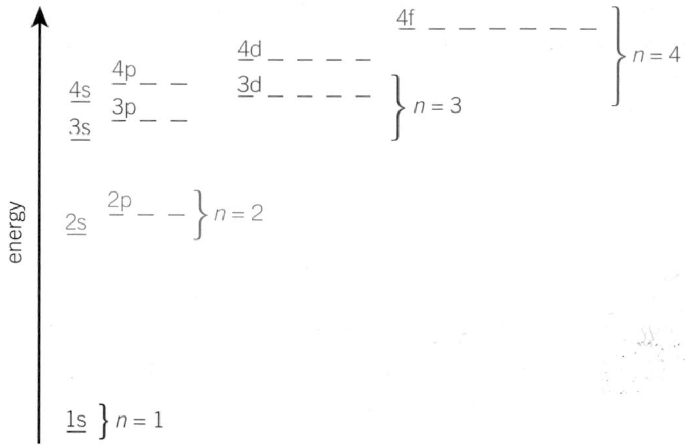

▲ **Figure 6** *Energies of electron sub-shells from n = 1 to n = 4 in a typical many-electron atom. The order shown in the diagram is correct for the elements in Period 3 and up to nickel in Period 4 (periods are the horizontal rows on the periodic table). After nickel, the 3d-sub-shell has lower energy than 4s*

▲ **Figure 7** *Energy levels of atomic orbitals in the n = 1 to n = 4 shells*

The position of an electron cannot be mapped exactly. For an electron in a given atomic orbital, is is the *probability* of finding the electron in any region that is known (Figure 8).

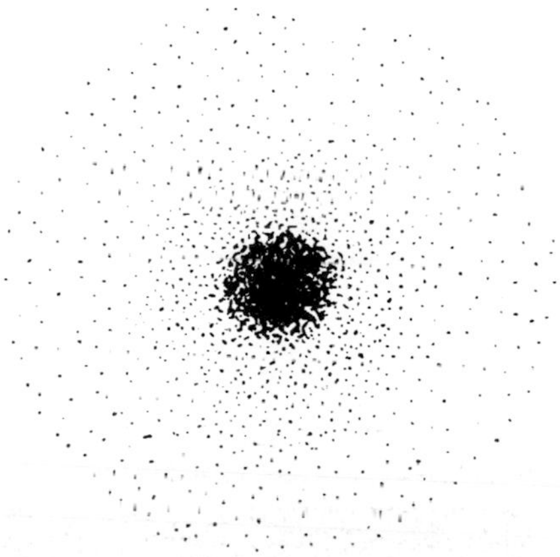

Each atomic orbital can hold a maximum of two electrons. Electrons in atoms have a *spin*, which is pictured as a spinning motion in one of two directions. Every electron spins at the same rate in either a clockwise ↑ or anticlockwise ↓ direction. Electrons can only occupy the same orbital if they have opposite, or paired, spins. They can be written as $\boxed{\uparrow\downarrow}$.

The box represents the atomic orbital and the arrows represent the electrons. Four pieces of information are needed when describing an electron:

- the electron shell it is in
- its sub-shell

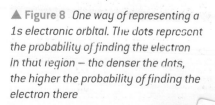

▲ **Figure 8** *One way of representing a 1s electronic orbital. The dots represent the probability of finding the electron in that region – the denser the dots, the higher the probability of finding the electron there*

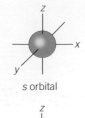

s orbital

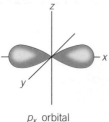

p_x orbital

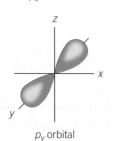

p_y orbital

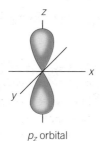

p_z orbital

▲ Figure 9 *Shapes of s-orbitals and p-orbitals*

- its orbital within the sub-shell
- its spin.

Atomic orbitals have a three dimensional shape that represents the volume of space where there is a high probability of finding up to two electrons (Figure 9).

Filling up atomic orbitals

The arrangement of electrons in shells and orbitals is called the **electronic configuration** of an atom. The orbitals are filled in a specific order to produce the lowest energy arrangement possible.

The orbitals are filled in order of increasing energy. Where there is more than one orbital with the same energy, these orbitals are first occupied singly by electrons. This keeps the electrons in an atom as far apart as possible. Only when every orbital is singly occupied do the electrons pair up in orbitals. For the lowest energy arrangement, electrons in singly occupied orbitals have parallel spins.

Figure 10 shows how the 11 electrons in a sodium atom are arranged in atomic orbitals. The electronic configuration of a sodium atom can also be represented as $1s^2 2s^2 2p^6 3s^1$. The large numbers show the principal quantum number of each shell, the letters show the sub-shells, and the small superscripts indicate the numbers of electrons in each sub-shell.

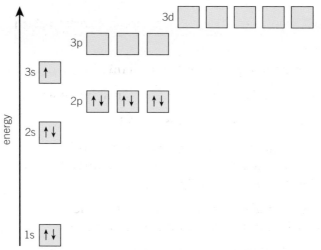

▲ Figure 10 *Arrangement of electrons in atomic orbitals in a ground state sodium atom*

Writing electronic configurations

Hydrogen is the simplest element, with atomic number $Z = 1$. It has one electron that occupies the s-orbital of the $n = 1$ shell.

1s ↑   H $1s^1$

The next element, helium ($Z = 2$) has two electrons that both occupy the 1s orbital with paired spins.

1s ↑↓   He $1s^2$

Lithium ($Z = 3$) has three electrons. The third electron cannot fit in the $n = 1$ shell, so occupies the next lowest orbital – the 2s orbital.

$$1s \boxed{\uparrow\downarrow} \quad 2s \boxed{\uparrow} \qquad \text{Li } 1s^2 2s^1$$

Nitrogen ($Z = 7$) has seven electrons.

$$1s \boxed{\uparrow\downarrow} \quad 2s \boxed{\uparrow\downarrow} \quad 2p \boxed{\uparrow}\boxed{\uparrow}\boxed{\uparrow} \qquad \text{N } 1s^2 2s^2 2p^3$$

The three electrons in the 2p-sub-shell occupy the three separate p-orbitals singly and their spins are parallel.

Oxygen ($Z = 8$) has eight electrons

$$1s \boxed{\uparrow\downarrow} \quad 2s \boxed{\uparrow\downarrow} \quad 2p \boxed{\uparrow\downarrow}\boxed{\uparrow}\boxed{\uparrow} \qquad \text{O } 1s^2 2s^2 2p^4$$

However, to write the electronic configuration of the element, scandium ($Z = 21$), look back to Figure 6. The energy level of the 3d-sub-shell lies just above that of the 4s-sub-shell but just below the 4p-sub-shell. This means that the 4s-orbital fills before the 3d-orbital. Once the 4s level is filled in calcium ($Z = 20$), scandium has the electronic structure $1s^2 2s^2 2p^6 3s^2 3p^6 \mathbf{3d^1 4s^2}$. The 3d-sub-shell continues to be filled across the period in the elements scandium to zinc. Zinc ($Z = 30$) has the electronic configuration $1s^2 2s^2 2p^6 3s^2 3p^6 \mathbf{3d^{10} 4s^2}$. The electronic configurations of elements with atomic numbers 1 to 36 are shown in Table 2.

> **Study tip**
>
> The 3d-sub-shell is written alongside the other $n = 3$ sub-shells even though it is filled after the 4s-sub-shell.

🖩 Worked example: Working out electronic configuration of an atom

What is the electronic configuration of vanadium?

Step 1: Identify the number of electrons in a vanadium atom.

From the periodic table, $Z = 23$

Step 2: Electrons go into lowest energy shells and sub-shells, using the appropriate numbers and letters, until they are full, so for vanadium $1s^2 2s^2 2p^6 3s^2 3p^6$

Step 3: The energy of the 4s-sub-shell is lower than the 3d-sub-shell so this is filled with two electrons first.

Step 4: This leaves the three remaining electrons to go into the 3d-sub-shell

Step 5: Group together the same-shell electrons. The final overall configuration is therefore $1s^2 2s^2 2p^6 3s^2 3p^6 3d^3 4s^2$

> **Study tip**
>
> You will have access to a periodic table in examinations.

> **Study tip**
>
> You need to be able to write the electron configuration of an atom from hydrogen to krypton, but not copper and chromium.

Electron configurations show how the periodic table is built up. Period 2 (lithium to neon) corresponds to the filling of the 2s- and 2p-orbitals, Period 3 (sodium to argon) corresponds to the filling of the 3s-and 3p-orbitals, and so on. You can also write electron configurations of ions, which you will do in Topic EL 4.

> **Synoptic link**
>
> You will find out more about copper and chromium in Chapter 10, Colour by design.

▼ Table 2 *Ground state electronic configurations of elements with atomic numbers 1–36*

Period	Atomic number / Z	Element	Electronic configuration	Period	Atomic number / Z	Element	Electronic configuration
$n = 1$	1	hydrogen	$1s^1$		19	potassium	$1s^2 2s^2 2p^6 3s^2 3p^6 4s^1$
	2	helium	$1s^2$		20	calcium	$1s^2 2s^2 2p^6 3s^2 3p^6 4s^2$
$n = 2$	3	lithium	$1s^2 2s^1$		21	scandium	$1s^2 2s^2 2p^6 3s^2 3p^6 3d^1 4s^2$
	4	beryllium	$1s^2 2s^2$		22	titanium	$1s^2 2s^2 2p^6 3s^2 3p^6 3d^2 4s^2$
	5	boron	$1s^2 2s^2 2p^1$		23	vanadium	$1s^2 2s^2 2p^6 3s^2 3p^6 3d^3 4s^2$
	6	carbon	$1s^2 2s^2 2p^2$		24	**chromium**	$\mathbf{1s^2 2s^2 2p^6 3s^2 3p^6 3d^5 4s^1}$
	7	nitrogen	$1s^2 2s^2 2p^3$		25	manganese	$1s^2 2s^2 2p^6 3s^2 3p^6 3d^5 4s^2$
	8	oxygen	$1s^2 2s^2 2p^4$		26	iron	$1s^2 2s^2 2p^6 3s^2 3p^6 3d^6 4s^2$
	9	fluorine	$1s^2 2s^2 2p^5$	$n = 4$	27	cobalt	$1s^2 2s^2 2p^6 3s^2 3p^6 3d^7 4s^2$
	10	neon	$1s^2 2s^2 2p^6$		28	nickel	$1s^2 2s^2 2p^6 3s^2 3p^6 3d^8 4s^2$
	11	sodium	$1s^2 2s^2 2p^6 3s^1$		29	**copper**	$\mathbf{1s^2 2s^2 2p^6 3s^2 3p^6 3d^{10} 4s^1}$
	12	magnesium	$1s^2 2s^2 2p^6 3s^2$		30	zinc	$1s^2 2s^2 2p^6 3s^2 3p^6 3d^{10} 4s^2$
	13	aluminium	$1s^2 2s^2 2p^6 3s^2 3p^1$		31	gallium	$1s^2 2s^2 2p^6 3s^2 3p^6 3d^{10} 4s^2 4p^1$
$n = 3$	14	silicon	$1s^2 2s^2 2p^6 3s^2 3p^2$		32	germanium	$1s^2 2s^2 2p^6 3s^2 3p^6 3d^{10} 4s^2 4p^2$
	15	phosphorous	$1s^2 2s^2 2p^6 3s^2 3p^3$		33	arsenic	$1s^2 2s^2 2p^6 3s^2 3p^6 3d^{10} 4s^2 4p^3$
	16	sulfur	$1s^2 2s^2 2p^6 3s^2 3p^4$		34	selenium	$1s^2 2s^2 2p^6 3s^2 3p^6 3d^{10} 4s^2 4p^4$
	17	chlorine	$1s^2 2s^2 2p^6 3s^2 3p^5$		35	bromine	$1s^2 2s^2 2p^6 3s^2 3p^6 3d^{10} 4s^2 4p^5$
	18	argon	$1s^2 2s^2 2p^6 3s^2 3p^6$		36	krypton	$1s^2 2s^2 2p^6 3s^2 3p^6 3d^{10} 4s^2 4p^6$

Summary questions

1 $1s^2$ is an example of a notation for electronic configuration.
   a What does the number 1 refer to?   (*1 mark*)
   b What does the s refer to?   (*1 mark*)
   c What does the superscript 2 refer to?   (*1 mark*)

2 Write out the electronic configurations of the following atoms.
   a boron ($Z = 5$)   (*1 mark*)   b phosphorus ($Z = 15$)   (*1 mark*)

3 The electronic configuration of the outermost shell of an atom of an element X is $3s^2 3p^4$. What is the atomic number and name of the element?   (*2 marks*)

4 Electronic configurations are sometimes abbreviated by labelling the core of filled inner shells as the electronic configuration of the appropriate noble gas. For example, the electronic configuration of neon is $1s^2 2s^2 2p^6$ and that of sodium is $1s^2 2s^2 2p^6 3s^1$ so you can write the electronic configuration of sodium as $[Ne]3s^1$. Name the elements from the electronic configurations.
   a $[Ne]3s^2 3p^5$ (*1 mark*)   c $[Ar]3d^2 4s^2$ (*1 mark*)
   b $[Ar]4s^1$ (*1 mark*)   d $[Kr]4d^{10} 5s^2 5p^2$ (*1 mark*)

EL 4 Organising the elements of life
Specification reference: EL(f), EL(m), EL(n)

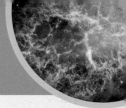

The periodic table

As elements were discovered and more was learnt about their properties, chemists looked for patterns so the elements could be grouped. Much of the work on finding patterns in the elements was done by Johann Döbereiner and Lothar Meyer in Germany, John Newlands in England, and Dmitri Mendeleev in Russia. These chemists looked at similarities in the chemical reactions of the elements they knew about, and also patterns in physical properties such as melting point, boiling point, and density.

Mendeleev arranged elements in order of increasing atomic mass and so that elements with similar properties were in the same vertical group. His grouping was seen as the most credible. Mendeleev's values for atomic masses were not accurate because the existence of isotopes was not known at that time.

▲ Figure 1 *Dimitri Mendeleev*

▼ Table 1 *A form of Mendeleev's periodic table – the asterisks denote elements that he thought were yet to be discovered*

	Group 1	Group 2	Group 3	Group 4	Group 5	Group 6	Group 7	Group 8
Period 1	H							
Period 2	Li	Be	B	C	N	O	F	
Period 3	Na	Mg	Al	Si	P	S	Cl	
Period 4	K Cu	Ca Zn	* *	Ti *	V As	Cr Se	Mn Br	Fe, Co, Ni
Period 5	Rb Ag	Sr Cd	Y In	Zr Sn	Nb Sb	Mo Te	* I	Ru, Rh, Pd

Making predictions using the periodic table

Mendeleev left gaps in his table of elements. These gaps corresponded to his predictions of elements that had not been discovered at the time. He was so confident of his table that he even made predictions about the properties of these undiscovered elements. When the element germanium was discovered in 1886, its properties were found to be in excellent agreement with the predictions Mendeleev had made using his table of elements.

Some elements do not exist naturally and are made synthetically in a laboratory. The first two elements to be made synthetically were neptunium ($Z = 93$) and plutonium ($Z = 94$). They were formed by bombarding uranium atoms with neutrons. The heaviest element synthesized had an atomic number of 118 but only existed for 200 microseconds!

The modern periodic table is based on the one originally drawn up by Mendeleev. It is one of the most amazingly compact stores of information ever produced – with a copy of the periodic table in front of you, and some knowledge of how it was put together, you have thousands of facts at your fingertips!

(1) 1	(2)												(3) 13	(4) 14	(5) 15	(6) 16	(7) 17	(0) 18
1 H 1.0	2		**key**															2 He 4.0
3 Li 6.9	4 Be 9.0		atomic number symbol relative atomic mass										5 B 10.8	6 C 12.0	7 N 14.0	8 O 16.0	9 F 19.0	10 Ne 20.2
11 Na 23.0	12 Mg 24.3	3	4	5	6	7	8	9	10	11	12		13 Al 27.0	14 Si 28.1	15 P 31.0	16 S 32.1	17 Cl 35.5	18 Ar 39.9
19 K 39.1	20 Ca 40.1	21 Sc 45.0	22 Ti 47.9	23 V 50.9	24 Cr 52.0	25 Mn 54.9	26 Fe 55.8	27 Co 58.9	28 Ni 58.7	29 Cu 63.5	30 Zn 65.4		31 Ga 69.7	32 Ge 72.6	33 As 74.9	34 Se 79.0	35 Br 79.9	36 Kr 83.8
37 Rb 85.5	38 Sr 87.6	39 Y 88.9	40 Zr 91.2	41 Nb 92.9	42 Mo 95.9	43 Tc 98]	44 Ru 101.1	45 Rh 102.9	46 Pd 106.4	47 Ag 107.9	48 Cd 112.4		49 In 114.8	50 Sn 118.7	51 Sb 121.8	52 Te 127.6	53 I 126.9	54 Xe 131.3
55 Cs 132.9	56 Ba 137.3	57–71	72 Hf 178.5	73 Ta 180.9	74 W 183.8	75 Re 186.2	76 Os 190.2	77 Ir 192.2	78 Pt 195.1	79 Au 197.0	80 Hg 200.6		81 Tl 204.4	82 Pb 207.2	83 Bi 209.0	84 Po [209]	85 At [210]	86 Rn [222]
Fr	Ra		Rf	Db	Sg	Bh	Hs	Mt	Ds	Rg	Cn			Fi		Lv		

57 La 138.9	58 Ce 140.1	59 Pr 140.9	60 Nd 144.2	61 Pm 144.9	62 Sm 150.4	63 Eu 152.0	64 Gd 157.2	65 Tb 158.9	66 Dy 162.5	67 Ho 164.9	68 Er 167.3	69 Tm 168.9	70 Yb 173.0	71 Lu 175
Ac	90 Th 232.0	Pa	92 U 238.1	Np	Pu	Am	Cm	Bk	Cf	Es	Fm	Md	No	Lr

▲ Figure 2 *A modern periodic table*

Activity EL 4.1

In this activity you will use the Internet to look for patterns in the properties of elements.

✚ Superheavy elements and the Island of Stability

The periodic table arranges elements in order by the number of protons residing in an element's nucleus. Neutrons and protons (collectively known as **nucleons**) reside in the nucleus. The protons are positively charged and repel each other, with the repulsion increasing as the protons get closer. However another force – the strong nuclear force – is stronger than this repulsion of charge, and holds the nucleus together. Neutrons don't have an overall charge, but they have some magnetic properties. Neutrons act as a buffer between protons. Usually, the more protons you have, the more neutrons you need. Beyond element 92, uranium, the strong nuclear force is insufficient to hold the nucleus of the element together and all isotopes are unstable, breaking down by radioactive decay. The Island of Stability follows from theoretical predictions that some superheavy elements with a total number of nucleons around 300 may hang around for a while longer than their lighter cousins, before radioactively decaying into other types of nuclei.

1 Look up the neutron to proton ratio for naturally occurring isotopes of the elements (the so called 'band of stability')
   a State the neutron/proton ratio for the first 20 elements.
   b Describe how this ratio changes as the atomic number increases beyond 20.
2 Explain, in terms of neutron/proton ratio and relative stability, why either alpha or beta radioactive decay might be expected to occur for elements above atomic number 83.

Chemical ideas: The periodic table 11.1

Periodicity

The modem periodic table (Figure 2) is based on one proposed by the Russian chemist Mendeleev in 1869. In Mendeleev's periodic table the elements were arranged in order of increasing relative atomic mass. At first glance, the same seems to be true of the modern version – but not quite. Some pairs of elements are out of order, for example, tellurium, Te ($A_r = 128$) comes before iodine, I ($A_r = 127$). It is atomic number – the number of protons in the nucleus – that actually determines the place of an element in the periodic table.

With the exception of hydrogen, the elements can be organised into four blocks labelled s, p, d, and f (Figure 3). Elements in the same block show general similarities. For example, all the non-metals are in the p-block – many of the reactive metals (like sodium, potassium, and strontium) are in the s-block.

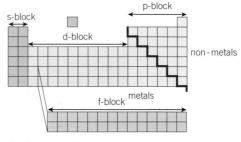
▲ Figure 3 Blocks in the periodic table

Vertical columns in the periodic table are called **groups**. Horizontal rows in the periodic table are called **periods**.

Physical properties

The elements in a group show patterns in their physical properties. There are *trends* in the properties as you go down a group. Because they cut across the groups there are fewer common features among the elements of a period but many properties vary in a fairly regular way as you move across a period from left to right. The pattern is then repeated as you go across the next period. The occurrence of periodic patterns is called **periodicity**.

One of the most obvious periodic patterns is the change from metals to non-metals as you go across the periods. The zig-zag line across the p-block in Figure 3 marks the change. The elements to the left of the zig-zag line display properties associated with metals, such as electrical conductivity. Elements to the right of the zig-zag line non-metals, displaying properties associated with them, for example, non-conductors of electricity. Other physical properties also show periodicity.

Melting points and boiling points

When elements are melted or boiled, the bonds between the individual atoms or molecules – the intermolecular forces – must be overcome. The strength of these bonds influences whether an element has a high or low melting point or boiling point.

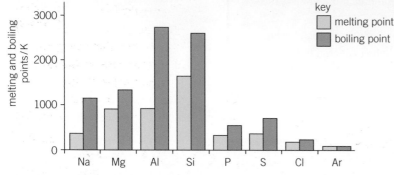

key
■ melting point
■ boiling point

In Period 3 (sodium to argon), both the melting and boiling points initially increase as you go across the period, and then fall dramatically from silicon, Si, to phosphorus, P (Figure 4). This means that the bonds between the particles in phosphorus must be much weaker and easier to overcome than those between the particles in silicon. Similar patterns in melting and boiling points are observed for the elements in other periods.

▲ Figure 4 *Melting and boiling points of elements in Period 3*

Electronic structure

The arrangement of elements by rows and columns in the periodic table is a direct result of the electronic structure of atoms. The number of outer shell electrons determines the group number. For example, lithium, sodium, and potassium all have one outer shell electron and are in Group 1. The noble gases all have filled outer shells – they have eight electrons in their outer shells, except helium which has two. They are in Group 0 (the zero refers to the shell beyond the filled outer one, which does not have any electrons in it yet).

The number of the shell that is being filled determines the period an element belongs to. From lithium to neon, the outer electrons are being placed in shell two, so these elements belong to Period 2.

Chemical properties

The chemical properties of an element are decided by the electrons in the incomplete outer shells – these are the electrons that are involved in chemical reaction.

Compare the electronic arrangement of the noble gases:

He	$1s^2$	Ar	$1s^2 2s^2 2p^6 3s^2 3p^6$
Ne	$1s^2 2s^2 2p^6$	Kr	$1s^2 2s^2 2p^6 3s^2 3p^6 3d^{10} 4s^2 4p^6$

All have sub-shells fully occupied by electrons. Such arrangements are called **closed shell** arrangements. These are particularly stable arrangements.

Now compare the electronic configurations of the elements in Group 1. They all have one electron in the outermost s-sub-shell and as a result show similar chemical properties. Group 2 elements all have two electrons in the outermost s-sub-shell. Groups 1 and 2 elements are known as s-block elements.

In Groups 3, 4, 5, 6, 7, and 0 the outermost p-sub-shell is being filled. These elements are known as p-block elements. The elements where a d-sub-shell is being filled are called d-block elements, and those where an f-sub-shell is being filled are called f-block elements (Figure 3). Dividing it up in this way is very useful to chemists since it groups together elements with similar electronic configurations and similar chemical properties.

Electronic configuration of s- and p-block ions

- Group 1 elements form +1 ions.
- Group 2 elements form +2 ions.
- Group 7 elements usually form −1 ions.
- Group 6 elements form −2 ions.
- Aluminium forms +3 ions.

Knowing this, you can write the electronic configurations of ions formed by the s- and p-block elements in the first four periods.

 Worked example: Electronic configuration of an anion

What is the electronic configuration of S^{2-}?

Step 1: Write out the electronic configuration of a sulfur atom.

$$1s^2 2s^2 2p^6 3s^2 3p^4$$

Step 2: Use the charge on the ion to decide how many electrons to add to give the required charge.

S^{2-} has a charge of −2 so need to add two electrons.

Step 3: Add the required number of electrons to the electronic configuration.

$$1s^2 2s^2 2p^6 3s^2 3p^6$$

 Worked example: Electronic configuration of a cation

Draw out the electronic configuration of Na^+, showing the sub-shells and atomic orbitals.

Step 1: Write out the electronic configuration of a sodium atom.

$$1s^2 \quad 2s^2 \quad 2p^6 \quad 3s^1$$

Step 2: Use the charge on the ion to decide how many electrons to take away to give the required charge.

Na^+ has a charge of +1 so need to take away one electron.

Step 3: Add the required number of electrons to the electronic configuration.

$$1s^2 \quad 2s^2 \quad 2p^6$$

Summary questions

1 The electron shell configuration for sodium can be written as 2.8.1. Use this notation to write down the electron shell configurations for:
   a lithium (*1 mark*)
   b phosphorus (*1 mark*)
   c calcium. (*1 mark*)

2 The electron shell configurations of unknown elements **A** to **E** are given below. Which of these elements are in the same group? (*2 marks*)

   A 2.8.2 **B** 2.6 **C** 2.8.8.2
   D 2.7 **E** 2.2

3 Copy and complete the following table that shows the electron shell configurations of some elements and the groups and periods to which they belong. (*3 marks*)

Electron shell configuration	Group	Period
2.8.7	7	3
2.3		
2.8.6		
	4	2
	4	3
2.1		
2.8.1		
2.8.8.1		

4 Classify the following elements as s-, p-, d-, or f-block elements.
   a $[Kr]5s^1$ (*1 mark*)
   b $1s^2 2s^2 2p^6 3s^2 3p^4$ (*1 mark*)
   c $[Ar]3d^{10}4s^2 4p^6$ (*1 mark*)
   d $[Xe]6s^2$ (*1 mark*)

5 Classify the following elements as s-, p-, d-, or f-block elements.
   a chromium (*1 mark*)
   b aluminium (*1 mark*)
   c uranium (*1 mark*)
   d strontium (*1 mark*)

EL 5 The molecules of life

Specification reference: EL(i), EL(j), EL(k)

Learning outcomes

Demonstrate and apply knowledge and understanding of:

→ chemical bonding in terms of electrostatic forces – simple electron dot-and-cross diagrams to describe the electron arrangements in ions and covalent and dative covalent bonds

→ the bonding in simple molecular structure types; the typical physical properties (melting point, solubility in water, electrical conductivity) characteristic of these structure types

→ use of the electron pair repulsion principle, based on dot-and-cross diagrams, to predict, explain, and name the shapes of simple molecules (such as $BeCl_2$, BF_3, CH_4, NH_3, H_2O, and SF_6) and ions (such as NH_4^+) with up to six outer pairs of electrons (any combination of bonding pairs and lone pairs) and assigning bond angles to these structures.

Humans are not made up of single atoms but rather molecules and some ions. What are the molecules of life and how did they come into existence?

Cold chemistry and the molecules of life

Although hydrogen is the most common element in space, its atoms are spread out. There is about one atom per cubic centimetre (cm^3) in the space between the stars, so there is almost no chance that hydrogen atoms will come together to form hydrogen molecules.

However, dense gas clouds, or molecular gas clouds, do exist between stars and may contain between 100 and 1×10^6 particles per cm^3. The gas clouds are made up of a mixture of atoms and molecules, mainly of hydrogen, together with the dust of solid material from the breakup of old stars. They have been detected by radio and infrared telescopes on Earth and by spectroscopic instruments carried by rockets.

Table 1 shows some of the chemical species found in these gas clouds. Many of the substances contain carbon atoms bonded to elements other than just oxygen – they are **organic species**. These elements are major constituents of the human body.

Monatomic (1 atom)	Diatomic (2 atoms)	Triatomic (3 atoms)	Tetra-atomic (4 atoms)	Penta-atomic (5 atoms)
C^+	H_2	H_2O	NH_3	HCOOH
Ca^{2+}	OH	H_2S	H_2CO	NH_2CN
H^+	CO	HCN	HNCO	HC_3N
	CN	HNC	HNCS	C_4H
	CS	SO_2	C_3N	CH_2NH
	NS	OCS		CH_4
	SO	N_2H^+		
	SiO	HCS^+		
	SiS	HCO^+		
	C_2	NaOH		
	CH^+			
	NO			

Hexa-atomic (6 atoms)	Hepta-atomic (7 atoms)	Octa-atomic (8 atoms)	Nona-atomic (9 atoms)	Others
CH_3OH	CH_3CHO	$HCOOCH_3$	CH_3CH_2OH	HC_9N
NH_2CHO	CH_3NH_2		CH_3OCH_3	$HC_{11}N$
CH_3CN	H_2CCHCN		CH_3CH_2CN	
CH_3SH	CH_3C_2H		HC_7N	
CH_2CCH				

▶ Table 1 Some chemical species in the dense gas clouds

Where did the molecules of life come from?

In 1950 the scientist Stanley Miller put methane, ammonia, carbon dioxide, and water – simple molecules like those present in the dense gas clouds – into a flask. He heated and subjected the mixture to an electrical discharge to simulate the effect of lightning. Miller found that some of the reaction mixture had been converted into amino acids – the building blocks of proteins. Proteins are a group of compounds needed for correct cell functioning. This is one of the experiments that suggests life on Earth could have originated from the molecules in the dense gas clouds in outer space.

Analysis of dust, collected from a comet, showed the presence of polycyclic aromatic hydrocarbons as well as organic compounds rich in oxygen and nitrogen. These molecules are of interest to astrobiologists as these compounds could play important roles in terrestrial biochemistry. Temperatures in gas clouds are too low for chemical reactions to occur naturally, but ultraviolet light may penetrate the clouds and provide the energy to break covalent bonds. The simple molecules may then go on to form larger molecules. The evidence is building but the jury is still out.

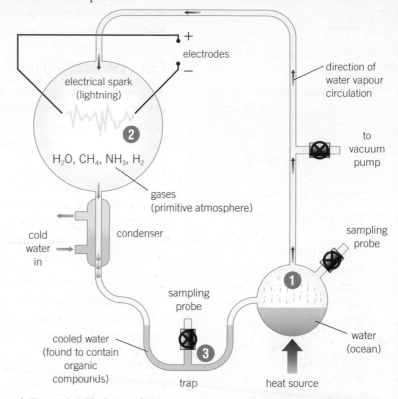

▲ Figure 1 *Miller's experiment*

> **Study tip**
>
> Hydrocarbons are molecules made of carbon and hydrogen.

Chemical ideas: Bonding, shapes, and sizes 3.1a

Chemical bonding

Covalent bonding

When scientists realised that the noble gases all have eight electrons in their outer shells (with the exception of helium) they linked this to the chemical stability of the noble gases. They suggested that other elements achieve eight outer shell electrons by losing or gaining electrons during reactions to form compounds. Furthermore, it seemed that some light elements achieve stability by reaching the helium configuration of two outer shell electrons. This was the basis for early ideas about why elements combine to form compounds. It is still useful today – though it is not the whole story.

> **Synoptic link**
>
> You will find out about ions and ionic bonding in Topic FL 7, Blood, sweat, and seas.

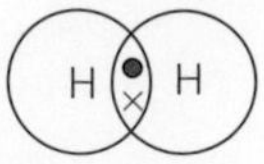

▲ **Figure 2** *Electron sharing in the hydrogen molecule, H$_2$*

Dot-and-cross diagram

Diagrams used to represent the way that atoms bond together. In these diagrams, the outer shell electrons of one atom are represented by dots, with crosses for the other.

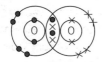

▲ **Figure 4** *Dot-and-cross diagram for oxygen, O$_2$*

Study tip

The dot-and-cross model is an over-simplified model for bonding but it is nevertheless useful. It needs to be extended before it can do more than explain simple situations. Don't be surprised if dot-and-cross diagrams sometimes don't work, or if some molecules seem to break the rules.

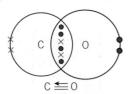

▲ **Figure 7** *Dative covalent bonding in carbon monoxide, CO*

Covalent bonds

When non-metallic elements react it is not energetically favourable to form ions – electrons are *shared* between the atoms of the elements in these compounds. This sharing of electrons is called **covalent bonding**. The resulting compound is more stable than the individual elements. Shared electrons count as part of the outer shell of *both* atoms in the bond. Dot-and-cross diagrams can be used to show the bonding in covalent compounds (Figure 2).

The dot-and-cross diagrams show how sharing electrons gives the atoms more stable electron structures, like noble gases. Electron pairs that form bonds are called bonding pairs. Pairs of electrons not involved in bonding are called **lone pairs**. Ammonia, NH$_3$, has one lone pair of electrons and water, H$_2$O, has two lone pairs of electrons (Figure 3).

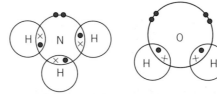

▲ **Figure 3** *Dot-and-cross diagrams for ammonia, NH$_3$, and water, H$_2$O*

When two pairs of electrons form a covalent bond, it is called a *double bond*. The bonds in molecular oxygen are double covalent bonds (Figure 4).

When three pairs of electrons form a bond, it is called a *triple bond*. The bonds in nitrogen, N$_2$, and between carbon and nitrogen in hydrogen cyanide, HCN, are examples (Figure 5).

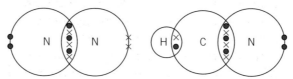

▲ **Figure 5** *Dot–and-cross diagrams for nitrogen, N$_2$, and hydrogen cyanide, HCN*

Dot-and-cross diagrams are useful for representing individual electrons in a chemical bond. A simpler way of drawing molecules is by using lines to represent a pair of electrons shared between two atoms. A single line represents a single covalent bond, whilst double and triple lines represent double and triple covalent bonds (Figure 6).

H — H   O = C = O   H — C ≡ N
hydrogen   *carbon dioxide*   *hydrogen cyanide*

O = O   N ≡ N
oxygen   *nitrogen*

▲ **Figure 6** *Covalently bonded molecules*

Dative covalent bonds

Carbon monoxide, CO, involves a triple bond. Two of the pairs of electrons are formed by the carbon and oxygen atoms each contributing one electron to the pair – these are ordinary covalent bonds. However both electrons in the third pair come from the oxygen atom – this is called a **dative covalent bond**. In a dative bond, *both* bonding electrons come from the *same* atom. A dative covalent bond is shown by an arrow, with the arrow pointing away from the atom that donates the pair of electrons (Figure 7).

Physical properties of covalently bonded simple molecules

The covalent intramolecular bonds *within* simple molecules such as water and ammonia are strong. However, the electrostatic attractions between different simple molecules are weak. This means relatively small amounts of energy are needed to separate one molecule from another. Therefore, elements and compounds with a simple molecular structure have relatively low melting and boiling points. Because there are no charged particles, covalent simple molecules do not conduct electricity. They mostly do not dissolve readily in water.

Chemical ideas: Bonding, shapes, and sizes 3.2

The shapes of molecules

Simple, covalently bonded molecules have a specific shape that can be determined using a theory called electron pair repulsion theory.

Electron pair repulsions

In methane, CH_4, there are *four groups* of electrons around the central carbon atom – the four covalent bonding pairs. Because similar charges repel, these groups of electrons arrange themselves so that they are as far apart as possible.

The farthest apart they can get is when the H—C—H bond angles are 109.5°. This corresponds to a tetrahedral shape, with the hydrogen atoms at the corners and the carbon atom in the centre. When drawing three-dimensional structures, bonds that lie in the plane of the paper are drawn as solid lines. Bonds that come towards you are represented by solid, wedged-shaped lines. Bonds that go away from you are shown as dashed wedges (Figure 8).

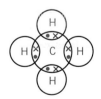

▲ Figure 8 *A dot-and-cross diagram showing the covalent bonding in methane (left), tetrahedral methane (middle), and a tetrahedron (right)*

All carbon atoms that are surrounded by four single bonds have a tetrahedral distribution of bonds, for example, in ethane both carbons have a tetrahedral arrangement (Figure 9). The same ideas apply to ions, for example, the ammonium ion, NH_4^+, is tetrahedral (Figure 10).

Lone pairs count too

Molecules such as ammonia and water have lone pairs in the outer shells of nitrogen and oxygen (Figure 3). These lone pairs repel

Synoptic link

Particles of simple molecular and lattice structures are covered in Topic EL 7, Blood, sweat, and seas.

Study tip

A group of electrons can be a single covalent bond (two electrons), a double covalent bond (four electrons), a triple covalent bond (six electrons), or a lone pair of electrons not involved in bonding.

Activity EL 5.1

Balloons can be used to model the formation of different molecular shapes. Try them out with this activity.

look along this bond

▲ Figure 9 *The tetrahedral bonding around the carbon atoms in ethane*

▲ Figure 10 *The tetrahedral arrangement of the ammonium ion*

Activity EL 5.2

In this activity you can predict the shapes of molecules and ions using dot-and-cross diagrams.

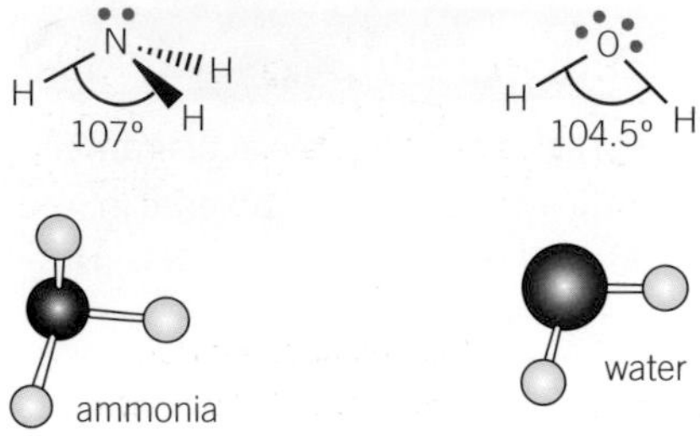

▲ Figure 11 *The shapes of molecules of ammonia (left) and water (right)*

the bonding pairs of electrons, as in methane. The electron pairs in ammonia and water adopt a tetrahedral shape, but one or more of the corners are occupied by lone pairs of electrons.

The H—N—H and H—O—H bond angles are both close to 109° (Table 2). Ammonia is said to have a *pyramidal* shape and water is described as *bent* (Figure 11).

The shapes of covalent molecules are decided by a simple rule – groups of electrons in the outer shell repel one another and move as far apart as possible. It doesn't matter if they are groups of bonding electrons or lone pairs – they all repel one another. Lone pairs of electrons repel more strongly than electrons involved in bonding. This explains the different bond angles observed in methane, ammonia, and water even though all three molecules have four groups of electrons around a central atom (Table 2).

▼ Table 2 *Summary of bond angles for compounds with different numbers of covalent bonds and lone pairs of electrons*

Number of groups of electrons	Types of groups	Example	Bond angle
4	4 × single covalent bonds	methane	109.5°
4	3 × single covalent bonds 1 × lone pair of electrons	ammonia	107.0°
4	2 × single covalent bonds 2 × lone pair of electrons	water	104.5°

Other shapes

There are several other types of shapes that molecule can have.

Linear molecules

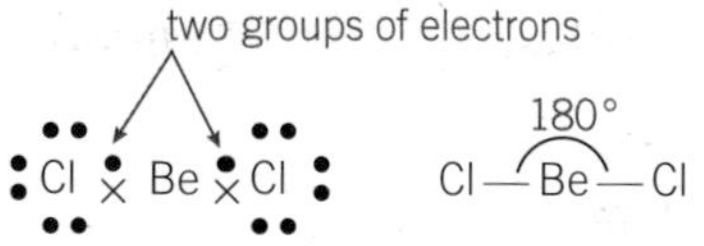

two groups of electrons

$$:Cl \overset{\times}{\bullet} Be \overset{\times}{\bullet} Cl: \qquad Cl—\overset{180°}{Be}—Cl$$

▲ Figure 12 *The shape of the BeCl₂ molecule*

In $BeCl_2$, there are two groups of electrons around the central atom (Figure 12). Because there are fewer groups of electrons than in methane, they can get further apart. The furthest apart they can get is at an angle of 180°, so $BeCl_2$ is linear with a Cl—Be—Cl angle of 180°.

$$O=C=O \qquad H—C\equiv C—H$$

▲ Figure 13 *Carbon dioxide(left) and ethyne (right) both have linear molecules*

Carbon dioxide and ethyne are other examples of linear molecules – there are only two groups of electrons around the central atoms in these molecules (Figure 13).

Planar molecules

In BF_3 there are *three*, not four, groups of electrons around the central atom. The F—B—F bond angle is 120°. BF_3 is flat and shaped like a triangle with the boron atom at the centre and the fluorine atoms at the corners (Figure 14). It is described as planar triangular.

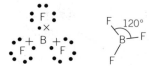

▲ Figure 14 *The shape of the BF₃ molecule*

Methanal and ethane are other examples of planar structures (Figure 15). In methanal there are three groups of electrons around the carbon atom, and in ethene *both* carbon atoms have three groups of electrons. Remember it is the number of *groups* of electrons that determine the shape – there are four pairs of electrons around the carbon atoms in these molecules, but the bonds are *not* tetrahedrally directed because there are only three *groups* of electrons.

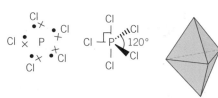

methanal

ethene

▲ **Figure 15** *Methanal and ethene have planar molecules*

Bipyramidal molecules

Five groups of electrons around a central atom give rise to a trigonal bipyramidal shape, with electrons at the five corners of the shape. An example of such a molecule is phosphorus pentachloride, PCl_5. Bond angles approximate to 120° or 90°, depending on their position within the molecule (Figure 16).

▲ **Figure 16** *The shape of the PCl_5 molecule. The lone pairs of electrons on chlorine have been omitted for clarity*

Octahedral molecules

Six groups of electrons around a central atom gives rise to an octahedral shape, where the electrons are directed to the six corners of an octahedron. Some molecules, such as SF_6, adopt this structure but it is more commonly found in the octahedral shapes of complexes of metal ions with six ligands (Figure 17).

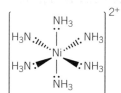

▲ **Figure 17** *Six groups of electrons around a central atom (left) or an ion (right) give an octahedral shape (because there are eight faces). The lone pairs of electrons on fluorine have been omitted for clarity*

Summary questions

1   Draw electron dot-and-cross diagrams for the following covalent substances.
   a   chlorine, Cl_2 *(1 mark)*
   b   hydrogen chloride, HCl *(1 mark)*
   c   methane, CH_4 *(1 mark)*
   d   hydrogen sulfide, H_2S *(1 mark)*
   e   aluminium bromide, $AlBr_3$ *(1 mark)*
   f   silicon chloride, $SiCl_4$ *(1 mark)*

2   Draw electron dot-and-cross diagrams for the following covalent substances:
   a   ethane *(1 mark)*
   b   ethyne *(1 mark)*
   c   methanol *(1 mark)*

3   Ammonia, NH_3, and boron trifluoride, BF_3, combine together to form the molecule NH_3BF_3.
   This molecule has a dative covalent bond between the nitrogen atom and the boron atom.
   Draw a dot-and-cross diagram for this molecule. *(1 mark)*

4   For each of the following molecules, draw an electron dot-and-cross diagram to show the bonding pairs and lone pairs of electrons in the molecule. Give the bond angles you would expect in each molecule.
   a SiH_4 *(1 mark)*
   b H_2S *(1 mark)*
   c PH_3 *(1 mark)*
   d CO_2 *(1 mark)*
   e SF_2 *(1 mark)*
   f BCl_3 *(1 mark)*
   g C_2H_2 *(1 mark)*

5   Draw a diagram to show the bonds and lone pairs of electrons in each of the following molecules. Write in the bond angles you would expect to find in each structure.
   a CH_3CH_3 *(2 marks)*
   b CH_3OH *(2 marks)*
   c CH_3NH_2 *(2 marks)*
   d $CH_2{=}CH_2$ *(2 marks)*
   e $CH_3C{\equiv}N$ *(2 marks)*
   f NH_2OH *(2 marks)*
   g $COCl_2$ *(2 marks)*

6   What shapes are the following molecules?
   a NF_3 *(1 mark)*
   b BF_3 *(1 mark)*
   c SF_6 *(1 mark)*

What are you made of? You could answer this biologically, and talk about organs and bones, or be more detailed and mention proteins, fats, and DNA. However, a chemist would talk about atoms and molecules or elements and compounds.

Elements and the body

A body is not really made up of a mixture of elements but rather a mixture of compounds, many of which appear quite complicated. The elements that make up these compounds are shown in Figure 1. You will be finding out more about some of these compounds later in this chapter. To begin with, however, you will look at the elements that are most likely to be in the compounds in your body.

Counting atoms of elements

Elements in the body are classified as major constituent elements, trace elements, and ultra-trace elements (Table 1). The ultra-trace elements are not given because their quantities are so small.

▼ **Table 1** *The major constituent elements in the human body*

Element	Mass in a 60 kg person / g	Percentage of atoms	Element classification
oxygen	38 800	25.50	major constituent
carbon	10 900	9.50	major constituent
hydrogen	5990	63.00	major constituent
nitrogen	1860	1.40	major constituent
calcium	1200	0.31	trace
phosphorus	650	0.22	trace
potassium	220	0.06	trace
sulfur	150	0.05	trace
chlorine	100	0.03	trace
sodium	70	0.03	trace
magnesium	20	0.01	trace

oxygen – 64.7%

potassium – 0.37%
hydrogen – 10%
phosphorus – 1.1%
chlorine – 0.17%
magnesium – 0.03%
calcium – 2.0%
nitrogen – 3.1%
iron – 0.006%
carbon – 18.2%

sulfur – 0.25%
sodium – 0.12%

▲ **Figure 1** *The elements a human body is made of (% is by mass)*

The data gives conflicting interpretations of the importance of different elements in the body. For example, there are more atoms of hydrogen in your body than atoms of any other element, but hydrogen contributes far less than carbon or oxygen to the mass of your body. To decide which category an element belongs to, scientists determine how many atoms of an element there are in a body.

The mass of an element can be converted into the number of atoms by using a unit called the mole. One mole is the amount of an element that contains the same number of atoms as 12 g of carbon. Because atoms have exceedingly small masses, the number of atoms in one mole of an element is large: 6×10^{23} atoms per mole. Once you know the total number of atoms in the body and the number of atoms of each element in a body, you can work out the percentage of each element in the body. The mole is an important concept in chemistry and can be applied to both elements and compounds.

Chemical ideas: Measuring amounts of substance 1.1

Amount of substance

Imagine a bag containing 10 golf balls and 20 table tennis balls. The golf balls would make up most of the mass, but the table tennis balls would make up most of the volume. The composition of water has a similar situation. Oxygen makes up nearly 90% of the mass of water, but $\frac{2}{3}$ of the atoms are hydrogen atoms. However, you cannot pick up and count the number of oxygen and hydrogen atoms in water.

The formula of water is H_2O – there are *two* hydrogen atoms combined with *one* oxygen atom in each water molecule. The relative atomic mass A_r of an element links the mass of an element with the number of atoms.

Relative atomic mass

The relative atomic mass of an element is the mass of its atom relative to carbon-12, which is assigned a relative atomic mass of exactly 12. A_r values have no units. The relative atomic masses of some elements are listed in Table 2.

▼ Table 2 *Some relative atomic masses*

Element	Symbol	Approximate relative atomic mass A_r
hydrogen	H	1
helium	He	4
carbon	C	12
nitrogen	N	14
oxygen	O	16
magnesium	Mg	24
sulfur	S	32
calcium	Ca	40
iron	Fe	56
copper	Cu	64
iodine	I	127
mercury	Hg	200

Study tip

In this topic the A_r values are rounded to whole numbers. You will usually use A_r values to one decimal place.

▲ **Figure 2** *Two carbon-12 atoms have the same mass as one magnesium atom. So the relative atomic mass of magnesium is 2 × 12 = 24*

Chemical quantities

If you had two bottles containing equal masses of copper (A_r = 64) and sulfur (A_r = 32) you would know that you had twice as many sulfur atoms as copper atoms because sulfur atoms have only half the mass of copper atoms. If you had a bottle containing mercury (A_r = 200) that was five times heavier than a similar bottle containing calcium (A_r = 40), you would know that both bottles contained equal numbers of atoms because each mercury atom has five times the mass of each calcium atom.

12 g of carbon, 1 g of hydrogen, and 16 g of oxygen all contain equal numbers of atoms because these masses are in the same ratio as their relative atomic masses. This amount of each of these elements is called a **mole**, or mol for short. The mole is a unit for measuring the amount of substance. One mole of a substance contains as many particles (e.g., atoms, molecules, groups of ions) as there are atoms in 12 g of carbon-12.

Chemical amounts are defined so that the mass of one mole (the **molar mass**) is equal to the relative atomic mass in grams. So, the molar mass of carbon is 12 g mol^{-1}. If you had 6 g of carbon you would have 0.5 mol of carbon atoms and 3 g of hydrogen would contain 3 mol of hydrogen atoms.

$$\text{amount in moles } n \text{ (mol)} = \frac{\text{mass } m \text{ (g)}}{\text{molar mass } M \text{ (g mol}^{-1})}$$

 Worked example: Calculating the amount of substance

Calculate the amount of substance in 4 g of oxygen.

Step 1: Identify the molar mass of oxygen. 16 g mol^{-1}

Step 2: Substitute the values into the equation to calculate the amount of substance, in moles.

$$\frac{4 \text{ g}}{16 \text{ g mol}^{-1}} = 0.25 \text{ mol}$$

Relative formula mass

You can use the mole to deal with compounds as well as elements. A molecule of methane, CH_4, is formed when one carbon atom combines with four hydrogen atoms. Therefore one mole of methane is formed when one mole of carbon atoms combines with four moles of hydrogen atoms.

The **relative formula mass** M_r (or relative molecular mass) of a substance is the sum of the relative atomic masses of the elements making it up. The M_r of a substance can be worked out by writing the chemical formula of the substance then adding together the relative atomic masses of each of the atoms in the formula.

 Worked example: Relative formula mass

Calculate the relative formula mass of methane.

Step 1: Write out the chemical formula of methane. CH_4

Step 2: Use a periodic table to identify the relative atomic mass of each of the elements.

$$A_r(C): 12 \qquad A_r(H): 1$$

Step 3: Add together the relative atomic mass for each atom in the chemical formula.

$$1\ C \qquad 4\ H$$
$$(1 \times 12) + (4 \times 1) = 16$$

Like relative atomic mass, relative formula masses have no units and are on the same scale – relative to $A_r(^{12}C) = 12.000$. When a substance is made of discrete molecules, such as methane, the relative formula mass is often called the relative molecular mass. This also has the symbol M_r.

Formula units

Substances are made up of formula units. They are the basic units or building blocks, and match the formulae of the substances. Formula units can be single atoms, molecules, or groups of ions.

For example, the formula unit in metal copper is simply a copper atom, and the formula unit in non-metal carbon is a carbon atom. The formula unit of most elements is a single atom so their relative formula mass is identical to their relative atomic mass. There are some exceptions – in oxygen gas, the formula unit is the O_2 molecule so the relative formula mass is *twice* the relative atomic mass of oxygen.

In many covalent compounds, the formula unit is a molecule, for example, in methane the formula unit is the CH_4 molecule. However in ionic compounds the formula unit is a *group of ions*. In calcium nitrate the formula unit (or repeating unit) is $Ca(NO_3)_2$ and contains a group of three ions – one Ca^{2+} and two NO_3^-. These groups of ions are not labelled with any special name and so you just use the general name formula unit when referring to ionic compounds.

Moles of formula units

The relative formula mass in grams is equal to the molar mass, so the molar mass of methane is 16 g and calcium nitrate is 164 g. This means you can calculate the amount in moles of formula units.

amount in moles of formula units = mass ÷ molar mass

So, 8 g of methane contains 0.5 mol of CH_4 formula units and 41 g of calcium nitrate contains 0.25 mol of $Ca(NO_3)_2$ formula units.

As well as describing 16 g of methane as consisting of one mole of formula units (or molecules) of CH_4, you can also say that it contains one mole of carbon atoms and $(1 \times 4 =)$ four moles of hydrogen atoms. Similarly, 164 g of calcium nitrate contains one mole of formula units of $Ca(NO_3)_2$ – it also contains one mole of Ca^{2+} two moles 2 mol of NO_3^- ions.

For elements that exist as diatomic gases, such as oxygen, you must be especially careful. 32 g of oxygen gas contains one mol of O_2 molecules, but two moles of oxygen atoms. It is therefore essential to give the formula unit you are referring to when using moles, for example, moles of oxygen atoms or moles of O_2 molecules.

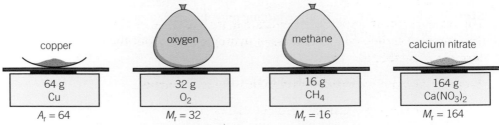

copper	oxygen	methane	calcium nitrate
64 g Cu	32 g O_2	16 g CH_4	164 g $Ca(NO_3)_2$
$A_r = 64$	$M_r = 32$	$M_r = 16$	$M_r = 164$

all contain the same number of formula units because their masses are in the same ratio as their relative formula masses

▲ **Figure 3** *The important thing to remember is that equimolar amounts of substances contain equal numbers of formula units*

The Avogadro constant

The number of formula units in one mole of a substance is a constant. It is called the Avogadro constant N_A, after the Italian scientist Amedeo Avogadro. The value of the Avogadro constant is 6.02×10^{23} formula units per mole. One mole of atoms, molecules, and electrons will all contain $6.02 \times 10^{23} \, mol^{-1}$. The number is so huge that it is difficult to comprehend (Figure 4).

To work with specific numbers of formula units, you only have to use moles – easily done by using molar mass and mass of substance.

Working out chemical formulae

Moles can be used to work out chemical formulae. If a known mass of magnesium is reacted with oxygen to form magnesium oxide, you can find out the mass of oxygen that combined with the magnesium.

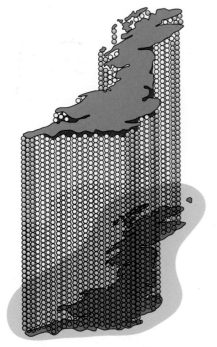

▲ **Figure 4** *If the 6.02×10^{23} atoms in 12 g of carbon were turned into marbles, the marbles could cover Great Britain to a depth of 1500 km*

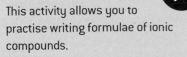

Activity EL 6.1

This activity allows you to practise writing formulae of ionic compounds.

 Worked example: Chemical formula using moles

In an experiment, 0.84 g of magnesium was burnt and combined with 0.56 g of oxygen. Using these results, work out the chemical formula of the magnesium oxide produced.

Step 1: Calculate the amount in moles of magnesium atoms in the reaction.
$$\frac{0.84 \, g}{24 \, g \, mol^{-1}} = 0.035 \, mol^{-1}$$

Step 2: Calculate the amount in moles of oxygen atoms in the reaction.
$$\frac{0.56 \, g}{16 \, g \, mol^{-1}} = 0.035 \, mol$$

Step 3: Calculate the ratio of moles of atoms of Mg : O in magnesium oxide. 1 : 1

Step 4: Write the formula of magnesium oxide. MgO

Another way of analysing a compound is to find the *percentage mass* of each element it contains.

 Worked example: Chemical formula using percentage mass

The results from an experiment indicate that methane contains 75% mass of carbon and 25% mass of hydrogen. Work out the formula of methane.

Step 1: Calculate the mass of carbon in 100 g of methane.

100 g of methane contains 75 g of carbon.

Step 2: Calculate the amount of carbon atoms in 75 g.

$$\frac{75\,g}{12\,g\,mol^{-1}} = 6.25\,mol$$

Step 3: Calculate the mass of hydrogen in 100 g of methane.

100 g methane contains 25 g of hydrogen.

Step 4: Calculate the amount of hydrogen atoms in 25 g.

$$\frac{25\,g}{1\,g\,mol^{-1}} = 25\,mol$$

Step 5: Calculate the ratio of moles of atoms of C : H in methane. 1 : 4

Step 6: Write the formula of methane. CH_4

In a methane molecule a central carbon atom is surrounded by four hydrogen atoms – the simple ratio of atoms is the same as the formula of the molecule. This isn't the case for all substances. The **molecular formula** tells you the actual numbers of different types of atom. The molecular formula of ethane is C_2H_6, but the simplest ratio for the moles of atoms of C : H is 1 : 3 – a calculation from percentage masses would give a formula CH_3. This is the **empirical formula**. Table 3 gives examples of the two types of formula.

▼ Table 3 *Some molecular formulae and empirical formulae*

Substance	Molecular formula	Empirical formula
ethene	C_2H_4	CH_2
benzene	C_6H_6	CH
butane	C_4H_{10}	C_2H_5
phosphorus(V) oxide	P_4O_{10}	P_2O_5
oxygen	O_2	O
bromine	Br_2	Br

Activity EL 6.2

This activity allows you to analyse Epsom salts, a hydrated salt.

▲ Figure 5 *When water is added to anhydrous copper(II) sulfate (white), hydrated copper(II) sulfate is formed (blue)*

Water of crystallisation

The crystals of some ionic lattices include molecules of water. These water molecules are fitted within the lattice in a regular manner, and are called water of crystallisation. The crystals are said to be hydrated, for example, $CuSO_4 \bullet 5H_2O$ is hydrated copper(II) sulfate and is blue coloured. When hydrated crystals are heated, the water of crystallisation is removed as steam leaving the anhydrous solid, for example, $CuSO_4$ is the anhydrous form of copper(II) sulfate and is white.

Calculations using compounds with a water of crystallisation

The formula of hydrated compounds can be determined using a simple experimental technique. The mass of a sample of the hydrated salt is measured. The salt is then heated, with the mass measured at regular intervals. When no more mass is lost, the formula can be determined.

Worked example: Calculating the formula of a hydrated ionic compounds

2.53 g of hydrated magnesium chloride, $MgCl_2 \bullet xH_2O$, was heated to constant mass. 1.17 g of solid remained. What is the formula of the hydrated compound?

Step 1: Calculate the mass of water in 2.53 g $MgCl_2$

$$2.53 - 1.17 = 1.36\,g$$

Step 2: Calculate the number of moles of water.

$$\frac{1.36}{18} = 0.076\ \text{moles}$$

Step 3: Calculate the number of moles of $MgCl_2$ in the anhydrous solid. ($M_r(MgCl_2) = 95.9$)

$$\frac{1.17}{95.9} = 0.012\ \text{moles}$$

Step 4: Calculate the ratio of moles of magnesium chloride to moles of water by dividing both vales by the number of moles of magnesium chloride.

$$\frac{0.012}{0.012} = 1 \qquad \frac{0.076}{0.012} = 6.3 \qquad 1:6.3$$

Step 5: Use the ratio from Step 4 to determine the formula of the hydrated magnesium chloride.

$$MgCl_2 \bullet 6H_2O$$

Working out percentage yield

The mole can also be used to calculate the expected amount of product in a reaction carried out under ideal conditions. This is called the *theoretical yield*. However, certain factors can reduce the amount of products produced:

- loss of products from reaction vessels, particularly if there are several stages to the reaction
- side-reactions occurring, producing unwanted by-products
- impurities in the reactants
- changes in temperature and pressure
- if the reaction is an equilibrium system.

The reduced amount of products (called the *experimental* yield) can be expressed as a percentage of the theoretical (maximum) yield. This is the percentage yield.

$$\text{percentage yield} = \frac{\text{experimental yield}}{\text{theoretical yield}} \times 100$$

The theoretical yield can be calculated from the balanced equation for the particular reaction.

Synoptic link

You will find out more about equilibrium systems in Chapter 3, Elements from the sea.

 Worked example: Calculating percentage yield

0.84 g of magnesium was burnt in excess oxygen, producing 1.1 g of magnesium oxide. Using these results, work out the percentage yield of the reaction.

Step 1: Write the balanced equation for the reaction.

$$2Mg(s) + O_2(g) \rightarrow 2MgO(s)$$

Step 2: Calculate the amount of substance, in moles, of magnesium.

$$\frac{0.84\,g}{24\,g\,mol^{-1}} = 0.035\,mol$$

Step 3: The balanced equation tells us that the ratio of Mg to MgO produced is 2:2 or as the lowest whole number ratio 1:1. This means the maximum amount of magnesium oxide that can be produced from this amount of magnesium is also 0.035 mol.

Step 4: Calculate the formula mass of MgO.

$$A_r(Mg) = 24; \; A_r(O) = 16 \quad M_r(MgO) = 24 + 16 = 40$$

Step 5: Calculate the maximum mass of MgO you can expect to produce.

$$0.035\,mol\;of\;MgO \quad so, \; 40 \times 0.035 = 1.4\,g$$

Step 6: The experiment produced 1.1 g. Calculate the percentage yield.

$$\frac{1.1\,g}{1.4\,g} \times 100 = 79\% \; (2\,s\,f)$$

Study tip

You will need to use a table of relative atomic masses or the periodic table to answer the questions that follow. From now on, you will always see A_r and M_r figures quoted to one decimal place.

In the example above you were given the necessary balanced chemical equation. However, there will be many times where you will need to write your own equation.

Chemical ideas: Measuring amounts of substance 1.2

Balanced equations

A balanced chemical equation tells you the reactants and products in a reaction, and the relative amounts involved. The equation is balanced so that there are equal numbers of each type of atom on both sides of the arrow.

$$CH_4(g) + 2O_2(g) \rightarrow CO_2(g) + 2H_2O_{(l)}$$

This equation tells you that one molecule of methane reacts with two molecules of oxygen to form one molecule of carbon dioxide and two molecules of water. These are also the amounts in moles of the substances involved in the reaction, that is, one mole of methane reacts with two moles of oxygen.

The number written in front of each formula in a balanced equation tells you the number of formula units involved in the reaction. Remember that a formula unit may be a molecule or another species such as an atom or an ion. The small subscript numbers are part of the formulae and cannot be changed.

Writing balanced equations

The only way to be sure of the balanced equation for a reaction is to do experiments to find out what is formed in the reaction and what quantities are involved. However, equations are used a lot in chemistry and it isn't possible to do experiments every time. Fortunately, if you know the reactants and products, usually the formulae can be worked out and a balanced equation predicted. Equations can only be balanced by putting numbers in front of the formulae. You cannot balance them by altering the formulae because that would create different substances.

> **Study tip**
>
> State symbols are included in chemical equations to show the physical state of the reactants and products.
>
> | State | Symbol |
> |---|---|
> | gas | (g) |
> | liquid | (l) |
> | solid | (s) |
> | aqueous solution | (aq) |

 Worked example: The steps for predicting balanced equations

What is the balanced equation for the reaction between calcium and water.

Step 1: Decide what the reactants and products are.

calcium + water → calcium hydroxide + hydrogen

Step 2: Write the formula for the substances involved, including state symbols.

$$Ca(s) + H_2O(l) \rightarrow Ca(OH)_2(aq) + H_2(g)$$

Step 3: Balance the equation so that there are the same numbers of each type of atom on each side.

$$Ca(s) + 2H_2O \rightarrow Ca(OH)_2(aq) + H_2(g)$$

Summary questions

1 One atom of element X is approximately 12 times heavier than one carbon atom.

  a What is the approximate relative atomic mass of this element? *(1 mark)*

  b Identify X. *(1 mark)*

2 Balance the following equations.

  a $Mg + O_2 \rightarrow MgO$ *(1 mark)*

  b $H_2 + O_2 \rightarrow H_2O$ *(1 mark)*

  c $CaCO_3 + HCl \rightarrow CaCl_2 + CO_2 + H_2O$ *(1 mark)*

  d $HCl + Ca(OH)_2 \rightarrow CaCl_2 + H_2O$ *(1 mark)*

  e $CH_3OH + O_2 \rightarrow CO_2 + H_2O$ *(1 mark)*

3 Write balanced equations, including state symbols for the following reactions.

  a zinc reacting with sulfuric acid, H_2SO_4, to form zinc sulfate, $ZnSO_4$, and hydrogen. *(2 marks)*

  b magnesium carbonate, $MgCO_3$, decomposing on heating to form magnesium oxide, MgO, and carbon dioxide. *(2 marks)*

  c barium oxide, BaO, reacting with hydrochloric acid to form barium chloride, $BaCl_2$, and water. *(2 marks)*

4 The empirical formula of a compound can be calculated when you know the masses of the elements in a sample of it.

The steps to calculate the empirical formula of a compound if a 16.7 g sample of it contains 12.7 g of iodine and 4.0 g of oxygen are shown in Table 1.

  a Why do we need to know the mass of the sample as well as the masses of the elements in it? *(1 mark)*

  b In step 2, what do the numbers 0.1 and 0.25 represent? *(1 mark)*

  c Why, in step 3, do we divide both 0.1 and 0.25 by 0.1? *(1 mark)*

  d Why have we doubled the numbers in moving from step 3 to step 4? *(1 mark)*

  e Write down three possibilities for the molecular formula of this compound based on its empirical formula. *(3 marks)*

  f What additional information do you need to work out the actual molecular formula of the compound? *(1 mark)*

5 How many moles of atoms are contained in:

  a 32.1 g of sulfur *(1 mark)*

  b 31.8 g of copper *(1 mark)*

  c Explain why they are different, even though each question involves approximately 32 g of an element. *(2 marks)*

6 Calculate how many moles of each substance are contained in

  a 88 g of carbon dioxide *(1 mark)*

  b 2.92 g of sulfur hexafluoride, SF_6 *(1 mark)*

  c 0.37 kg of calcium hydroxide, $Ca(OH)_2$ *(2 marks)*

  d 18 tonnes of water (1 tonne = 1×10^6 g) *(2 marks)*

▼ Table 1 *Steps involved in working out an empirical formula. Units have been omitted*

	Iodine	Oxygen
Step 1	12.7	4.0
Step 2	$\dfrac{12.7}{126.9} = 0.1$	$\dfrac{4.0}{16.0} = 0.25$
Step 3	$\dfrac{0.1}{0.1} = 1$	$\dfrac{0.25}{0.1} = 2.5$
Step 4	2	5
Step 5	The empirical formula of the compound is I_2O_5	

Learning outcomes

Demonstrate and apply knowledge and understanding of:

→ balanced ionic chemical equations, including state symbols

→ the bonding in giant lattice (metallic, ionic, covalent network and simple molecular) structure types; the typical physical properties (melting point, solubility in water, electrical conductivity) characteristic of these structure types

→ structures of compounds that have a sodium chloride type lattice

→ the relationship between the position of an element in the s- or p-block of the Periodic Table and the charge on its ion; the names and formulae of NO_3^-, SO_4^{2-}, CO_3^{2-}, OH^-, NH_4^+, HCO_3^-, Cu^{2+}, Zn^{2+}, Pb^{2+}, Fe^{2+}, Fe^{3+}; formulae and names for compounds formed between these ions and other given anions and cations

→ the solubility of compounds formed between the following cations and anions: Li^+, Na^+, K^+, Ca^{2+}, Ba^{2+}, Cu^{2+}, Fe^{2+}, Fe^{3+}, Ag^+, Pb^{2+}, Zn^{2+}, Al^{3+}, NH_4^+, CO_3^{2-}, SO_4^{2-}, Cl^-, Br^-, I^-, OH^-, NO_3^-; colours of any precipitates formed; use of these ions as tests, for example, Ba^{2+} as a test for SO_4^{2-}; a sequence of tests leading to the identification of a salt containing the ions above

→ techniques and procedures for making soluble salts by reacting acids and bases and insoluble salts by precipitation reactions.

Other than water, what connects blood, sweat, and seas? An important link is ions.

Salts

Sea water contains dissolved salts. Salts are ionic compounds and many of the ions found present in sea water are important for a healthy body. For example, sodium and potassium ions are present in both blood and sea water, but in *very* different amounts (Table 1). Na^+ ions are lost in sweat and these must be replaced. Natural sea salt sometimes contains iodide ions, I^-, which are also needed for a healthy life. However, evidence shows that too much salt (in everyday language, salt usually means sodium chloride) can be harmful.

Salts get into our seas and rivers by dissolving out of rocks. Occasionally shallow seas form in hot climates, particularly over geological time, and evaporation of the water leaves large deposits of solid salts, called evaporites (Figure 1).

▼ **Table 1** *Concentration of sodium and potassium ions in blood and sea water*

Substance	Concentration / mmol dm^{-3}	
	Na$^+$	K$^+$
sea water	470	10
blood cells	10	150

Salts of other metallic elements are also important in the body. Compounds of calcium are vital for healthy bones and teeth and these salts are insoluble. Calcium ions are precipitated from sea water as calcium carbonate, and make the shells and skeletons of many marine animals (Figure 2).

▲ **Figure 1** *Lithium salt mining in Atacama desert, Chile*

▲ **Figure 2** *Calcareous fossils called foraminifera*

Formation of salts

A consequence of the increase in atmospheric carbon dioxide is that more of the gas dissolves in sea water. This lowers the pH of the sea and could result in the calcium carbonate shells of marine animals dissolving, with disastrous consequences to the marine ecosystem.

The reaction of carbonate ions with acids to form a salt is an example of a neutralisation reaction. Table 2 gives some examples of other salts and their formation and uses.

▼ Table 2 *Some salts and their uses*

Salt	Use	Possible method (with reactants) to make salt in the lab		Formula of salt
		Reactant 1	Reactant 2	
magnesium sulfate	bath salts 	sulfuric acid acid	magnesium metal metal	$MgSO_4$
lithium chloride	lithium batteries 	hydrochloric acid acid	lithium oxide metal oxide	$LiCl$
barium sulfate	'barium meal' shows up soft tissue on X-ray 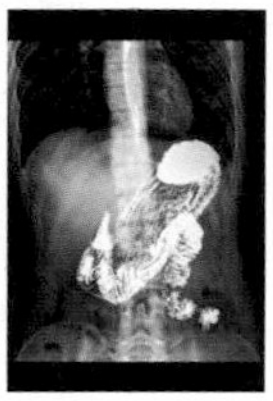	barium chloride $Ba^{2+}(aq)$	magnesium sulfate $SO_4^{2-}(aq)$	$BaSO_4$ this salt is insoluble and made by precipitation
sodium ethanoate	hand warmers 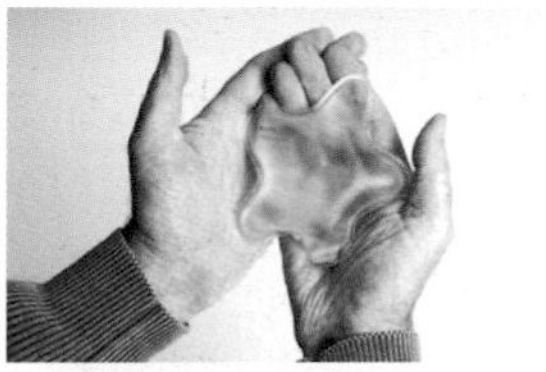	ethanoic acid acid	sodium carbonate carbonate	CH_3COONa
ammonium nitrate	fertiliser 	nitric acid acid	ammonium hydroxide (ammonia solution) hydroxide	NH_4NO_3

Acids

In almost all the methods in Table 2, one of the reactants is always an acid. Acids produce hydrogen ions, H^+, in solution. Some common acids are:

- sulfuric acid, H_2SO_4
- hydrochloric acid, HCl
- nitric acid, HNO_3
- ethanoic (or acetic) acid, CH_3COOH.

Chemical bonding

Formation of ions and electrostatic bonds

The elements in Groups 1 and 2 of the periodic table have only one or two outer shell electrons. They *lose* these electrons to form positively charged ions, called **cations**.

Most non-metal atoms have more than three outer shell electrons. They are able to *gain* electrons to form negatively charged ions, called **anions**.

Gaining one electron gives an anion with a single negative charge. A second electron is repelled by this anion because their charges are the same, so making a doubly charged anion is much harder. Getting a anion with a −3 charge is very difficult and does not often happen. It is also hard to remove three or more electrons from atoms – cations with a +4 charge are almost unknown.

- Group 1 elements form +1 ions
- Group 2 elements form +2 ions
- Group 6 elements form −2 ions
- Group 7 elements form −1 ions

Ionic bonding

When *metals* react with *non-metals* in a chemical reaction, ions are only formed if the *overall* energy change for the reaction is favourable. Electrons are *transferred* from the metal atoms to the non-metal atoms, often giving both the metal and the non-metal a stable electronic structure like that of a noble gas. The cations and anions formed are held together by their opposite charges in an **electrostatic bond**.

Figure 3 shows dot-and-cross diagrams for the formation of sodium chloride and magnesium fluoride. Each sodium atom loses one electron and each chlorine atom gains one electron, so the compound formed has the formula NaCl. Each magnesium atom loses two electrons but each fluorine atom gains only one electron, so the formula for magnesium fluoride is MgF_2.

Activity EL 7.1

In this activity you can check your understanding of why atoms form ions.

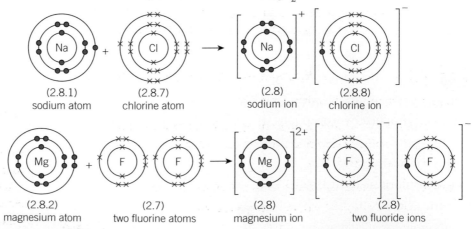

(2.8.1)
sodium atom

(2.8.7)
chlorine atom

(2.8)
sodium ion

(2.8.8)
chlorine ion

(2.8.2)
magnesium atom

(2.7)
two fluorine atoms

(2.8)
magnesium ion

(2.8)
two fluoride ions

▲ Figure 3 *Dot-and-cross diagrams for sodium chloride (top) and magnesium fluoride (bottom). The numbers show the arrangement of electrons in shells and only the outer electrons are shown*

The oppositely charged ions attract each other strongly in an ionic bond. In the solid compound, each ion attracts many others of opposite charge and the ions build up into a giant lattice (Figure 3).

Writing chemical formulae of ionic compounds

Table 3 lists some common cations and anions.

▼ Table 3 *The names and formulae of some common ions. For elements where there is more than one common ion, the oxidation state is given*

Cations (positive ions)			Anions (negative ions)	
+1	+2	+3	−1	−2
H^+ hydrogen	Mg^{2+} magnesium	Al^{3+} aluminum	F^- fluoride	O^{2-} oxide
Li^+ lithium	Ca^{2+} calcium	Fe^{3+} iron(III)	Cl^- chloride	CO_3^{2-} carbonate
Na^+ sodium	Ba^{2+} barium		Br^- bromide	SO_4^{2-} sulfate
K^+ potassium	Fe^{2+} iron(II)		I^- iodide	
NH_4^+ ammonium	Cu^{2+} copper(II)		OH^- hydroxide	
H^+ hydrogen	Zn^{2+} zinc		NO_3^- nitrate(V) or nitrate	
	Pb^{2+} lead(II)		HCO_3^- hydrogen carbonate	

Some ions contain more than one type of atom, such as hydroxide, OH^-, and sulfate, SO_4^{2-}. These are complex ions and consist of a group of atoms held together by covalent bonds, but the whole group carries an electric charge. The dot-and-cross diagram for sodium hydroxide is shown in Figure 4.

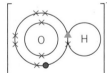

▲ Figure 4 *Dot-and-cross diagram for sodium hydroxide*

Once you know the formulae of the ions in an ionic compound, it is easy to write the formula of the compound. You need to make sure that the total positive charge is equal to the total negative charge – when you add all the charges on the ions, they add up to zero.

 Worked example: Writing formulae for ionic compounds

Write the formula for sodium chloride.

Step 1: Identify the atoms and charges a compound is made of – sodium chloride is made up of sodium ions, Na^+, and chloride ions, Cl^-.

Step 2: Work out the formula so the charges balance.

$(+1) + (-1) = 0$, so the formula is NaCl.

Write the formula for chromium(III) hydroxide.

Step 1: Identify the atoms and charges a compound is made of – chromium(III) hydroxide is made up of chromium(III) ions (Cr^{3+}) and hydroxide ions (OH^-).

Step 2: Work out the formula so the charges balance.

To balance the +3 charge you need three −1 charges. The formula is $Cr(OH)_3$.

Making ionic salts

The salts in Table 2 are of ionic compounds. There are several ways of making salts. You should already be familiar with these:

- acid + alkali → salt + water
- acid + base → salt + water
- acid + carbonate → salt + water + carbon dioxide
- acid + metal → salt + hydrogen

Chemical ideas: Structure and properties 5.1

Ionic substances in solution

Many ionic substances dissolve readily in water. There are however some ionic compounds that are not soluble and these include:

- barium, calcium, lead, and silver sulfates
- silver and lead halides (chlorides, bromides, and iodides)
- all metal carbonates
- metal hydroxides (except Group 1 hydroxides and ammonium hydroxide).

When ionic substances dissolve, the ions become surrounded by water molecules and spread out through the solution. Once they are separated they behave independently of each other. This presence of hydrated ions in solution explains why aqueous solutions of salts can conduct electricity.

Ionic equations

When ions are in solution, the reactions of an ionic substance, such as sodium chloride, quite often involve only one of the two types of ion – the other ion does not get involved in the reaction. For example, if you add a solution of sodium chloride to silver nitrate solution, you get a white precipitate of silver chloride. Silver ions, Ag^+, and chloride ions, Cl^-, have come together to form insoluble silver chloride, which precipitates out. You can write an equation for this reaction showing only the ions that take part in the reaction.

$$Ag^+(aq) + Cl^-(aq) \rightarrow AgCl(s)$$

Synoptic link

You will find out more about neutralisation in EL 9, How salty?

Synoptic link

You can find out more about making soluble and insoluble salts in Techniques and procedures.

▲ **Figure 5** *Lead(II) chromate – also known as the pigment chrome yellow – precipitating from mixing lead nitrate and potassium chromate solutions*

Study tip

It is worth remembering that all Group 1 compounds, ammonium compounds, and all nitrates are soluble in water.

Activity EL 7.2

This activity allows you to apply your understanding of precipitation reactions.

The $Na^+(aq)$ and $NO_3^-(aq)$ ions are not involved in the reaction so can be left out of the equation – they are **spectator ions**. This is an **ionic equation** – it only shows the ions that take part in the reaction and excludes the spectator ions.

State symbols are very important in ionic equations. They help to identify that the above reaction involves ionic **precipitation** – a suspension of solid particles is produced by a chemical reaction in solution.

Sometimes precipitation reactions can be used to identify certain metal cations in solution (Table 5).

▼ Table 4 *State symbols*

State	State symbol
gas	(g)
liquid	(l)
solid	(s)
aqueous solution	(aq)

▼ Table 5 *Precipitation tests for some cations and anions*

Cation or anion being tested for	Solution added	Precipitate formed	Colour of precipitate	Overall ionic equation
Cu^{2+}	sodium hydroxide	copper hydroxide	blue	$Cu^{2+}(aq) + 2OH^-(aq) \rightarrow Cu(OH)_2(s)$
Fe^{2+}	sodium hydroxide	iron(II) hydroxide	'dirty' green	$Fe^{2+}(aq) + 2OH^-(aq) \rightarrow Fe(OH)_2(s)$
Fe^{3+}	sodium hydroxide	iron(III) hydroxide	orange/brown	$Fe^{3+}(aq) + 3OH^-(aq) \rightarrow Fe(OH)_3(s)$
Pb^{2+}	potassium iodide	lead iodide	bright yellow	$Pb^{2+}(aq) + 2I^-(aq) \rightarrow PbI_2(s)$
Cl^-	silver nitrate	silver chloride	white	$Ag^+(aq) + Cl^-(aq) \rightarrow AgCl(s)$
Br^-	silver nitrate	silver bromide	cream	$Ag^+(aq) + Br^-(aq) \rightarrow AgBr(s)$
I^-	silver nitrate	silver iodide	pale yellow	$Ag^+(aq) + I^-(aq) \rightarrow AgI(s)$
SO_4^{2-}	barium chloride	barium sulfate	white	$Ba^{2+}(aq) + SO_4^{2-}(aq) \rightarrow BaSO_4(s)$

Precipitation reactions have lots of practical applications, including water treatment, production of coloured pigments for paints and dyes, and in the identification of certain metal ions in solution.

Chemical ideas: Structure and properties 5.2

Bonding, structure, and properties

Ionic bonding

Ionic compounds are typically solids at room temperature and pressure and have lattice structures consisting of repeating positive and negative ions in all three dimensions. Because of this regular arrangement of ions, ionic compounds often form regularly shaped crystals (Figure 6).

chloride ion, Cl^-

sodium ion, Na^+

▲ Figure 6 *The sodium chloride lattice, built up from oppositely charged sodium ions and chloride ions*

The electrostatic attraction of oppositely charged ions overcomes any repulsion between ions of the same charge, holding together the ionic lattice structures (including those of hydrated salts). This is called ionic bonding. The electrostatic attractions are strong. This means a large amount of energy is needed to pull the ions apart. This is the reason why ionic compounds have high melting points. Once melted the ions are free to move and this is why molten ionic compounds conduct electricity.

Metallic bonding

Metals also have a lattice structure. One simple model that accounts for some of the most important characteristics of metals is the electron-sea model. In this model the metal is pictured as a giant lattice structure of metal cations in a 'sea' of delocalised valence electrons. These electrons are free to move, so account for the flow of electricity in metals. The whole structure is held together by the attraction of the metal cations to the delocalised electrons (Figure 7). These electrostatic attractions are strong, so account for the relatively high melting and boiling points of metals.

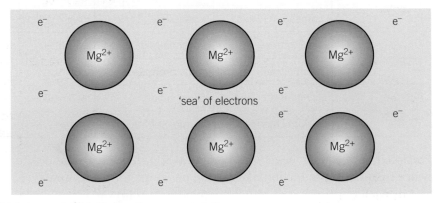

▲ Figure 7 *Metallic bonding – a sea of electrons*

Covalent networks

Some elements and compounds, such as carbon in the form of diamond (Figure 8) and silicon dioxide found in quartz, form large covalently bonded networks. These have strong covalent bonds between the atoms in the network. Because it takes a lot of energy to break these intramolecular electrostatic attractions, these networks have very high melting and boiling points. These networks are also insoluble in water and do not conduct electricity (graphite is an exception, due to its unique structure).

▲ Figure 8 *The structure of diamond. Each circle represents a carbon atom*

Summary of bonding, structure, and properties

How does the type of the structure and bonding in a substance affect its properties? Table 6 summarises the similarities and differences, including covalent bonding in simple molecules.

▼ Table 6 *A summary of the various types of structures and bonding*

	Giant lattice			Covalent molecular	
	Ionic	Covalent network	Metallic	Simple molecular	Macromolecular
What substances have this type of structure?	compounds of metals with non-metals	some elements in Group 4 and some of their compounds	metals	some non-metal elements and some non-metal/ non-metal compounds	polymers
Examples	sodium chloride, NaCl, and calcium oxide, CaO	diamond and graphite (both C) and silica, SiO_2	sodium, Na, copper, Cu, and iron, Fe	carbon dioxide, CO_2, chlorine, Cl_2, and water, H_2O	poly(ethene), nylon, proteins, DNA
What type of particles does it contain?	ions	atoms	positive ions surrounded by delocalised electrons	small molecules	long-chain molecules
How are the particles bonded together?	strong ionic bonds; attraction between oppositely charged ions	strong covalent bonds – attraction of atoms' nuclei for shared electrons	strong metallic bonds – attraction of atoms' nuclei for delocalised electrons Figure 7	weak intermolecular bonds between molecules – strong covalent bonds between the atoms within molecules	weak intermolecular bonds between molecules – strong covalent bonds between the atoms within molecules
What are the typical properties? Melting point and boiling point	high	very high	generally high (except mercury)	low	moderate – often decompose on heating
Hardness	hard but brittle	very hard (if three-dimensional)	hard but malleable (except mercury)	soft	variable; many are soft but often flexible
Electrical conductivity	electrolytes conduct when molten or dissolved in water	do not normally conduct (except graphite)	conduct when solid or liquid	do not conduct	do not normally conduct
Solubility in water	often soluble	insoluble	insoluble (but some react)	usually insoluble, unless molecules contain groups which can hydrogen bond with water	usually insoluble
Solubility in non-polar solvents (e.g., hexane)	insoluble	insoluble	insoluble	usually soluble	sometimes soluble

Activity EL 7.3

In this activity you can check your understanding of the links between properties and structure.

Summary questions

1 Draw electron dot-and-cross diagrams for the following ionic compounds.
   a lithium hydride, LiH   (2 marks)
   b potassium fluoride, KF   (2 marks)
   c magnesium oxide, MgO   (2 marks)
   d calcium sulfide, CaS   (2 marks)

2 Which type of structure would you expect each of these substances to have?
   a A white solid which starts to soften at 200 °C and can be drawn into fibres.   (1 mark)
   b A white solid which melts at −190 °C.   (1 mark)
   c A white solid which melts at 770 °C and conducts electricity when molten, but not in the solid state.   (1 mark)

3 Draw electron dot-and-cross diagrams for the following ionic compounds.
   a calcium chloride, $CaCl_2$ (2 marks)
   b sodium sulfide, Na_2S (2 marks)

4 An ammonia molecule, NH_3, forms a dative bond with a hydrogen ion, H^+, to produce an ammonium ion, NH_4^+. The other three hydrogen atoms are held to the nitrogen atom by conventional covalent bonds.
   a What is the essential difference between a dative bond and a covalent bond?   (2 marks)
   b Draw dot-and-cross diagrams for the ammonia molecule and the ammonium ion.   (2 marks)

5 Which type of structure would you expect each of the following substances to have?
   a A hard grey solid which conducts electricity and melts at 3410 °C.   (1 mark)
   b A liquid which conducts electricity and solidifies at −39 °C.   (1 mark)

6 Draw electron dot-and-cross diagrams for the following ionic compounds.
   a sodium nitride, Na_3N (2 marks)
   b aluminium fluoride, AlF_3 (2 marks)

EL 8 Spectacular metals

Specification reference: EL(p), EL(q), EL(r), EL(u)

The metals of Groups 1 and 2 are the most reactive metals in the periodic table. Some of their reactions can be impressive (Figure 1).

Uses of Group 1 and 2 metals

Group 1 and 2 metals have some interesting uses. The **alloy** of the Group 1 metals sodium and potassium remains liquid at room temperature and is used in the nuclear industry as a heat transfer fluid.

The Group 2 metal beryllium is used in mirrors in satellites and magnesium uses range from pencil sharpeners to parts of car engines. The *compounds* of Group 2 find even more extensive use in our everyday lives (Table 1).

▼ Table 1 *Uses of Group 2 compounds*

Group 2 compound	Property	Use
beryllium oxide, BeO	high melting point	nose-cones of rockets
magnesium hydroxide, $Mg(OH)_2$	basic oxide	some indigestion tablets
calcium carbonate, $CaCO_3$	insoluble and reacts with acids	agriculture to neutralise acid soils
calcium sulfate, hydrated $CaSO_4$	insoluble	plaster to keep broken limbs in place
strontium carbonate, $SrCO_3$	red colour when heated	fireworks
barium sulfate, $BaSO_4$	opaque to X-rays and insoluble	barium meal to allow X-raying of the digestive tract
radium chloride, $RaCl_2$	radioactive	treat certain cancers

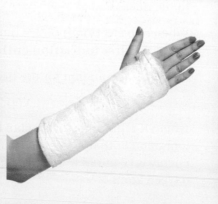

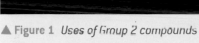

▲ Figure 1 *Uses of Group 2 compounds*

The s-block: Groups 1 and 2

The s-block contains two groups of reactive metals. Group 1 metals (Figure 3) are also called the alkali metals. Group 2 metals (Figure 4) are also called the alkaline earth metals. Groups 1 and 2 display two trends visible in other groups in the periodic table.

- Elements become *more* metallic *down* a group, for example, they more readily form cations in ionic compounds. For this reason, the most reactive metals in Groups 1 and 2 are found at the bottom of each group.

- Elements become *less* metallic *across* a period from left to right. For this reason, the Group 1 metals are more reactive than the Group 2 metals in the same period.

The Group 1 and 2 metals are not as widely used (in elemental form) as the familiar metals of the d-block, such as iron, copper, and chromium. The s-block metals tend to be soft, weak metals with low melting points, and the metals themselves are too reactive with water and oxygen to have many uses. However, the *compounds* of the s-block elements are very important (for examples, see Table 1).

Chemical reactivity

As Group 1 and 2 metals are all very reactive, they are never found naturally in their elemental form. However, compounds of s-block metals are very common throughout nature.

Like all groups in the periodic table, Groups 1 and 2 show patterns of reactivity down the group. There are *similarities* between the reactions of the elements within a group, and *differences* that show up as patterns, or trends. The *similarities* happen because the elements in a particular group all have similar arrangements of electrons in their atoms. The *differences* happen because the size of the atom increases down the group.

First ionisation enthalpy

If sufficient energy is given to an atom, an electron is lost and the atom becomes a positive ion – ionisation has taken place. An input of energy is *always* needed to remove electrons because they are attracted to the nucleus.

When one electron is pulled out of an atom, the energy required is called the **first ionisation enthalpy** (or first ionisation energy).

The general equation for the first ionisation process is

$$X(g) \rightarrow X^+(g) + e^-$$

where X represents the symbol for the element. For oxygen, the first ionisation enthalpy is $+1320\,kJ\,mol^{-1}$. This means that $1320\,kJ$ of energy is required to remove one electron from one mole of gaseous oxygen atoms.

$$O(g) \rightarrow O^+(g) + e^-$$

▲ **Figure 2** *Caesium reacting with water– an indicator has been added to the water to show how the solution becomes alkaline*

6.9 Li lithium 3	9.0 Be beryllium 4
23.0 Na sodium 11	24.3 Mg magnesium 12
39.1 K potassium 19	40.1 Ca calcium 20
85.5 Rb rubidium 37	87.6 Sr strontium 38
132.9 Cs caesium 55	137.3 Ba barium 56
[223] Fr francium 87	[226] Ra radium 88

▲ **Figure 3** *The elements of Group 1*   ▲ **Figure 4** *The elements of Group 2*

First ionisation enthalpy

The first ionisation enthalpy of an element is the energy needed to remove one electron from every atom in one mole of isolated gaseous atoms of the element – a mole of gaseous ions with one positive charge are formed.

The first ionisation removes the most loosely held electron. This will be one of the outer shell electrons since they are furthest from the nucleus.

Figure 5 shows the first ionisation enthalpy for elements 1–56. The elements at the peaks are all in Group 0 (the noble gases). These elements have high first ionisation enthalpies – they are difficult to ionise and are very unreactive. The elements at the troughs are all in Group 1 (the alkali metals). These elements, with only one outer shell electron, have low ionisation enthalpies – they are easy to ionise and are very reactive.

This pattern provides chemists with data to support the idea of electron shells, for example, the outermost filled electron shell for neon is n = 2. But the outermost electron in sodium is in the n = 3 shell. In sodium, there is more shielding of the outermost electrons from the nucleus and therefore less energy needed to remove an outer electron from sodium than from neon. Trends in Period 2 and Period 3 are similar for similar reasons.

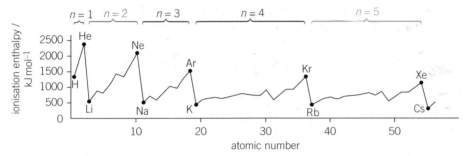

▲ **Figure 5** *Variation of first ionisation enthalpy for elements 1–56*

Figure 6 shows the first ionisation enthalpies for elements 1–20 in more detail. There is a general trend that as you go across a period in the periodic table it becomes more difficult to remove an electron. Across a period, electrons are being added to the outer shell but, at the same time, protons are being added to the nucleus. As the nuclear charge becomes more positive, the electrons will be held more tightly and so it gets harder to pull one from the outer shell. This is an example of periodicity.

Ionisation energies and electron shells

The pattern of first ionisation enthalpies can be used as evidence to support the existence of sub-shells as well as shells. Look at Figure 6 for Period 2 (n = 2). Although the general trend across the period is an increase, there is some variation.

Between beryllium and boron there is a decrease. The electronic configuration of beryllium is $1s^2 2s^2$ and boron is $1s^2 2s^2 2p^1$. The s-sub-shell is lower in energy than the p-sub-shell, therefore less energy is needed to remove the outer electron of boron, in spite of increased nuclear charge.

Between nitrogen and oxygen there is a decrease in first ionisation enthalpies.

N is 2p ☐ ☐ ☐   and   O is 2p ☐ ☐ ☐

    2s ☐   2s ☐

    1s ☐   1s ☐

The extra repulsion from the paired electron sub-shell in oxygen means less energy is needed to remove one of the paired electrons, despite the increased number of protons on going from nitrogen to oxygen.

The observed data (ionisation enthalpies) supports the theory of electron sub-shells.

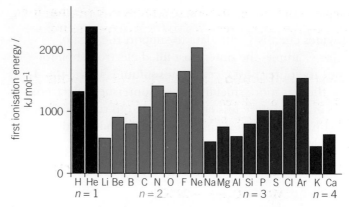

▲ Figure 6 *First ionisation enthalpies for elements 1–20*

Going down a group in the periodic table the first ionisation enthalpies *decrease*. This is because the attraction between the nucleus and the outermost electron decreases. There are more filled shells of electrons between the nucleus and the outermost electron and these shield the positively charged nucleus from the outermost electron, reducing the attraction the electron experiences. It is therefore easier for the outermost electron to be removed. The metals of Groups 1 and 2 react by losing their outer electrons, so the first ionisation energy trends also correspond to an increase in reactivity down the group.

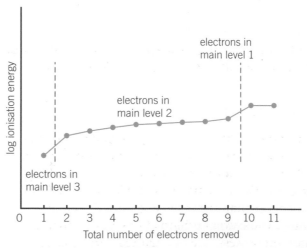

▲ Figure 7 *Successive ionisation enthalpies for sodium – a logarithmic scale is used for the ionisation enthalpy*

Successive ionisation enthalpies

More than one electron can be removed from an atom (except from a hydrogen atom, of course). So there are second, third, and fourth ionisation enthalpies, and so on. This is the energy required to

Activity EL 8.1

In this activity you will look at the reactions of Group 2 metals and their compounds.

remove further electrons. The general equations for these ionisation processes are:

first ionisation: $\quad Ca(g) \quad \rightarrow \quad Ca^+(g) + e^-$

second ionisation: $\quad Ca^+(g) \quad \rightarrow \quad Ca^{2+}(g) + e^-$

Each of the second and subsequent ionisation processes involves the removal of an electron from a *positive ion*, for example, the second ionisation enthalpy is the energy needed to remove one electron from an $X^+(g)$ ion. It is *not* the energy required to remove two electrons from an $X(g)$ atom.

Chemical properties of Group 2 elements and compounds

The elements of Group 2 are all reactive. They form compounds containing ions with a +2 charge, such as Mg^{2+} and Ca^{2+}.

Reactions with oxygen
All the Group 2 elements react with oxygen to produce the metal oxide.

$$2M(s) + O_2(g) \rightarrow 2MO(s)$$

Reactions with water
All the elements react with water to form hydroxides and hydrogen, with an increase in reactivity down the group. Magnesium (third period) reacts slowly, even when the water is heated. Barium (sixth period) reacts rapidly, giving a steady stream of hydrogen. The general equation, using M to represent a typical Group 2 metal reacting with water, is:

$$M(s) + 2H_2O(l) \rightarrow M(OH)_2(aq) + H_2(g)$$

The Group 2 metals do not react with water as vigorously as the Group 1 metals do.

Effect of heating carbonates
The general formula of Group 2 carbonates is MCO_3. When carbonates are heated, they decompose forming the oxide and releasing carbon dioxide.

$$MCO_3(s) \rightarrow MO(s) + CO_2(g)$$

The carbonates become more difficult to decompose down the group. For example, magnesium carbonate is easily decomposed by heating in a test tube over a Bunsen burner flame. Calcium carbonate needs much stronger heating directly in the flame before it will decompose. The **thermal stability** of calcium carbonate is greater than that of magnesium carbonate. The decomposition of calcium carbonate (limestone) is an important process, used to manufacture calcium oxide (quicklime).

The change in thermal stability down the group can be explained in terms of the **charge density** of the cations. Charge density is a measure of the concentration of charge on the cation. The smaller the +2 ion the higher the charge density. Cations with a higher charge density (those at the top of Group 2) can distort or **polarise** the negative charge cloud

▲ Figure 8 *Magnesium metal burning in air*

Activity EL 8.2
Here you will be able to investigation the decomposition of Group 2 carbonates.

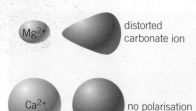

Mg²⁺ — distorted carbonate ion

Ca²⁺ — no polarisation

▲ Figure 9 *Effect of charge density on distortion of a carbonate ion*

around the carbonate ion making it less stable and easier to break up on heating (Figure 9).

Oxides and hydroxides

The general formula of the Group 2 oxides is MO, and that of the hydroxides is $M(OH)_2$. In water, the oxides and hydroxides form alkaline solutions, although they are not very soluble. Forming alkaline solutions is typical of metal oxides and hydroxides, in contrast to non-metals whose oxides are usually acidic. The most strongly alkaline oxides and hydroxides are those at the bottom of the group.

The oxides and hydroxides react with acids to form salts.

$$MO(s) + 2HCl(aq) \rightarrow MCl_2(aq) + H_2O(l)$$

$$M(OH)_2(s) + H_2SO_4(aq) \rightarrow MSO_4(aq) + 2H_2O(l)$$

This neutralising effect is used by farmers when they put lime (calcium hydroxide) on their fields to neutralise soil acidity.

Summary

The trends in properties of some compounds of Group 2 elements are summarised in Table 2.

▼ Table 2 *Summary of the reactivity trends of Group 2 compounds*

Element	Trend in reactivity with water	Trend in thermal stability of carbonate	Trend in pH of hydroxide in water	Trend in solubility of hydroxide	Trend in solubility of carbonate
Mg ↓ Ba ↓	increasing reactivity ↓	decomposes at increasingly higher temperature ↓	increasing pH ↓	increasing solubility ↓	decreasing solubility ↓

Summary questions

1 State the two factors that affect the charge density of an ion. (*2 marks*)

2 Write a word equation and a balanced chemical equation (with state symbols) for each of the following reactions.
   a the action of heated magnesium on steam (*2 marks*)
   b the neutralisation of hydrochloric acid with calcium oxide (*2 marks*)
   c the thermal decomposition of beryllium carbonate (*2 marks*)
   d the action of sulfuric acid on barium hydroxide. (*2 marks*)

3 The first, second, and third ionisation enthalpies of calcium are $+596 \, kJ \, mol^{-1}$, $+1160 \, kJ \, mol^{-1}$, and $+4930 \, kJ \, mol^{-1}$ respectively.
   a Write equations corresponding to each of these three ionisation enthalpies. (*3 marks*)
   b Explain why the second ionisation enthalpy of calcium is larger than its first ionisation enthalpy. (*2 marks*)
   c Explain why there is a very sharp rise between the second and third ionisation enthalpies of calcium. (*2 marks*)

EL 9 How salty?
Specification reference: EL(b), EL(c), EL(t)

You have already looked at the different types of salts, where they are found, and how they are made. But how can you find out how much salt is in the sea, in our blood, or our sweat?

Calculating quantities of salts

The mole and certain practical techniques can be used to calculate the amounts of products and reactants formed and used up in reactions.

Titrations are used to calculate quantities or concentrations of substances in solution. In titrations a solution of known concentration (a standard solution) is reacted with another solution of unknown concentration. Along with the balanced equation for the reaction, it is possible to calculate the concentration of the unknown solution.

There are many applications of titrations, from calculating the amount of acid in acid rain and determining the amount of active ingredient in a medicine, to determining the amount of a pollutant ion in waste from mining. In the industries that do lots of titrations, automatic titration systems are used. These enable one operator to set up and do many titrations in a short time.

> **Chemical ideas**: Measuring amount of substances 1.3

Using equations to work out reacting masses

A balanced chemical equation tells you the reactants and products in a reaction, and the relative amounts involved. The equation is balanced so that there are equal numbers of each type of atom on both sides. For example, in the combustion of methanol the theoretical equation is:

$$2CH_3OH(g) + 3O_2(g) \rightarrow 2CO_2(g) + 4H_2O(g)$$

The number written in front of each formula in a balanced equation tells you the number of formula units involved in the reaction. Remember that a formula unit may be a molecule or another species such as an atom or an ion. The small subscript numbers are part of the formulae and cannot be changed. Atoms are not created or destroyed in chemical reactions, they are simply rearranged, so equations must balance and the total mass on each side of the equation must always be the same. Chemists can use equations to work out the masses of reactants and products involved in a reaction, without having to do an experiment.

▲ Figure 1 *An automatic titration device being used to perform quality controls on a pharmaceutical product*

Study tip

The steps for working out reacting masses

Step 1: Write a balanced equation.

Step 2: In words, state what the equation tells you about the amount in moles of the substances you are interested in.

Step 3: Change amounts in moles to masses in grams.

Step 4: Scale the masses to the ones in the question.

 Worked example: Calculating masses from balanced equations

Calculate the mass of calcium oxide and carbon dioxide that would be produced if 100.0 g of calcium carbonate is heated.

Step 1: Write a balanced equation for the reaction.

$$CaCO_3(s) \rightarrow CaO(s) + CO_2(g)$$

Step 2: From the equation, work out the moles of reactants and moles of products.

1 mole of $CaCO_3$ produces 1 mole of CaO and 1 mole of CO_2.

Step 3: Calculate the molecular mass of the reactants and products from the atomic mass of the atoms that make up the substances.

$$A_r(Ca) = 40.0 \qquad A_r(C) = 12.0 \qquad A_r(O) = 16.0$$

$$M_r(CaCO_3) = 40.0\,g + 12.0\,g + (3 \times 16.0\,g) = 100.0\,g$$

$$M_r(CaO) = 40.0\,g + 16.0\,g = 56.0\,g$$

$$M_r(CO_2) = 12.0\,g + (2 \times 16.0\,g) = 44.0\,g$$

Step 4: Work out the masses of the products from the mass that was reacted

100.0 g of calcium carbonate gives 56.0 g of calcium oxide and 44.0 g carbon dioxide.

Chemical ideas: Acids and bases 8.1

Acids, bases, alkalis, and neutralisation

A key property of the Group 2 oxides and hydroxides is that they are bases – they react with acids to form salts.

What do we mean by acid and base?

Hydrochloric acid is an **acid** because of its properties. It turns litmus red, reacts with carbonates to give carbon dioxde, and is neutralised by bases – all properties that are expected of acids.

Chemists explain properties in terms of what goes on at the level of atoms, molecules, and ions. The properties of acids are due to the ability to transfer H^+ ions to something else. The substance which accepts the H^+ ion is called a **base**.

Take the reaction of hydrogen chloride with ammonia. The reaction forms a white salt, ammonium chloride

$$HCl(g) + NH_3(g) \rightarrow NH_4Cl(s)$$

Hydrochloric acid donates a proton, H^+, to the ammonia – it acts as an acid. The ammonia accepts a proton – it acts as a base. The general

Acid

A compound that dissociates in water to produce hydrogen ions.

Base

A compound that reacts with an acid – is a proton acceptor – to produce water (and a salt).

definition of an acid is a substance that donates H^+ in a chemical reaction. The substance that accepts the H^+ is a base. The reaction is an acid–base reaction. Since a hydrogen atom consists of only a proton and an electron, a H^+ ion corresponds to just one proton. Acids are sometimes referred to as *proton donors* and bases as *proton acceptors*. This theory of H^+ transfer is known as the Brønsted-Lowry theory of acids and bases.

Solutions of acids and bases

Hydrogen chloride is a gas containing HCl molecules. Water is almost totally made up of H_2O molecules. But hydrochloric acid, a solution of hydrogen chloride in water, readily conducts electricity so it must contain ions. A reaction between the hydrogen chloride molecules and the water molecules produces these ions.

$$HCl(aq) + H_2O(l) \rightarrow H_3O^+ (aq) + Cl^-(aq)$$
$$\text{acid} \qquad \text{base} \qquad \text{oxonium ion}$$

In this reaction, water behaves as a base. The H_3O^+ ion is called the oxonium ion. It is a very common ion and is present in every solution of an acid in water – it occurs in every *acidic* solution. The acid donates H^+ to H_2O to form H_3O^+ (Figure 2).

▲ Figure 2 *The bonding in the H_3O^+ ion. A lone pair on the oxygen atom forms a dative covalent bond to H^+. (Once the bond has formed it is indistinguishable from the two other O—H bonds)*

The H_3O^+ ion can itself act as an acid, donating H^+ and becoming an H_2O molecule. The familiar properties of acidic solutions are all properties of the H_3O^+ ion. You will often see the formula $H_3O^+(aq)$ shortened to $H^+(aq)$ and the dissociation of HCl(aq) into ions represented by

$$HCl(aq) \rightarrow H^+(aq) + Cl^-(aq)$$

When an acid dissolves in water, $H^+(aq)$ ions form in solution.

An **alkali** is a base that dissolves in water to produce hydroxide ions, $OH^-(aq)$ (Figure 3).

Some alkalis, such as sodium hydroxide and potassium hydroxide, already contain hydroxide ions, whereas others, such as sodium carbonate and ammonia, form $OH^-(aq)$ ions when they react with water.

$$CO_3^{2-}(aq) + H_2O(l) \rightleftharpoons HCO_3^-(aq) + OH^-(aq)$$

$$NH_3(aq) + H_2O(l) \rightleftharpoons NH_4^+(aq) + OH^-(aq)$$

When an alkali reacts with an acid a salt is formed. This is a **neutralisation** reaction. The reaction involves the hydrogen ions in the acidic solution and the hydroxide ions in the alkali, so the ionic equation for a neutralization reaction is:

$$H^+(aq) + OH^- \rightarrow H_2O(l)$$

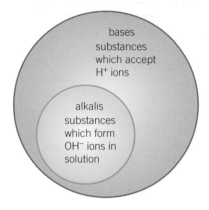

▲ Figure 3 *The relationship between alkalis and bases*

Alkali

A base that dissolves in water to produce hydroxide, OH^-, ions

Synoptic link

You will develop your understanding of acids in Chapter 8, Oceans.

Synoptic link

You can find out how to weigh solids and measure volumes of liquids in Techniques and procedures.

Activity EL 9.1

This activity gives you the opportunity to make a standard solution.

Synoptic link

You can find out more about these three practical skills in Techniques and procedures.

Activity EL 9.2

In this activity you will be able to find the concentration of an acid solution.

Activity EL 9.3

In this activity you will practise your titration technique and carry out accurate dilution of solutions.

Activity EL 9.4

In this activity you will make a compound that can be used in a fertiliser and calculate your percentage yield.

Study tip

You may see $mol\,dm^{-3}$ abbreviated to M. This abbreviation was once widely used, but now $mol\,dm^{-3}$ is used.

Ionic equations can also be written for insoluble salts formed by precipitation reactions.

$$Ba^{2+}(aq) + SO_4{}^{2-}(aq) \rightarrow BaSO_4(s)$$

Analysing acids and bases

It is often useful to know exactly when an acid and a base in solution have neutralised each other. From this, the amount of acid or base in the original solution can be calculated. By using a known concentration of one of the reactants – either acid or base – the exact reacting volume at neutralisation can be found out.

In order to successfully do this you will need to be able to do three things:

- dilute a solution
- make up a standard solution
- Carry out an acid-base titration.

The volume and concentration of the standard sodium hydroxide solution required for neutralisation can be used to calculate the concentration of the hydrochloric acid.

Chemical ideas: Measuring amounts of substance 1.4

Concentrations of solutions

Reactions are often carried out in solution. When using a solution of a substance, it is important to know how much of the substance is dissolved in a particular volume of solution. Concentrations are sometimes measured in $g\,dm^{-3}$. A solution containing 80 g of sodium hydroxide made up to $1\,dm^3$ of solution has a concentration of $80\,g\,dm^{-3}$.

However, chemists usually prefer to measure quantities in moles rather than in grams, because working in moles tells gives the number of particles present. The preferred unit for measuring concentration is $mol\,dm^{-3}$. To convert $g\,cm^{-3}$ to $mol\,dm^{-3}$, you need the molar mass of the substance involved. For example, the molar mass of sodium hydroxide, NaOH, is $40.0\,g\,mol^{-1}$, so a solution containing $80\,g\,dm^{-3}$ has a concentration of:

$$\frac{80.0\,g\,dm^{-3}}{40.0\,g\,mol^{-1}} = 2.0\,mol\,dm^{-3}$$

In general:

$$\text{concentration } (mol\,dm^{-3}) = \frac{\text{concentration } (g\,dm^{-3})}{\text{molar mass } (g\,mol^{-1})}$$

When a solution is made, its concentration will depend on:

- the amount of solute
- the final volume of the solution (Figure 6).

With the concentration of a solution, you can work out the amount of solute in a particular volume. In general:

amount (mol) = concentration of solution ($mol\,dm^{-3}$) × volume of solution

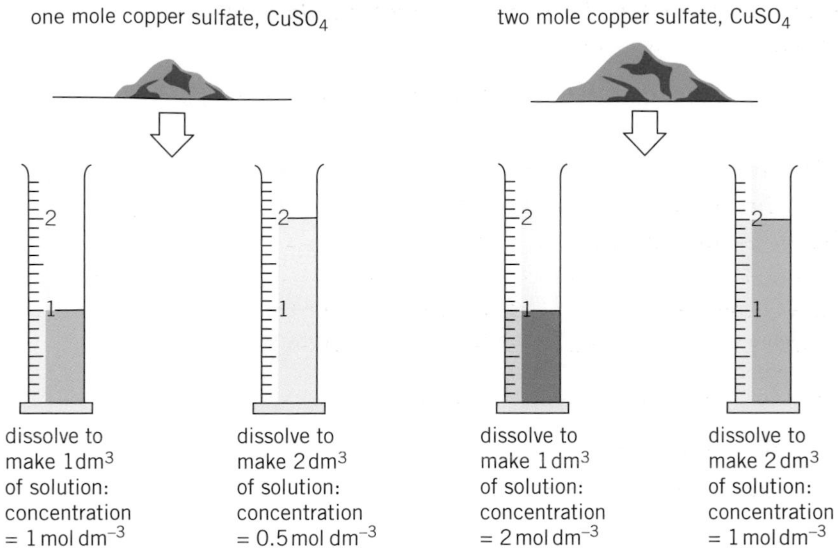

one mole copper sulfate, $CuSO_4$

two mole copper sulfate, $CuSO_4$

dissolve to make $1\,dm^3$ of solution: concentration = $1\,mol\,dm^{-3}$

dissolve to make $2\,dm^3$ of solution: concentration = $0.5\,mol\,dm^{-3}$

dissolve to make $1\,dm^3$ of solution: concentration = $2\,mol\,dm^{-3}$

dissolve to make $2\,dm^3$ of solution: concentration = $1\,mol\,dm^{-3}$

▲ **Figure 4** *The concentration of a solution depends on the amount of solute and the final volume of the solution*

Worked example: The amount of a solute in a solution

Calculate how many moles of sodium hydroxide are in a $250\,cm^3$ solution of sodium hydroxide with a concentration of $2\,mol\,dm^{-3}$.

Step 1: Convert $250\,cm^3$ to dm^3.

$$\frac{250}{1\,000} = 0.25\,dm^3$$

Step 2: Calculate the moles in $0.25\,dm^3$ of solution.

$$2\,mol\,dm^{-3} \times 0.25\,dm^3 = 0.5\,mol\text{ of sodium hydroxide}$$

Using concentrations in calculations

When carrying out a chemical reaction in solution, and you know the equation for the reaction, you can use the concentrations of the reacting solutions to predict the volumes you will need.

$$\text{concentration } c \text{ (mol\,dm}^{-3}) = \frac{\text{amount } n \text{ (moles)}}{\text{volume } V \text{ (dm}^3)} \qquad \textbf{Equation 1}$$

This equation can be rearranged easily using Figure 7. The horizontal line represent dividing n by c or V (Equation 1 and 2). The vertical lines represents multiplying c and V (Equation 3).

$$\text{volume } V \text{ (dm}^3) = \frac{\text{amount } n \text{ (moles)}}{\text{concentration } c \text{ (mol\,dm}^{-3})} \qquad \textbf{Equation 2}$$

$$\text{amount } n \text{ (moles)} = \text{concentration } c \text{ (mol\,dm}^{-3}) \times \text{volume } V \text{ (dm}^3) \qquad \textbf{Equation 3}$$

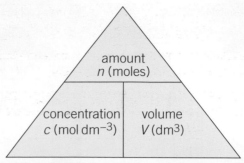

▲ Figure 5 *A memory aid showing the relationship between concentration, the amount, and the volume of solution*

In general, the steps for working out the reacting volumes of solutions are as follows:

1   Write a balanced equation.

2   Write down what the equation tells you about the amount in moles of the substances you are interested in.

3   Use the known concentrations of the solutions to change amounts in moles to volumes of solutions.

4   Scale the volumes of solutions to the ones in the question.

 Worked example: Predicting volumes from chemical equations

Consider the reaction of sodium hydroxide with hydrochloric acid. The concentrations of both solutions are $2.0\,mol\,dm^{-3}$. If you have $0.25\,dm^3$ of sodium hydroxide solution, what volume of hydrochloric acid would be needed to neutralise it?

Step 1: Write the equation for the reaction.

$$NaOH(aq) + HCl(aq) \rightarrow NaCl(aq) + H_2O(l)$$

Step 2: Calculate the moles of sodium hydroxide.

$$2.0\,mol\,dm^{-3} \times 0.25\,dm^3 = 0.50\,mol$$

Step 3: Calculate the amount of hydrochloric acid needed.

From the equation, one mole of NaOH reacts with one mole of HCl, therefore $0.50\,mol$ of HCl needed.

Step 4: Calculate the volume of hydrochloric acid needed.

$$\text{volume of HCl needed} = \frac{0.50\,mol}{2.0\,mol\,dm^{-3}} = 0.25\,dm^3$$

Summary questions

1 Complete the following table that shows the concentrations of ions in water from a sample of Dead Sea water. *(7 marks)*

Ion	Concentration / g dm^{-3}	Concentration / mol dm^{-3}
Cl$^-$	183.0	
Mg^{2+}	36.2	
Na$^+$		1.37
Ca^{2+}		0.355
K$^+$	6.8	
Br$^-$	5.2	
SO$_4{}^{2-}$		0.00625

2 Calculate the mass of solute needed to make up the following solutions.
  a 1 dm^3 of a 2 mol dm^{-3} solution of NaCl *(2 marks)*
  b 250 cm^3 of a 0.1 mol dm^{-3} solution of KMnO$_4$ *(2 marks)*
  c 50 cm^3 of a 0.5 mol dm^{-3} solution of KOH *(2 marks)*
  d 15 dm^3 of a 2 mol dm^{-3} solution of Pb(NO$_3$)$_2$ *(2 marks)*
  e 10 cm^3 of a 0.01 mol dm^{-3} solution of LiOH *(2 marks)*

3 In a titration, 25.00 cm^3 of a sodium hydroxide solution were pipetted into a conical flask. A 0.10 mol dm^{-3} solution of sulfuric acid was run from a burette into the flask. An indicator in the flask changed colour at an average of 22.00 cm^3 of the acid added.

$$H_2SO_4(aq) + 2NaOH(aq) \longrightarrow Na_2SO_4(aq) + 2H_2O(l)$$

  a What is the concentration of the sodium hydroxide solution in mol dm^{-3}? *(2 marks)*
  b The student washes the conical flask with water between titrations and does not dry it. Explain the effect, if any, on the titre of the next titration. *(1 mark)*

4 A standard solution of sodium hydroxide cannot be made by direct weighing of the solid.
  a Explain the meaning of the term *standard solution*. *(1 mark)*
  b Suggest why sodium hydroxide cannot be made into a standard solution by direct weighing of the solid. *(2 marks)*

Practice questions

1   How many hydrogen atoms are there in a mole of methanol, CH_3OH?

   A 4   **C** 1.8×10^{24}

   B 6×10^{23} **D** 2.4×10^{24} (*1 mark*)

2   What mass of aluminium is contained in 5.10 g of aluminium oxide?

   A 1.35 g   **C** 2.70 g

   B 4.26 g   **D** 3.66 g   (*1 mark*)

3   Which is the correct method for making a pure dry sample of barium sulfate?

   A Mix barium carbonate with sulfuric acid, filter, and evaporate the filtrate.

   B Mix barium chloride solution with sodium sulfate solution. Filter the precipitate and allow it to dry.

   C Mix barium nitrate solution with sulfuric acid. Filter, wash and dry the precipitate.

   D Add barium to sulfuric acid. Evaporate the solution.   (*1 mark*)

4   The change in structure and bonding in the elements across a period of the periodic table follows the following pattern:

   A metallic to giant covalent to small molecules

   B metallic to ionic to giant covalent

   C ionic to giant covalent to small molecules

   D giant covalent to small molecules to ionic.   (*1 mark*)

5   In the Geiger and Marsden experiment, the scientists fired alpha particles at a thin piece of gold foil. Which of the following happened to the alpha particles?

   A Most of the particles were deflected.

   B Some particles came almost straight back.

   C The particles were mostly absorbed by the foil.

   D The foil emitted neutrons.   (*1 mark*)

6   Which row of the table below contains two correct formulae for ionic substances?

A	$CaOH_2$	$CaSO_4$
B	$Fe_2(SO_4)_3$	$Fe(NO_3)_2$
C	AIN	AIS
D	KCO_3	KNO_3

   (*1 mark*)

7   Which is a correct equation for a reaction from the chemistry of barium?

   A $BaCl_2(aq) + 2H_2O(l) \rightarrow$
   $$Ba(OH)_2(aq) + 2HCl(aq)$$

   B $Ba(s) + H_2O(l) \rightarrow BaO(s) + H_2(g)$

   C $Ba(OH)_2(s) + 2HCl(aq) \rightarrow$
   $$BaCl_2(aq) + 2H_2O(l)$$

   D $BaCO_3(s) + H_2SO_4(aq) \rightarrow$
   $$BaSO_4(aq) + CO_2(g) + H_2O(l)$$
   (*1 mark*)

8   Which of the following is/are true about atomic emission spectra?

   1 They consist of bright lines on a dark background.

   2 They are caused by electrons being excited to higher energy levels.

   3 The lines get closer at higher wavelength.

   A 1, 2, and 3 correct

   B 1 and 2 are correct

   C 2 and 3 are correct

   D Only 1 is correct   (*1 mark*)

9   Which of the following is/are true about a sample of solid sodium chloride?

   1 There are equal numbers of sodium and chloride ions.

   2 The ions are arranged in a lattice.

   3 The ions cannot move through the lattice.

   A 1, 2, and 3 correct

   B 1 and 2 are correct

   C 2 and 3 are correct

   D Only 1 is correct   (*1 mark*)

10   Which of the following is/are reasons why the bond angle in NH_3 is smaller than the bond angle in CH_4?

   1 Lone pair – lone pair repulsion is greater than lone pair – bond pair repulsion.

   2 The hydrogen atoms repel each other less in ammonia.

   3 Methane has more lone pairs than ammonia.

   A 1, 2, and 3 correct

   B 1 and 2 are correct

   C 2 and 3 are correct

   D Only 1 is correct   (*1 mark*)

11 Helium is made in the Sun by reactions such as that shown below:

$$^2_1H + ^3_1H \rightarrow ^4_2He + ^1_0n$$

a   Name this type of reaction that involves light nuclei joining to form heavier ones.   (*1 mark*)

b   Explain why ^2_1H and ^3_1H are described as *isotopes*. Give their similarities and differences in terms of nuclear particles.   (*2 marks*)

c   Characteristic lines in a star's absorption spectrum show the presence of hydrogen gas in the gas surrounding the star.

   (i)   Describe the appearance of an absorption spectrum.   (*1 mark*)

   (ii)   Why do hydrogen atoms give an absorption spectrum whereas hydrogen nuclei do not?   (*1 mark*)

   (iii)   Draw a diagram of the energy levels in a hydrogen atom. Draw arrows on this diagram to show the origin of two lines in the hydrogen absorption spectrum.   (*2 marks*)

d   Using a dot-and-cross diagram for a hydrogen molecule, explain how the atoms are held together.   (*2 marks*)

e   (i)   Write the equation to represent the first ionisation enthalpy of hydrogen.   (*1 mark*)

   (ii)   Explain how the value of this first ionisation enthalpy would compare with that of lithium.   (*2 marks*)

12 Calcium carbonate rocks give off carbon dioxide when strongly heated. This can dissolve in water from underground springs making it fizzy. The fizzy water is sometimes sold as 'naturally carbonated spring water'.

a   Write the equation for the decomposition of calcium carbonate, showing state symbols.   (*1 mark*)

b   Calcium hydroxide can be made from one of the products of the reaction. Describe tests that could be done to identify each of the ions in aqueous calcium hydroxide.   (*2 marks*)

c   (i)   Calculate the volume of carbon dioxide that is given off when 0.35 g of calcium carbonate is fully decomposed. (*2 marks*)

   (ii)   Calculate the mass of barium carbonate that would be needed to produce the same volume of gas.   (*1 mark*)

   (ii)   Draw a diagram of an apparatus you could use to check your answers to (i) and (ii). Explain how you could use your apparatus to show that barium carbonate was more thermally stable than calcium carbonate.   (*6 marks*)

   (iii)   Explain in terms of the ions involved, why barium carbonate is more thermally stable than calcium carbonate.   (*2 marks*)

d   Draw a dot-and-cross diagram for carbon dioxide and use it to explain the shape and bond angle of the molecule.   (*3 marks*)

13 Some students set out to make pure copper sulfate crystals from copper carbonate and sulfuric acid.

$$CuCO_3(s) + H_2SO_4(aq) \rightarrow$$
$$CuSO_4(aq) + CO_2(g) + H_2O(l)$$

a   They plan to use the following methods:

   Student A: React excess copper carbonate with sulfuric acid. Evaporate and allow to crystallise.

   Student B: React excess sulfuric acid with copper carbonate. Filter, wash, and dry.

   Student C: React excess copper carbonate with sulfuric acid. Filter. Evaporate the filtrate to dryness.

   (i)   None of these students would end up with pure crystals of copper sulfate? In each case, say what *would* be left at the end.   (*3 marks*)

   (ii)   Outline the method that should be used.   (*2 marks*)

   (iii)   Calculate the mass of copper carbonate that would react exactly with 20 cm^3 of 2.0 mol dm^{-3} sulfuric acid.   (*2 marks*)

   (iv)   How could the students test for the presence of sulfate in their crystals once they have made them? (*2 marks*)

b   Draw a dot-and-cross diagram for the carbonate ion, CO$_3^{2-}$, using the minus symbol for the extra two electrons. Give the bond angle in the ion.   (*2 marks*)

Chapter 1 Summary

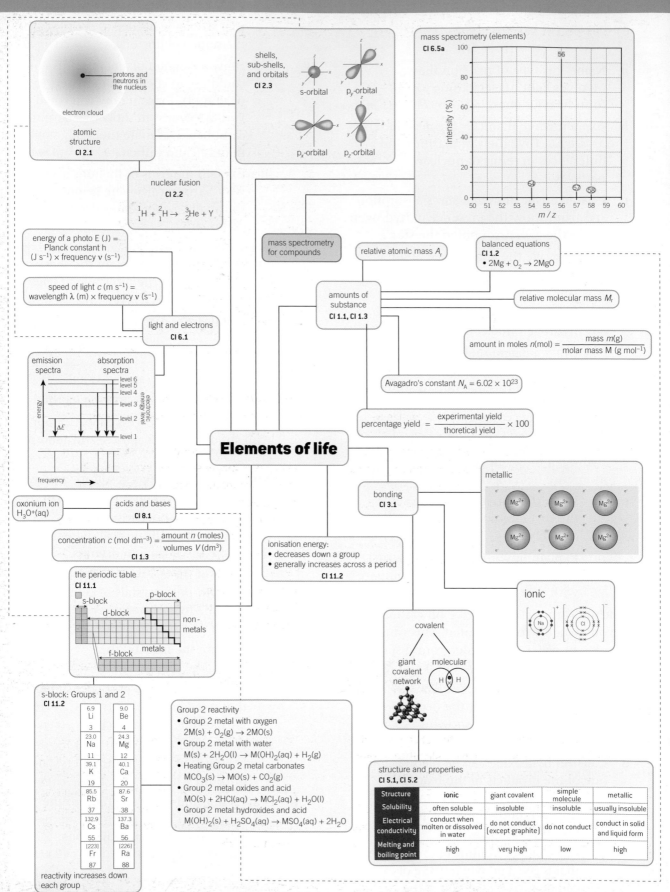

atomic structure CI 2.1 — protons and neutrons in the nucleus, electron cloud

shells, sub-shells, and orbitals CI 2.3 — s-orbital, p_y-orbital, p_x-orbital, p_z-orbital

mass spectrometry (elements) CI 6.5a

nuclear fusion CI 2.2
$$^1_1H + ^2_1H \rightarrow ^3_2He + Y$$

energy of a photo E (J) = Planck constant h (J s^{-1}) × frequency ν (s^{-1})

speed of light c (m s^{-1}) = wavelength λ (m) × frequency ν (s^{-1})

light and electrons CI 6.1

emission spectra — absorption spectra — level 6, level 5, level 4, level 3, level 2, level 1, electronic energy level, ΔE, energy, frequency

mass spectrometry for compounds

relative atomic mass A_r

balanced equations CI 1.2
• $2Mg + O_2 \rightarrow 2MgO$

amounts of substance CI 1.1, CI 1.3

relative molecular mass M_r

$$\text{amount in moles } n(\text{mol}) = \frac{\text{mass } m(g)}{\text{molar mass M (g mol}^{-1})}$$

Avagadro's constant $N_A = 6.02 \times 10^{23}$

$$\text{percentage yield} = \frac{\text{experimental yield}}{\text{thoretical yield}} \times 100$$

Elements of life

oxonium ion $H_3O^+(aq)$

acids and bases CI 8.1

$$\text{concentration } c \text{ (mol dm}^{-3}) = \frac{\text{amount } n \text{ (moles)}}{\text{volumes } V \text{ (dm}^3)}$$
CI 1.3

the periodic table CI 11.1 — s-block, p-block, d-block, non-metals, metals, f-block

ionisation energy:
• decreases down a group
• generally increases across a period
CI 11.2

bonding CI 3.1

metallic — Mg^{2+}

ionic — Na, Cl

covalent — giant covalent network, molecular — H H

s-block: Groups 1 and 2 CI 11.2

6.9 Li 3	9.0 Be 4
23.0 Na 11	24.3 Mg 12
39.1 K 19	40.1 Ca 20
85.5 Rb 37	87.6 Sr 38
132.9 Cs 55	137.3 Ba 56
[223] Fr 87	[226] Ra 88

reactivity increases down each group

Group 2 reactivity
• Group 2 metal with oxygen
  $2M(s) + O_2(g) \rightarrow 2MO(s)$
• Group 2 metal with water
  $M(s) + 2H_2O(l) \rightarrow M(OH)_2(aq) + H_2(g)$
• Heating Group 2 metal carbonates
  $MCO_3(s) \rightarrow MO(s) + CO_2(g)$
• Group 2 metal oxides and acid
  $MO(s) + 2HCl(aq) \rightarrow MCl_2(aq) + H_2O(l)$
• Group 2 metal hydroxides and acid
  $M(OH)_2(s) + H_2SO_4(aq) \rightarrow MSO_4(aq) + 2H_2O$

structure and properties CI 5.1, CI 5.2

Structure	ionic	giant covalent	simple molecule	metallic
Solubility	often soluble	insoluble	insoluble	usually insoluble
Electrical conductivity	conduct when molten or dissolved in water	do not conduct (except graphite)	do not conduct	conduct in solid and liquid form
Melting and boiling point	high	very high	low	high

The transport of oxygen

One of the important elements of life is iron as it is an important component of haemoglobin – the molecule that carries oxygen in the blood. Haemoglobin is a complex protein, but part of its structure contains 'heme' groups (Figure 1).

▲ Figure 2 *A horseshoe crab (top) and a sea cucumber (bottom)*

▲ Figure 1 heme B, $C_{34}H_{32}O_4N_4Fe$, M_r = 616 g mol^{-1}. One type of heme

Horseshoe crabs' blood contains a similar molecule called haemocyanin, which contains copper instead of iron. Sea cucumbers' blood contains vanabins, which contain vanadium, although the oxygen-carrying function of vanabins is uncertain as sea cucumbers also have haemoglobin. Haemocyanin and vanabins cause the blood to be blue and green respectively.

Other molecules with related structures include vitamin B-12 (cobalt-based) and chlorophyll (magnesium-based).

1 Write down the electron configurations, using s, p, d, and f notation, of magnesium, vanadium, iron, and cobalt.

2 Sketch a graph showing successive ionisation energies of magnesium and explain why it is evidence for the electron configuration of magnesium.

3 Explain why the bond angle in the CH_3 groups in heme B is 109 °.

4 Show that the percentage by mass of carbon in heme B is 66.2%. Calculate the percentage by mass of iron in heme B.

5 Describe and write ionic equations for the precipitation tests for Fe^{2+} and Cu^{2+} ions in solution.

Extension

1 Research the electron configuration of copper and chromium. Explain why they have slightly different configurations to other first row transition metals.

2 Prepare a summary of how to deduce shapes of molecules. For the molecules you choose, included details of the covalent bonds, repulsion of the groups of electrons and the shapes of the molecules. Give a wide range of examples.

3 On 12 November 2014 the *Philae* robotic lander detached from the European Space Agency *Rosetta* spacecraft and landed on the surface of comet 67P, orbiting near Jupiter, with the intention of studying the chemical composition of the comet. Research the findings of the mission and the instrumentation on board the lander.

CHAPTER 2
Developing fuels

Topics in this chapter

Why a chapter on Developing fuels?

This chapter tells the story of fuels including petrol and diesel – what they are, how they are made, and the use of food as fuels for our bodies. It describes the work of chemists on improving fuels for motor vehicles, and in developing alternative fuels and sustainable energy sources for the future. Important ideas about vehicle pollutants and their control are also covered.

Some fundamental chemistry is introduced to achieve this. There are two main areas. First, it is important to understand where the energy comes from when a fuel burns. This leads to a study of enthalpy changes

in chemical reactions, the use of energy cycles and the relationship between energy changes and the making and breaking of chemical bonds. Second, the module provides an introduction to organic chemistry. Alkanes and alkenes are studied in detail and other homologous series, such as alcohols and haloalkanes are introduced.

Isomerism is looked, and simple ideas about catalysis arise out of the use of catalytic converters to control exhaust emissions. All these topics will be developed and used in later modules.

Knowledge and understanding checklist

From your Key Stage 4 study you will have studied the following. Work through each point, using your Key Stage 4 notes and the support available on Kerboodle.

☐ Simple organic chemistry and homologous series.

☐ Useful products from crude oil.

☐ Combustion of alkanes.

☐ Exothermic and endothermic reactions.

☐ Addition polymerization.

☐ Sources of atmospheric pollutants.

☐ Catalysis.

You will learn more about some ideas introduced in earlier chapters:

☐ moles (**Elements of life**)

☐ empirical and molecular formulae (**Elements of life**)

☐ covalent bonding (**Elements of life**)

☐ polar bonds (**Elements of life**)

☐ molecular shape (**Elements of life**).

Maths skills checklist

In this chapter, you will need to use the following maths skills. You can find support for these skills on Kerboodle and through MyMaths.

☐ Recognise and make use of appropriate units in calculation.

☐ Use appropriate numbers of significant figures.

☐ Change to subject of an equation.

☐ Substitute numerical values into algebraic equations using appropriate units for physical quantities.

☐ Solve algebraic equations.

☐ Visualise and represent 2D and 3D forms including 2D representations of 3D objects.

☐ Understand the symmetry of 2D and 3D shapes.

☐ Plot data, lines of best fit, and extrapolate.

MyMaths.co.uk
Bringing Maths Alive

Learning outcomes

Demonstrate and apply knowledge and understanding of:

→ the terms – exothermic, endothermic, standard conditions, (standard) enthalpy change of reaction $\Delta_r H$, (standard) enthalpy change of combustion $\Delta_c H$, (standard) enthalpy change of formation $\Delta_f H$, (standard) enthalpy change of neutralisation $\Delta_{neut} H$

→ techniques and procedures for measuring the energy transferred when reactions occur in solution (or solids reacting with solutions) or when flammable liquids burn; the calculation of enthalpy changes from experimental results.

▲ **Figure 1** *A 1939 Mercedes (top) and a modern Mercedes (bottom)*

Fuel economy, or fuel efficiency, is the relationship between the distance travelled and the amount of fuel consumed by a vehicle. In the UK it is often measured in miles per gallon. Figure 1 shows a 1939 Mercedes car and a modern Mercedes car. The modern car has a fuel economy of nearly twice that of the 1939 vehicle. This shows the improvement in technology between 1939 and the present day. There are also hybrid and electric vehicles with even better consumption, but petrol-only or diesel-only engines will most likely be with us for many years yet.

The two cars in Figure 1 are petrol driven. Petrol is a highly concentrated energy source, meaning that the amount of energy released per gram of petrol is high when compared to many other fuels. To answer questions like 'How much energy can you get from a fuel?' and 'Which fuels store the most energy?' you have to know about thermochemistry.

Thermochemistry

Thermochemistry is the study of the energy and heat associated with chemical reactions. Different fuels give out different amounts of energy when one mole is burnt. Compare six important fuels (Figure 2).

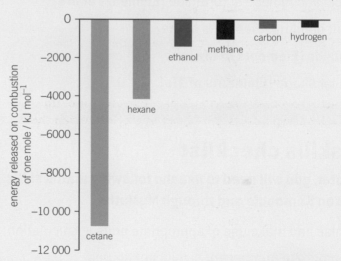

▲ **Figure 2** *The energy released on combustion of one mole of some important fuels*

The values vary widely. Why and what decides how much energy you get when you burn a mole of a particular fuel?

Fuels can be thought of as energy sources, but they can't release any energy until they have combined with oxygen. Therefore, the fuel–oxygen systems should be thought of as the energy sources. This is discussed in Topic DF 4. The following section looks at the energy changes during chemical reactions, such as combustion.

Chemical ideas: Energy changes and chemical reactions 4.1

Energy out, energy in

Energy changes are a characteristic feature of chemical reactions. Many chemical reactions give out energy and some take energy in. A reaction that gives out energy and heats the surroundings is described as **exothermic**. A reaction that takes in energy and cools the surroundings is **endothermic**.

During an exothermic reaction the chemical reactants are losing energy to their surroundings. This energy is used to heat the surroundings, for example, the air, the test tube, the laboratory, the car engine. The products end up with less energy than the reactants had but the surroundings end up with more, and get hotter. The energy transferred to and from the surroundings is measured as **enthalpy change**, ΔH. Enthalpy changes can be shown on an **enthalpy level diagram**, also called an energy level diagram (Figure 3).

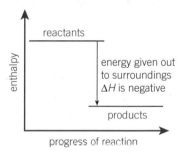

▲ **Figure 3** *Enthalpy level diagram for an exothermic reaction such as burning methane:*
$CH_4(g) + 2O_2(g) \rightarrow CO_2(g) + 2H_2O(l)$

In an endothermic reaction, the reactants take in energy from the surroundings leaving the products at a higher energy level than the reactants (Figure 4).

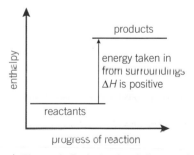

▲ **Figure 4** *Enthalpy level diagram for an endothermic reaction such as decomposing calcium carbonate:*
$CaCO_3(s) \rightarrow CaO(s) + CO_2(g)$

Enthalpy change

You cannot measure the enthalpy H of a substance. What you can measure is the change in enthalpy when a reaction occurs. This is represented ΔH (pronounced 'delta H')

$$\Delta H = H_{products} - H_{reactants}$$

The enthalpy change in a chemical reaction gives the quantity of energy transferred to or from the surroundings, when the reaction is carried out in an open container.

For an exothermic reaction, ΔH is negative. This is because, from the point of view of the chemical reactants, energy has been lost to the surroundings. Conversely, for an endothermic reaction ΔH is positive – energy has been gained from the surroundings. Enthalpy changes are measured in kilojoules per mole ($kJ\,mol^{-1}$).

 ## Worked example: Making sense of ΔH

Example 1

The equation for the reaction of methane with oxygen (Figure 3) is:

$$CH_4(g) + 2O_2(g) \rightarrow CO_2(g) + 2H_2O(l) \qquad \Delta H = -890\,kJ\,mol^{-1}$$

How much energy would be released to the surroundings when two moles of methane undergo compete combustion?

Step 1: Calculate the energy transferred to heat the surroundings for one mole of methane.

From the equation $\Delta H = -890\,kJ$

Step 2: Multiply energy transfer for one mole by the number of moles in the question (in this case, two moles).

$$\Delta H \text{ for two moles} = 2 \times -890 = -1780\,kJ$$

This assumes that all the methane is converted into products and that none is left unburnt.

Example 2

When calcium carbonate is heated, it decomposes. Energy is taken in – it is an endothermic reaction. How much energy would be taken in if 0.1 moles of calcium carbonate decomposed?

$$CaCO_3(s) \rightarrow CaO(s) + CO_2(g) \qquad \Delta H = +179\,kJ\,mol^{-1}$$

Step 1: Calculate the energy transferred to heat the surroundings for one mole of calcium carbonate.

From the equation $= +179\,kJ$

Step 2: Multiply energy transfer for one mole by the number of moles in the question (in this case, 0.1 moles).

$$\Delta H \text{ for 0.1 moles} = 0.1 \times +179 = +17.9\,kJ$$

System or surroundings?

When chemists talk about enthalpy changes, they often refer to the system. This means the reactants and the products of the reaction they are interested in. The system may lose or gain enthalpy as a result of the reaction. The surroundings means the rest of the world – the test tube, the air, and so on.

Standard conditions

Like many physical and chemical quantities, ΔH varies according to the conditions. In particular, ΔH is affected by temperature, pressure, and the concentration of solutions. As such, certain **standard conditions** are chosen to compare enthalpy changes. Under standard conditions elements exist in their standard states.

If ΔH refers to these standard conditions, it is written as $\Delta H^{\ominus}_{298}$. You can use data sheets to look up $\Delta H^{\ominus}$ values.

A word about temperature

There are two scales of temperature used in science and you should be familiar with both. The kelvin K is the unit of absolute temperature and 0 K is called absolute zero. The kelvin is the **SI (International System)** unit of temperature and should always be used in calculations involving temperature. However, you will usually measure temperature using a thermometer marked in degrees Celsius °C. You can convert temperatures from the Celsius scale to the kelvin scale by adding 273 to the Celsius reading. Similarly, you can convert from the Kelvin scale to the Celsius scale by subtracting 273 from the Kelvin reading.

Worked example: Converting between Celsius and Kelvin

A liquid boils at 100 °C. What is this in K?

Step 1: Identify the conversion.

Converting from °C to K, so add 273.

Step 2: Add 273 to the Celsius value.

$$100 + 273 = 373\,\text{K}$$

A solid melts at 478 K. What is this in °C?

Step 1: Identify the converstion.

Converting from K to °C, so take away 273.

Step 2: Take away 273 from the kelvin value.

$$478 - 273 = 205\,°\text{C}$$

Different kinds of enthalpy change

There are several different kinds of enthalpy change that you need to be familiar with.

Standard enthalpy change for a reaction

The **standard enthalpy change for a** *reaction* is the enthalpy change when molar quantities of reactants, as stated in the equation, react together under standard conditions. This means at 1 atm (pressure) and 298 K (temperature), with all the substances in their standard states. The symbol for the standard enthalpy change for a reaction is $\Delta_{r}H^{\ominus}_{298}$.

Standard conditions

Set conditions to allow us to compare enthalpy changes:

- a specified temperature normally chosen as 298 K (25 °C).

- a standard pressure of 1 atm (equivalent to $1.01 \times 10^{5}\,\text{N m}^{-2}$)

- a standard concentration of 1 mol dm^{-3} for solutions.

Standard states

The physical state of a substance under standard conditions. This may be a pure solid, liquid, or gas.

Standard enthalpy change for a reaction $\Delta_{r}H^{\ominus}_{298}$

The enthalpy change when molar quantities of reactants as stated in the equation react together under standard conditions.

Worked example: Determining the standard enthalpy change of reaction

The equation for the reaction of hydrogen and oxygen is

$$2H_2(g) + O_2(g) \rightarrow 2H_2O(l) \qquad \textbf{Equation 1}$$

Determine the standard enthalpy of reaction given that

$$H_2(g) + \frac{1}{2}O_2(g) \rightarrow H_2O(l) \quad \Delta_r H^{\ominus}_{298} = -286\,\text{kJ}\,\text{mol}^{-1} \quad \textbf{Equation 2}$$

Step 1: Identify the difference in moles of the reaction the standard enthalpy value is given in (Equation 2) and the reaction you are calculating a value for (Equation 1).

$$2H_2(g) \quad + \quad O_2(g) \quad \rightarrow \quad 2H_2O(l) \qquad \textbf{Equation 1}$$

$$H_2(g) \quad + \quad \frac{1}{2}O_2(g) \quad \rightarrow \quad H_2O(l) \qquad \textbf{Equation 2}$$

$$2 - 1 = 1 \qquad 1 - \frac{1}{2} = \frac{1}{2} \qquad 2 - 1 = 1$$

Equation 1 is double the number of moles compared to Equation 2.

Step 2: Calculate how much energy is transferred under the reaction conditions, Equation 1.

Since 286 kJ are transferred to the surroundings when one mole of hydrogen reacts with oxygen and the equation you are interested in involves the reaction of two moles of hydrogen, the enthalpy change must also be doubled:

$$286 \times 2 = -572\,\text{kJ}$$

So, $2H_2(g) + O_2(g) \rightarrow 2H_2O(l) \qquad \Delta_r H^{\ominus}_{298} = -572\,\text{kJ}\,\text{mol}^{-1}$

The following kinds of enthalpy change are particularly important and are given special names.

Measuring enthalpy changes

Many enthalpy changes can be measured in the laboratory by arranging for the energy involved in a reaction to be transferred to or from water surrounding the reaction vessel. If it is an exothermic reaction, the water gets hotter. If it is endothermic, the water gets cooler. By measuring the temperature change of the water, knowing the mass, and knowing the **specific heat capacity** of water $(4.18\,\text{J}\,\text{g}^{-1}\,\text{K}^{-1})$, you can calculate the amount of energy that was transferred to or from the water during the chemical reaction. To do this, you need to use the relationship:

$$\begin{matrix} \text{energy} \\ \text{transferred} \\ q\ \text{(kJ)} \end{matrix} = \begin{matrix} \text{specific heat} \\ \text{capacity} \\ c\ (\text{J}\ \text{g}^{-1}\ \text{K}^{-1}) \end{matrix} \times \begin{matrix} \text{mass} \\ m\ \text{(g)} \end{matrix} \times \begin{matrix} \text{temperature} \\ \text{change}\ \Delta T\ \text{(K)} \end{matrix}$$

Using a bomb calorimeter to accurately measure energy changes

In a bomb calorimeter, the fuel is ignited electrically and burns in oxygen inside the pressurised vessel. Energy is transferred to the surrounding water, where the temperature rise is measured. The experiment is done at constant volume in a closed container. Enthalpy changes are for reactions carried out at constant pressure, so the result needs to be modified accordingly.

1 Suggest why the sample is burnt in oxygen under pressure.
2 Suggest why the temperature measured is more accurate than when using a simple calorimeter.
3 Suggest why the heat transferred at constant volume might be different from that transferred at constant pressure.

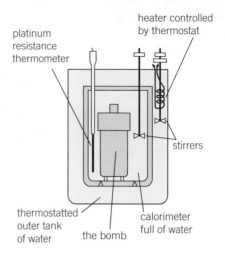

▲ **Figure 3** A bomb calorimeter

Standard enthalpy change of combustion

The **standard enthalpy change of** *combustion* is the enthalpy change that occurs when one mole of a substance is burnt completely in oxygen. In theory, the substance needs to be burnt under standard conditions – 1 atm and 298 K. In practice this is impossible, so the substance is burnt and then adjustments are made to allow for the non-standard conditions. The symbol for the standard enthalpy of combustion is $\Delta_c H^{\ominus}_{298}$.

For example, the enthalpy change of combustion of heptane, one of the alkanes found in petrol, is $-4187\,kJ\,mol^{-1}$. This is much bigger than for methane ($-890\,kJ\,mol^{-1}$) because burning heptane involves breaking and making more bonds than burning methane (see DF 4). Note that if no temperature is given with $\Delta_c H^{\ominus}$ then assume that the value refers to 298 K.

The equations for the combustion of methane and heptane are shown below.

$$CH_4(g) + 2O_2(g) \rightarrow CO_2(g) + 2H_2O(l) \qquad \Delta_c H^{\ominus}_{298} = -890\,kJ\,mol^{-1}$$

$$C_7H_{16}(l) + 11O_2(g) \rightarrow 7CO_2(g) + 8H_2O(l) \qquad \Delta_c H^{\ominus}_{298} = -4187\,kJ\,mol^{-1}$$

When writing an equation to represent an enthalpy change of combustion, the equation must always balance and show one mole of the substance reacting, even if this means having half a mole of oxygen molecules in the equation. You should always include state symbols.

Standard enthalpy change of combustion $\Delta_c H^{\ominus}_{298}$

The enthalpy change that occurs when one mole of a substance is burnt completely in oxygen under standard conditions in standard states.

Study tip

All combustion reactions are exothermic, so $\Delta_c H^{\ominus}_{298}$ is *always* negative.

Activity DF 1.1

In this activity you compare the energy given out by burning hexane and methanol, both compounds found in some types of petrol/

Standard enthalpy change of formation $\Delta_f H^\ominus_{298}$

The enthalpy change when one mole of a compound is formed from its elements under standard conditions in standard states.

Study tip

$\Delta_f H^\ominus_{298}$ may be positive or negative.

Study tip

Remember that, by definition, the standard enthalpy change of formation of a pure element in its standard state is zero.

Standard enthalpy change of neutralisation $\Delta_{neut} H^\ominus_{298}$

The enthalpy change when one mole of hydrogen ions react with one mole of hydroxide ions to form one mole of water under standard conditions and in solutions containing 1 mol dm^{-3}.

Synoptic link

You have studied neutralisation reactions in Chapter 1, Elements of life.

Activity DF 1.2

This activity allows you to measure and calculate the enthalpy change of neutralisation.

Standard enthalpy change of formation

The **standard enthalpy change of *formation*** is the enthalpy change when one mole of a compound is formed from its elements – again with both the compound and its elements being in their standard states under standard conditions. The symbol for standard enthalpy of formation is $\Delta_f H^\ominus_{298}$.

For example, the enthalpy change of formation of water, $H_2O(l)$, is -286 kJ mol^{-1}. When you make one mole of water from hydrogen and oxygen, 286 kJ are transferred to the surroundings. This is summed up as:

$$H_2(g) + \frac{1}{2} O_2(g) \rightarrow H_2O(l) \quad \Delta_f H^\ominus_{298} = -286 \text{ kJ mol}^{-1}$$

The equation refers to one mole of H_2O, so only $\frac{1}{2}$ mole of O_2 is needed in the equation.

It is often impossible to measure enthalpy changes of formation directly. For example, the standard enthalpy change of formation of methane is -75 kJ mol^{-1}. This refers to the reaction

$$C(s) + 2H_2(g) \rightarrow CH_4(g) \quad \Delta_f H^\ominus_{298} = -75 \text{ kJ mol}^{-1}$$

However, this reaction doesn't actually occur under standard conditions. So how did anyone manage to measure the value of $\Delta_f H^\ominus_{298}$? It has to be done indirectly, making use of quantities that can be measured and incorporating these into an **enthalpy cycle**. You can find out about enthalpy cycles in Topic DF 2.

Standard enthalpy change of neutralisation

The **standard enthalpy change of *neutralisation*** can be measured from the energy given out when acids react with alkalis in aqueous solution. The symbol for this change is $\Delta_{neut} H^\ominus_{298}$.

The enthalpy changes can be calculated from these measurements:

$NaOH(aq) + HCl(aq) \rightarrow NaCl(aq) + H_2O$	$\Delta_{neut} H^\ominus_{298} = -58 \text{ kJ mol}^{-1}$	**Equation 3**
$NaOH(aq) + HNO_3(aq) \rightarrow NaNO_3(aq) + H_2O$	$\Delta_{neut} H^\ominus_{298} = -58 \text{ kJ mol}^{-1}$	**Equation 4**
$NaOH(aq) + H_2SO_4(aq) \rightarrow NaHSO_4(aq) + H_2O$	$\Delta_{neut} H^\ominus_{298} = -58 \text{ kJ mol}^{-1}$	**Equation 5**
$2NaOH(aq) + H_2SO_4(aq) \rightarrow Na_2SO_4(aq) + 2H_2O$	$\Delta_r H^\ominus_{298} = -115 \text{ kJ mol}^{-1}$	**Equation 6**

Notice that many of these values are the same. This is because the reaction that is occurring in Equations 3–5 can be represented as:

$H^+(aq) + OH^- \rightarrow H_2O$	$\Delta H^\ominus = -58 \text{ kJ mol}^{-1}$	**Equation 7**

So the enthalpy change of neutralisation is defined per mole of H_2O formed.

If one mole of sodium hydroxide, NaOH, reacts with one mole of sulfuric acid, H_2SO_4, the reaction in Equation 7 occurs and the enthalpy change is -58 kJ mol^{-1}. If two moles of sodium hydroxide are available for each mole of sulfuric acid, then $\Delta_{neut} H^\ominus_{298}$ is virtually doubled. However, the enthalpy change of neutralisation (measured per mole of water produced) is virtually the same.

Summary questions

You will need to look up values for standard enthalpy changes when doing these problems.

1 Define the following enthalpy changes.
   a standard enthalpy change of combustion   (*1 mark*)
   b standard enthalpy change of formation   (*1 mark*)

2 Explain why standard enthalpy changes of formation may have a positive sign, but standard enthalpy changes of combustion are always negative.   (*2 marks*)

3 The standard enthalpy change of formation of hydrogen chloride is $-92.3\,kJ\,mol^{-1}$ and that of hydrogen iodide is $+26.5\,kJ\,mol^{-1}$. Draw labelled enthalpy level diagrams to represent the reactions which occur when each of these compounds is formed from its elements.   (*4 marks*)

4 The standard enthalpy change of combustion of carbon is equal to the standard enthalpy change of formation of carbon dioxide. Explain why this is so by referring to the equations for the two reactions.   (*3 marks*)

5 a Write the equation to represent the formation of one mole of water from its elements in their standard states.   (*2 marks*)
   b Look up and write down the standard enthalpy change of formation of water.   (*1 mark*)
   c Calculate the enthalpy change when 1.0 g of hydrogen burns in oxygen. What assumptions have you made?   (*3 marks*)
   d What is the standard enthalpy change for the following reaction?

$$H_2O(l) \longrightarrow H_2(g) + \frac{1}{2}O_2(g)$$ (*2 marks*)

6 Use Equations 3–7 to help you answer these questions.
   a $20\,cm^3$ of $0.10\,mol\,dm^{-3}$ NaOH reacts with $20\,cm^3$ $0.10\,mol\,dm^{-3}$ HCl. Calculate the temperature rise.   (*2 marks*)
   b Calculate the temperature rise if both the concentrations were doubled in (a).   (*2 marks*)
   c Calculate the temperature rise if $10\,cm^3$ of $0.10\,mol\,dm^{-3}$ NaOH is reacted with $20\,cm^3$ $0.10\,mol\,dm^{-3}$ HCl.   (*2 marks*)
   d Estimate the temperature rise if $20\,cm^3$ of $0.10\,mol\,dm^{-3}$ NaOH is reacted with $20\,cm^3$ $0.10\,mol\,dm^{-3}$ HBr (another 'strong' acid, like HCl, that is present entirely as its ions in solution). Explain your reasoning.   (*3 marks*)
   e Calculate the temperature rise when $20\,cm^3$ of $0.10\,mol\,dm^{-3}$ NaOH is reacted with $20\,cm^3$ $0.10\,mol\,dm^{-3}$ H_2SO_4.   (*2 marks*)
   f Calculate the temperature rise when $40\,cm^3$ of $0.10\,mol\,dm^{-3}$ NaOH is reacted with $20\,cm^3$ $0.10\,mol\,dm^{-3}$ H_2SO_4.   (*2 marks*)

Learning outcomes

Demonstrate and apply knowledge and understanding of:

→ the determination of enthalpy changes of reaction from enthalpy cycles and enthalpy level diagrams based on Hess' law

→ calculations involving enthalpy changes.

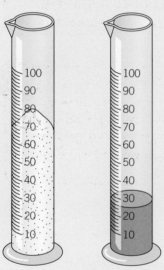

▲ **Figure 1** *Fats and oils are more energy-rich than carbohydrates – the oil on the right will provide the same quantity of energy as the solid glucose on the left*

| 1 single measure of spirits | 1 glass of wine | $\frac{1}{2}$ pint of beer or lager |

▲ **Figure 2** *Alcohol can be fattening: each of these drinks provides about 300 kJ of energy (about 70 Calories), equivalent to $1\frac{1}{2}$ slices of bread*

Important news for slimmers

Just as cars need fuels, so you need fuel for your body. When you eat too much of an energy-rich food, the excess energy gets stored in your body as fat. The more energy-rich the food, the more fattening it is.

Compare a carbohydrate such as glucose, $C_6H_{12}O_6$, with a fat such as glycerol trioleate, $C_{57}H_{104}O_6$, the main component of olive oil.

For each carbon atom, glucose has more oxygen atoms than olive oil, so glucose is much less energy-rich. From burning 1 g of a carbohydrate such as glucose you can get about 17 kJ. From burning 1 g of a fat such as olive oil you can get about 39 kJ. Gram for gram, fats contain twice as much energy as carbohydrates (Figure 1).

Alcohol is neither a fat nor a carbohydrate. In fact, there is a whole series of related compounds called alcohols, and the particular alcohol present in drinks is ethanol, C_2H_5OH. The same substance is used as an alternative to petrol for cars in some countries. It burns in the car engine releasing energy – and it also releases energy when metabolised in the body (Figure 2).

How is this energy measured? For some reactions, the enthalpy changes can be measured practically, but some cannot be measured directly and need to be calculated. The introductory activity in Topic DF 1 involved using a simple calorimeter to measure two enthalpy changes of combustion.

Chemical ideas: Energy changes and chemical reactions 4.2

Enthalpy cycles

For some reactions, measuring ΔH is very straightforward, for example, the burning of methane. For other reactions it is less simple. Decomposing calcium carbonate, $CaCO_3$, needs a temperature of over 800 °C. In cases like this, enthalpy changes can be measured indirectly, using enthalpy cycles.

Figure 4 shows an **enthalpy cycle**, also known as an energy cycle. There is both a direct and an indirect way to turn graphite, C, and hydrogen, H_2, into methane, CH_4. The enthalpy change for the direct route cannot be measured. The indirect route goes via carbon dioxide and water and involves two enthalpy changes both of which both can be measured. Since most organic compounds burn easily, cycles such as this can often be used, based on enthalpy changes of combustion, to work out indirectly the enthalpy change of an organic reaction.

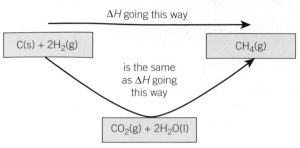

▲ **Figure 4** *An enthalpy cycle for finding the enthalpy change of formation of methane, CH_4*

The same reaction can be represented using an enthalpy level diagram (Figure 5) instead of an enthalpy cycle.

In Hess' law, the enthalpy change for any chemical reaction is independent of the intermediate stages, so long as the initial and final conditions are the same for each route.

This is an alternative way of representing the same processes as in Figure 4, and the numerical values can be processed in exactly the same way as the worked examples that follow.

Using enthalpy changes of combustion in enthalpy cycles

The key idea is that the total enthalpy change for the indirect route is the same as the enthalpy change via the direct route. Energy cannot be created or destroyed – this is the law of conservation of energy. So as long as your starting and finishing points are the same, the enthalpy change will always be the same, irrespective of how you get from start to finish. This is one way of stating **Hess' law**, and an enthalpy cycle, like the one in Figure 5, is called a Hess' cycle or a thermochemical cycle.

If you know the enthalpy changes involved in two parts of the cycle, you can work out the enthalpy change in the third. So, referring to Figure 6, if you can measure ΔH_2 and ΔH_3, ΔH_1 can be calculated, which is the enthalpy change that cannot be measured directly.

$$\Delta H_1 = \Delta H_2 - \Delta H_3$$

It has to be minus ΔH_3 because the reaction to which ΔH_3 applies actually goes in the opposite direction to the way you want it to go in order to produce methane. To find the enthalpy change for the reverse reaction, you must reverse the sign.

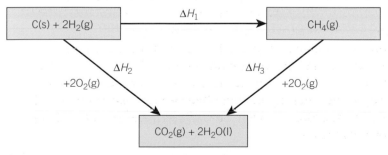

▲ **Figure 6** *Using an enthalpy cycle to find ΔH_1*
$$\Delta H_1 = \Delta H_2 - \Delta H_3$$

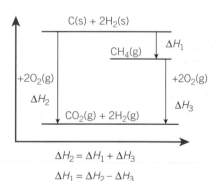

$$\Delta H_2 = \Delta H_1 + \Delta H_3$$
$$\Delta H_1 = \Delta H_2 - \Delta H_3$$

▲ **Figure 5** *An enthalpy level diagram for the enthalpy of formation of methane*

Study tip

Calculations look cluttered if $\Delta H^{\ominus}_{298}$ is written in full each time it occurs. So just $\Delta H^{\ominus}$ is often used for the standard enthalpy change at 298 K.

Study tip

Using enthalpy cycles enables you to find a value for an enthalpy change which you could not find directly. You will find these cycles very useful in other parts of your chemistry course.

 Worked example: Using Hess' cycles

Use the following standard enthalpy values to calculate the enthalpy of change of formation of methane (Figure 5).

$$\Delta_c H^{\ominus}(C) = -393 \, kJ \, mol^{-1}$$

$$\Delta_c H^{\ominus}(H_2) = -286 \, kJ \, mol^{-1}$$

$$\Delta_c H^{\ominus}(CH_4) = -890 \, kJ \, mol^{-1}$$

Step 1: Identify the reactions involved from the Hess' cycle (Figure 6).

ΔH_2 is the sum of the enthalpy changes of combustion of one mole of carbon and two moles of hydrogen.

$$\Delta H_2 = \Delta_c H^{\ominus}(C) + \Delta_c H^{\ominus}(H_2)$$

ΔH_3 is the enthalpy change of combustion of methane.

$$\Delta H_3 = \Delta_c H^{\ominus}(CH_4)$$

ΔH_1 is the enthalpy change of formation of methane – the quantity you are trying to find.

Step 2: Write the equation to calculate the ΔH_1 value.

$$\Delta H_1 = \Delta H_2 - \Delta H_3$$

This is the same as:

$$\Delta H_1 = \Delta_c H^{\ominus}(C) + 2\Delta_c H^{\ominus}(H_2) - \Delta_c H^{\ominus}(CH_4)$$

Step 3: Substitute in the standard enthalpy values.

$$= -393 + 2(-286) - (-890)$$

$$= -75 \, kJ \, mol^{-1}$$

Using enthalpy changes of formation in enthalpy cycles

Enthalpy changes for the formation of many compounds from their elements in their standard states are available in data books. Some of these have been measured directly and others indirectly, as shown above for methane. These data are very useful as they can be used to calculate the enthalpy change for a reaction, rather than doing experiments yourself.

 Worked example: Using enthalpy changes of formation to calculate an enthalpy change of reaction

Calculate the enthalpy change of reaction for the following process.

$$NH_3(g) + HCl(g) \rightarrow NH_4Cl(s)$$

Use the following data for the standard enthalpy of formation.

$$\Delta_f H^{\ominus}(NH_3) = -46.1 \, kJ \, mol^{-1}$$

$$\Delta_f H^{\ominus}(HCl) = -92.3 \, kJ \, mol^{-1}$$

$$\Delta_f H^{\ominus}(NH_4Cl) = -315 \, kJ \, mol^{-1}$$

Step 1: Draw a Hess' cycle for the reaction.

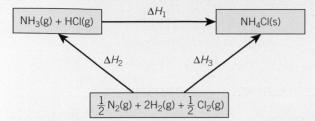

Step 2: Identify the reactions involved from the Hess' cycle.

ΔH_2 is the sum of the enthalpy changes of formation of one mole of ammonia and hydrogen chloride.

$$\Delta H_2 = \Delta_f H^{\ominus}(NH_3) + \Delta_f H^{\ominus}(HCl)$$

ΔH_3 is the enthalpy change of formation of ammonium chloride.

$$\Delta H_3 = \Delta_f H^{\ominus}(NH_4Cl)$$

ΔH_1 is the enthalpy change of reaction of ammonia and hydrogen chloride – the quantity you are trying to find.

Step 3: Write the equation to calculate the ΔH_1 value.

$$\Delta H_1 = -\Delta H_2 + \Delta H_3$$

This is the same as:

$$\Delta H_1 = -(\Delta_f H^{\ominus}(NH_3) - \Delta_f H^{\ominus}(HCl)) + \Delta_f H^{\ominus}(NH_4Cl)$$

Step 4: Substitute in the standard enthalpy of formation values.

$$= -(-46.1) - (-92.3) + (-315)$$
$$= -176.6 \, kJ \, mol^{-1}$$

You may find the following shortcut useful:

$\Delta_r H^{\ominus}$	$=$	$\sum \Delta_r H^{\ominus}{}_{(products)}$	$-$	$\Delta_r H^{\ominus}{}_{(reactants)}$
enthalpy change of reaction	$=$	sum of enthalpy changes of formation of products	$-$	sum of enthalpy changes of formation of reactants

Study tip

Remember, $\sum$ is a symbol for sum of.

Summary questions

1 You have been asked to measure the enthalpy change of combustion of methane. You have been given a gas cooker and a saucepan.
  a What other equipment would you need? *(1 mark)*
  b What measurements would you make? *(1 mark)*
  c What would be the main sources of error? *(1 mark)*

2 a Write an equation for the formation of one mole of butane, $C_4H_{10}(g)$, from its elements in their standard states. *(2 marks)*
  b Draw an enthalpy cycle to show the relationship between the formation of butane from carbon and hydrogen, and the combustion of these elements to give carbon dioxide and water (see Figure 6 if you need help). *(3 marks)*

c Use your enthalpy cycle to calculate a value for the standard enthalpy change of formation of butane.   (3 marks)

You will need the following data:

$$\Delta_c H^{\ominus}(C) = -393\,\text{kJ mol}^{-1}$$
$$\Delta_c H^{\ominus}(H_2) = -286\,\text{kJ mol}^{-1}$$
$$\Delta_c H^{\ominus}(C_4H_{10}) = -2877\,\text{kJ mol}^{-1}$$

3 A student measures the enthalpy change of the reaction:

$$\begin{array}{ccccc} CuSO_4(s) & + & 5H_2O & \rightarrow & CuSO_4 \cdot 5H_2O \\ \text{anhydrous} & & & & \text{hydrated} \\ \text{copper sulfate} & & & & \text{copper sulfate} \end{array}$$

The student uses two experimental steps and applies Hess' law to work out the answer.

Experiment 1

Dissolve 3.99 g (0.0250 mol) of anhydrous copper sulfate in 50 cm^3 water in a plastic cup. The temperature rose from 19.52 °C to 27.40 °C.

Experiment 2

Add 6.24 g (0.0250 mol) of hydrated copper sulfate to 48 cm^3 water in a plastic cup (to allow for the '5H$_2$O' making up the volume to 50 cm^3). The temperature changed from 19.56 °C to 18.28 °C.

a Why can't the enthalpy change of the reaction be measured directly?   (1 mark)

b Show, by calculation, that 48 cm^3 was the correct volume to use for the second experiment.   (2 marks)

c Calculate the energy change in each experiment in kJ mol^{-1}. (The specific heat capacity of water is 4.18 J g^{-1} K^{-1})   (2 marks)

d Explain any assumptions you made in these calculations.   (1 mark)

e Draw a Hess' cycle which connects the enthalpy changes in the two experiments with the enthalpy change in the equation.   (3 marks)

f Calculate a value for the enthalpy change of the reaction in the equation.   (2 marks)

g Making suitable assumptions about the apparatus used, calculate the percentage uncertainty in each value in part **c**. Work these out as actual ($\pm$) uncertainty values. Add these uncertainties to give the overall uncertainty in your answer to part **f**.   (3 marks)

h Which piece of apparatus generates the greatest percentage uncertainty?   (1 mark)

i The accepted value for the enthalpy change is −77.4 kJ mol^{-1}. Comment on this compared with your value and uncertainty limits.   (2 marks)

DF 3 What's in your tank?

Specification reference: DF(l), DF(m), DF(r)

What are petrol and diesel?

Petrol and diesel are both complex mixtures of many different compounds, carefully blended to give the right properties. The compounds are obtained from crude oil in several ways.

Crude oil is a mixture of many hundreds of hydrocarbons. It is a thick black liquid but dissolved in it are gases and solids. Oil from the North Sea is pumped along pipes on the seabed to UK refineries, and special tankers bring crude oil from distant oilfields, such as those in the Middle East and Alaska. These refineries are either close to the shore (such as at Fawley, near Southampton) or the oil is off-loaded into a pipeline leading to a refinery (such as from Finnart, on the west coast of Scotland, to the refinery at Grangemouth, near Edinburgh).

At the refinery the crude oil is heated to vaporise it and the vapour passes into a distillation column. This process is known as **fractional distillation**. The oil is separated into fractions, each having a specific boiling point range. The fractions do not have an exact boiling point because they are mixtures of many different hydrocarbons. For example, the gasoline fraction is a mixture of liquids, mostly alkanes with between five and seven carbon atoms, boiling in the range 25–75 °C.

The gasoline and gas oil fractions are the sources of petrol components. Another important fraction, naphtha, is also converted into high-grade petrol as well as being used in the manufacture of many organic chemicals. Table 1 contains information on some other fractions produced in the distillation of crude oil.

▲ **Figure 1** *Grangemouth refinery*

▼ **Table 1** *Fractions obtained from the fractional distillation of crude oil*

Name of fraction	Boiling point range / °C	Composition	% of crude oil	Use(s)
refinery gas	<25	$C_1–C_4$	1–2	liquid petroleum gas (propane, butane), blending in petrol, feedstock for organic chemicals
gasoline	25–75	$C_5–C_7$		car petrol
naphtha	75–190	$C_6–C_{10}$	20–40	production of organic chemicals, converted to petrol
kerosene	190–250	$C_{10}–C_{16}$	10–15	jet fuel, heating fuel (paraffin)
gas oil	250–350	$C_{14}–C_{20}$	15–20	diesel fuel, central heating fuel, converted to petrol
residue	>350	$>C_{20}$	40–50	fuel oil (e.g., power stations, ships), lubricating oils and waxes, bitumen or asphalt for roads and roofing

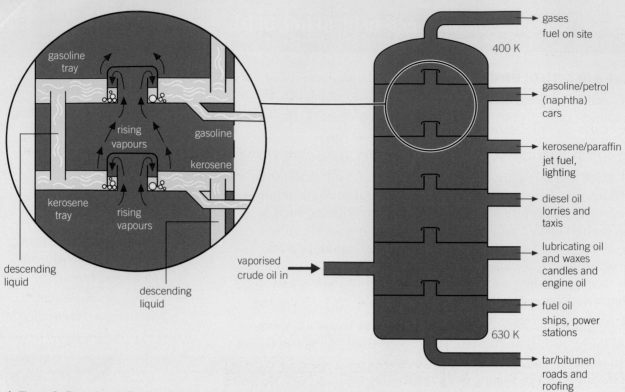

▲ **Figure 2** *The primary fractional distillation of crude oil is a continuous process. Vapour rises up through the column and liquids condense and are run off at different levels, depending on their volatility*

Chemical ideas: Organic chemistry: frameworks 12.1a

Organic chemistry – alkanes

Structure and naming of alkanes

Many carbon compounds are found in living organisms, which is why their study got the name **organic chemistry**. Today, organic chemistry includes all carbon compounds whatever their origin – except carbon monoxide, CO, carbon dioxide, CO_2, and the carbonates, which are traditionally included in inorganic chemistry studies.

Only carbon can form the diverse range of compounds necessary to produce the individuality of living things.

Why carbon?

Carbon's electron structure is shown in Figure 3. This electron structure makes it the first member of Group 4 in the centre of the periodic table, and is responsible for its special properties.

A carbon atom has four electrons in its outer shell. It could achieve stability by losing or gaining four electrons but this is too many electrons to lose or gain. The resulting carbon ions would have charges of +4 or −4 respectively, and would be too highly charged. So when carbon forms compounds the bonds are covalent rather than ionic.

carbon (2,4)

▲ **Figure 3** *The electron structure of carbon*

In methane, CH_4, the carbon atom achieves stability by sharing its outer electrons with four hydrogen atoms (Figure 4) forming four carbon–hydrogen covalent bonds.

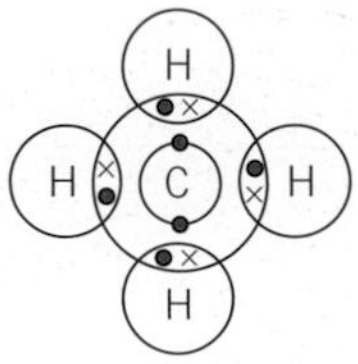

Carbon forms strong covalent bonds with itself to give chains and rings of its atoms, joined by carbon–carbon covalent bonds. This property is called **catenation** and leads to the limitless variety of organic compounds possible.

Each carbon atom can form four covalent bonds, so the chains may be straight or branched, and can have other atoms or groups substituted on them.

methane, CH_4

▲ **Figure 4** *A dot-and-cross diagram showing covalent bonds in methane*

Hydrocarbons

Chemists cope with the vast number of organic compounds by dividing them into groups of related compounds. **Hydrocarbons** are compounds containing *only* carbon atoms and hydrogen atoms. They are represented by the general molecular formula C_xH_y. There are different types of hydrocarbons (Table 2).

▼ **Table 2** *Some common hydrocarbons*

Name	Formula	Shape	Type of compound
methane	CH_4	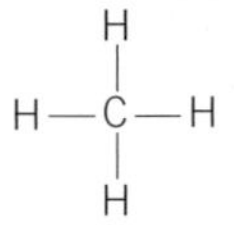 $H-\overset{\displaystyle H}{\underset{\displaystyle H}{C}}-H$	alkane
ethene	C_2H_4	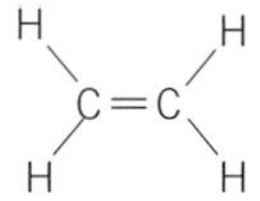 $C=C$ (with H atoms)	alkene
benzene	C_6H_6	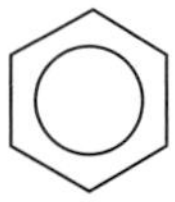 (hexagon with circle)	arene
cyclohexane	C_6H_{12}	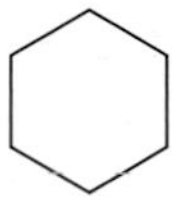 (hexagon)	cycloalkane

The ring of six carbon atoms in benzene has special properties. Compounds that contain a benzene ring are called **aromatic compounds**. These compounds are called arenes. Compounds that do not contain a benzene ring are called **aliphatic compounds**.

Hydrocarbons are relatively unreactive – this is particularly true of alkanes and arenes. They form the unreactive framework of organic compounds. But when you attach other groups to the hydrocarbon framework its properties are modified. So you can think of organic compounds as having hydrocarbon frameworks, with modifiers attached.

A modifier such as the hydroxyl group, –OH, is also called a **functional group**. Compounds with the –OH group are called **alcohols**. Petrol usually contains alcohols in the blend and you will find out more about them later.

Aromatic compounds

Compounds that contain one or more benzene rings.

Aliphatic compounds

Compounds that do not contain any benzene rings.

Functional group

Modifiers that are responsible for the characteristic chemical reactions of molecules.

The double bond in an **alkene**, C=C, is much more reactive than a carbon–carbon single bond. It is often regarded as a functional group, even though it is part of the hydrocarbon framework.

Alkanes

Alkanes are **saturated** hydrocarbons. Saturated means that they contain the maximum number of hydrogen atoms possible, with no double or triple bonds between carbon atoms.

The general molecular formula of the alkanes is C_nH_{2n+2} where n is the number of carbon atoms and can be any whole number. The names of all the alkanes end in –ane. Table 3 shows the names and formulae of the first 10 alkanes.

Look at the names in Table 3. The first part of the name of an alkane indicates the number of carbon atoms in each molecule. The second part, –ane, indicates that the compounds are part of the class of compounds called alkanes. The names are irregular up to butane – they do not use the normal prefix associated with 1, 2, 3, and 4 – but after butane they are more predictable. You need to learn the prefix for 1–10 carbons (Table 3).

A series of compounds related to each other in this way is called an **homologous series**. All the members of the series have the same general molecular formula, and each member differs from the next by a –CH$_2$– unit. All the compounds in a series have similar chemical properties, so chemists can study the properties of the series rather than those of individual compounds. However, physical properties such as melting point, boiling point, and density do change gradually in the series as the number of carbon atoms in the molecules increases.

Structure of alkanes

Figure 4 shows a dot-and-cross formula for methane. It shows all the outer electrons in each atom and how electrons are shared to form the covalent bonds. For larger molecules, dot-and-cross formulae can be complicated and the shared electron pairs can be replaced by lines representing the covalent bonds. Figure 5 is called a full structural formula – it shows all the atoms and all the bonds in the molecule.

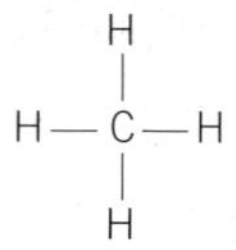

You can also write an abbreviated version known as a shortened structural formula. Take heptane, for example. Its full structural formula is shown in Figure 6.

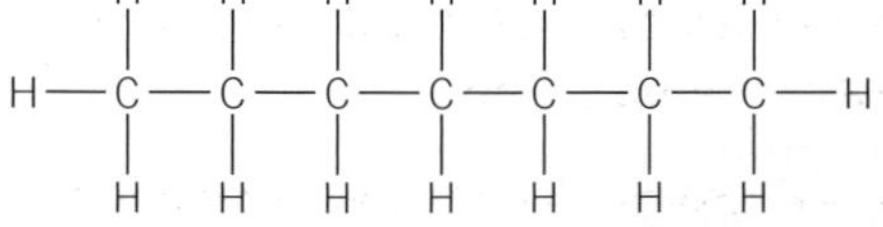

Saturated

Hydrocarbons containing the maximum number of hydrogen atoms possible, no carbon–carbon double or triple bonds.

▼ **Table 3** The first 10 alkanes

n	Molecular formula	Prefix	Name
1	CH_4	meth-	methane
2	C_2H_6	eth-	ethane
3	C_3H_8	prop-	propane
4	C_4H_{10}	but-	butane
5	C_5H_{12}	pent-	pentane
6	C_6H_{14}	hex-	hexane
7	C_7H_{16}	hept-	heptane
8	C_8H_{18}	oct-	octane
9	C_9H_{20}	non-	nonane
10	$C_{10}H_{22}$	dec-	decane

Homologous series

A series of compounds in which all members have the same general molecular formula

▲ **Figure 5** A full structural formula uses lines to represent covalent bonds

▲ **Figure 6** The full structural formula for heptane, C_7H_{16}

Its shortened structural formula is

$$CH_3-CH_2-CH_2-CH_2-CH_2-CH_2-CH_3$$

This can be further shortened to

$$CH_3CH_2CH_2CH_2CH_2CH_2CH_3$$

Table 4 gives the full structural formulae and shortened formulae for some alkanes. You will see that each carbon atom is bonded to four other atoms. Each hydrogen atom is bonded to only one other atom.

▼ **Table 4** *Structural formulae of alkanes*

Name	Molecular formula	Full structural formula	Shortened structural formula	Further shortened to
methane	CH_4	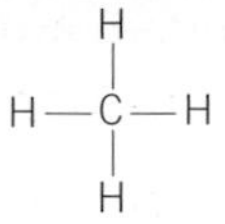	CH_4	
ethane	C_2H_6	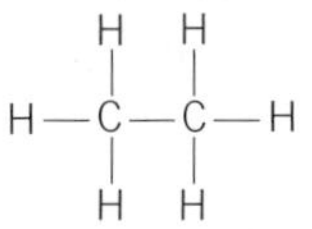	$CH_3—CH_3$	CH_3CH_3
propane	C_3H_8	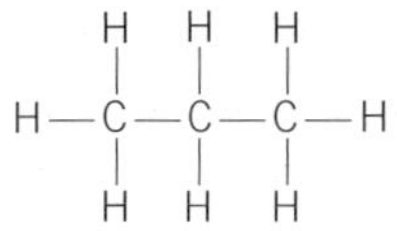	$CH_3—CH_2—CH_3$	$CH_3CH_2CH_3$

Branched alkanes

Alkanes may have straight or branched chains. It is often possible to draw more than one structural formula for a given molecular formula. There is often a straight-chain compound and one or more branched-chain compounds with the same molecular formula.

For a compound with the molecular formula C_4H_{10} there are two possible structural formulae:

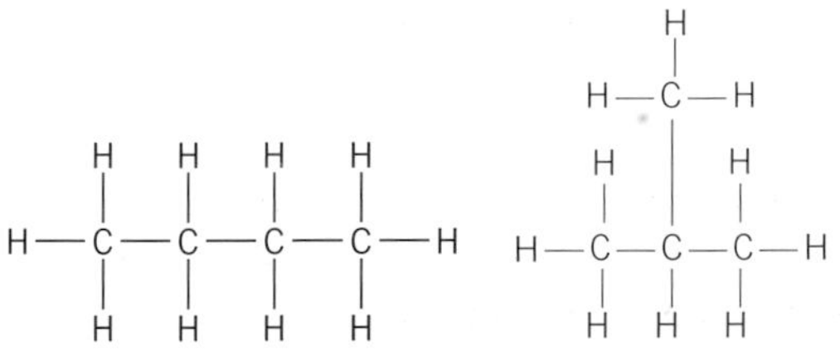

◀ **Figure 7** *The full structural formulae of butane (left) and methylpropane (right). Both have the molecular formula C_4H_{10}*

These two compounds are **structural isomers** because they have the same molecular formulae but different structural formulae. There is more about this type of isomerism later in this section.

The branched-chain isomer is regarded as being formed from a straight-chain alkane (propane) with a $-CH_3$ group attached to the second carbon atom. The $-CH_3$ group is just methane with a hydrogen atom removed so that it can join to another atom. It is called a methyl group. The isomer is therefore called methylpropane.

Side groups of this kind are called alkyl groups. They have the general formula C_nH_{2n+1} (Table 5) and are often represented by the symbol R.

Butane and methylpropane are **systematic names**. Every organic compound can be given a systematic name derived from an internationally agreed set of rules. Many compounds also have common names. The systematic name is important because it allows for the full structural formula to be determined. The systematic name can also be determined from the full structural formula.

▼ **Table 5** *Some common alkyl groups*

Alkyl group	Formula
methyl	$CH_3–$
ethyl	$CH_3CH_2–$
propyl	$CH_3CH_2CH_2–$
butyl	$CH_3CH_2CH_2CH_2–$
pentyl	$CH_3CH_2CH_2CH_2CH_2–$

Activity DF 3.1

In this activity you will practice naming alkanes and cycloalkanes.

Cycloalkanes

As well as open-chain alkanes, it is also possible for alkane molecules with cyclic structures to exist. These molecules are called cycloalkanes and have the general formula C_nH_{2n}. They have two fewer hydrogen atoms than the corresponding alkane, because there are no $-CH_3$ groups at the ends of the chain.

Table 6 shows some different ways of representing cycloalkanes. The skeletal formula shows only the shape of the carbon framework. Each line represents a carbon–carbon bond. The carbon atoms are at the corners. The carbon–hydrogen bonds are not shown but it is easy to work out how many there are – in saturated compounds, carbon always forms four covalent bonds

▼ **Table 6** *Cycloalkanes – the general formula is C_nH_{2n}*

Cycloalkane	Shortened structural formula	Skeletal formula
cyclopropane, C_3H_6	CH_2 at top, H_2C—CH_2 below (triangle ring)	triangle
cyclobutane, C_4H_8	H_2C—CH_2 / H_2C—CH_2 (square ring)	square
cyclohexane, C_6H_{12}	CH_2, H_2C CH_2, H_2C CH_2, CH_2 (hexagon ring)	hexagon

Naming alkanes

Alkanes can be either straight chain or branched.

They have the general formula C_2H_{2n+2}. The rules for naming the first 10 straight chain alkanes are given in Table 3.

Naming branched alkanes

When alkanes are branched, the following rules apply.

1   Find the longest chain of carbons in the molecule and name the chain, for example, a chain five carbon atoms long would be pentane.

2   Identify any side chains off the main chain.

3   Name these side chains as substituents on the main chain by adding –yl to the appropriate prefix.

4   State the location of any side chains by prefixing the name of the side chain with the number of the carbon atom to which the side chain is attached.

5   Keep the number as low as possible, for example:

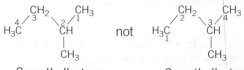

2-methylbutane   not   3-methylbutane

6   If there is more than one side chain, the rules still apply. Keep
     the numbering as low as possible, but when naming state the side
     chains in alphabetical order. If there are more than one of the
     same type of side group, use the prefixes di- (if there are two), tri-
     (if there are three), tetra- (if there are four).

7   Name the compound by starting the side chains and their location
     first, then the 'parent' (longest chain). Use hyphens between
     numbers and letters and commas between numbers.

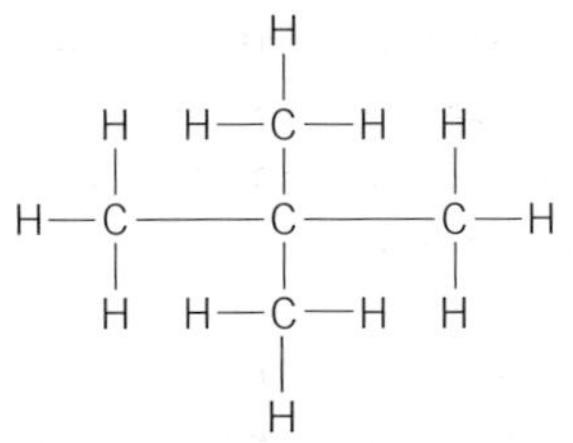

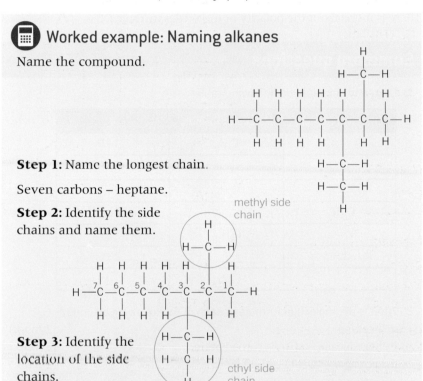

2,2-dimethyl propane

Worked example: Naming alkanes

Name the compound.

Step 1: Name the longest chain.

Seven carbons – heptane.

Step 2: Identify the side
chains and name them.

Step 3: Identify the
location of the side
chains.

Numbering the heptane chain from right to left gives the lowest
numbering with the methyl group on carbon 2 and the ethyl
group on carbon 3.

Step 4: List the two side chains in alphabetical order.
Therefore, 3-ethyl-2-methylheptane.

Naming cycloalkanes

Cycloalkanes have the general formula C_nH_{2n}. They can be names
using the following rules:

1   Count the number of carbons in the cyclic structure and identify
     the appropriate prefix (Table 3) to name the alkane.

2 Add the alkane name to the end of cyclo.

There are five carbons so this is cyclopentane.

What are alkanes like?

Whether an alkane is solid, liquid, or gas at room temperature depends on the size of its molecules. The first four members of the series ($n = 1-4$) are colourless gases (Figure 8). Higher members ($n = 5-16$) are colourless liquids and the larger alkanes ($n = 17+$) are white waxy solids (Figure 9).

Alkanes mix well with each other but do not mix with water. The alkanes and water form two separate layers. This is because alkanes contain non-polar molecules but liquids such as water contain polar molecules that attract each other and prevent the alkane molecules mixing with them. There is more about the polarity of molecules in Topic DF 6.

▲ **Figure 8** Propane, C_3H_8, gas cylinders

▲ **Figure 9** The hydrocarbon $C_{31}H_{64}$ is a typical component of paraffin wax

Summary questions

1 Copy and complete the following table.

Empirical formula	Molecular formula	M_r
	C_3H_8	44.0
CH_2		168.0
	C_6H_6	
$C_{10}H_{21}$		282.0
	C_5H_{10}	
CH		26.0
	$C_{10}H_8$	

2 What is the molecular formula of each of the following alkanes?
  a heptane (1 mark)
  b hexadecane – sixteen carbon atoms (1 mark)
  c eicosane – twenty carbon atoms (1 mark)

3 Draw dot-and-cross diagrams for:
  a ethane (1 mark)
  b ethene (1 mark)
  c propane (1 mark)

4 A hydrocarbon contains 85.7% of carbon and 14.3% of hydrogen by mass. Its relative molecular mass is 28.0.
  a Find its empirical formula. (1 mark)
  b Suggest a molecular formula for this compound. (1 mark)

5 A hydrocarbon contains 82.8% by mass of carbon.
  a Work out its empirical formula. (1 mark)
  b Suggest its molecular formula, explaining your reasons. (1 mark)

DF 4 Where does the energy come from?

Specification reference: DF(e)

Carrying fuels around

The enthalpy change of combustion may not be the most important thing to consider for a practical fuel. What really matters is the **energy density** – how much energy you get per kilogram of fuel. This can be worked out from the enthalpy change of combustion using the relative molecular mass (Table 1).

▲ **Figure 1** *The more energy this tanker can carry, the more cost effective the journey is*

Learning outcomes

Demonstrate and apply knowledge and understanding of:

→ the term average bond enthalpy and the relation of bond enthalpy to the length and strength of a bond; bond-breaking as an endothermic process and bond-making as exothermic; the relation of these processes to the overall enthalpy change for a reaction.

Why do different fuels release different amounts of energy? All chemical reactions involve breaking and making chemical bonds.

▼ **Table 1** *Energy densities of some important fuels*

Fuel	Formula	Standard enthalpy change of combustion $\Delta_c H^{\ominus}$ / kJ mol^{-1}	Relative molecular mass	Energy density – energy transferred on burning 1 kg of fuel / kJ kg^{-1}
hexadecane (cetane)	$C_{16}H_{34}(l)$	−10 700	226	−47 300
hexane	$C_6H_{14}(l)$	−4163	86	−48 400
methane	$CH_4(g)$	−890	16	−55 600
ethanol	$C_2H_5OH(l)$	−1367	46	−29 700
carbon	$C(s)$	−393	12	−32 800
hydrogen	$H_2(g)$	−286	2	−143 000

Bonds break in the reactants and new bonds form in the products. The energy changes in chemical reactions come from the energy changes that happen when bonds are broken and made.

Chemical ideas: Energy changes and chemical reactions 4.3

Bond enthalpies

A chemical bond is basically electrical attraction between atoms or ions. Breaking a bond involves overcoming these attractive forces. To break the bond completely, the atoms or ions need (theoretically) to be an infinite distance apart. Figure 2 illustrates this for the hydrogen–hydrogen bond in a molecule of hydrogen, H_2.

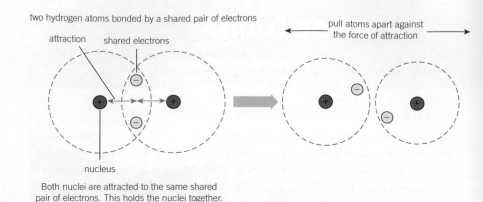

two hydrogen atoms bonded by a shared pair of electrons

attraction   shared electrons

pull atoms apart against
the force of attraction

nucleus

Both nuclei are attracted to the same shared
pair of electrons. This holds the nuclei together.

▲ **Figure 2** *Breaking a bond involves using energy to overcome the forces of attraction*

The quantity of energy needed to break a particular bond in a molecule is called the bond dissociation enthalpy, or **bond enthalpy**.

For the bond shown in Figure 2, the process involved is

$$H_2(g) \rightarrow 2H(g) \qquad \Delta H = +436\,\text{kJ}\,\text{mol}^{-1}$$

The bond enthalpy of the H—H bond is $+436\,\text{kJ}\,\text{mol}^{-1}$. ΔH has a positive value because breaking a bond is an endothermic process – it needs energy. Bond enthalpies are very useful because they indicate how strong bonds are. The stronger a bond, the more energy is needed to break it and the higher its bond enthalpy.

Bond enthalpy and bond length

When a bond like the one in Figure 2 forms, the atoms move together because of the attractive forces between nuclei and electrons. There are also repulsive forces between the nuclei of the two atoms. These get bigger as the atoms approach until the atoms stop moving together. The distance between them is now the equilibrium bond length (Figure 3). The shorter the bond length, the stronger the attraction between the atoms.

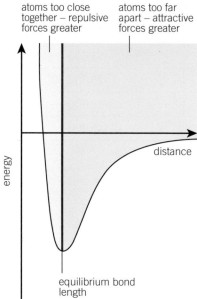

atoms too close together – repulsive forces greater

atoms too far apart – attractive forces greater

energy

distance

equilibrium bond length

▲ **Figure 3** *In a chemical bond there is a balance between attractive and repulsive forces*

▼ **Table 2** *Average bond enthalpies and bond lengths from a range of compounds (*except for C=O where the data is for carbon dioxide)*

Bond	Average bond enthalpy / kJ mol^{-1}	Bond length / nm
C—C	+347	0.154
C=C	+612	0.134
C≡C	+838	0.120
C—H	+413	0.108
O—H	+464	0.096
C—O	+358	0.143
C=O*	+805	0.116
O=O	+498	0.121
N≡N	+945	0.110

Table 2 gives some values for bond enthalpies and bond lengths. These are all **average bond enthalpies** because the exact value of a bond enthalpy actually depends on the particular compound in which the bond is found. From the table you can see that:

- Double bonds have much higher bond enthalpies than single bonds. Triple bond enthalpies are even higher.

- In general, the higher the bond enthalpy, the shorter the bond – you can see this if you compare the lengths of the single, double, and triple bonds between carbon atoms. This is because there are more electrons between the atoms being attracted to the positive nuclei. More attraction makes shorter bonds.

Average bond enthalpy
The average quantity of energy needed to break a particular bond.

Measuring bond enthalpies

It isn't easy to measure bond enthalpies because there is often more than one type of bond in a compound. It is also very difficult to make measurements when everything is in the gaseous state. For this reason, bond enthalpies are measured indirectly using enthalpy cycles.

Breaking and making bonds in a chemical reaction

Let's look again at the reaction that occurs when methane burns:

$$CH_4(g) + 2O_2(g) \rightarrow CO_2(g) + 2H_2O(l) \qquad \Delta H = -890\,kJ\,mol^{-1}$$

The reaction involves both breaking bonds and making new bonds. By looking at the structures of both reactants and products it is possible to work out which bonds have been broken and which have been formed. In this case, four bonds between carbon and hydrogen in a methane molecule and a bond between oxygen atoms in two oxygen molecules have been broken. This bond-breaking requires energy. Once the bonds have been broken, the atoms can join together again to form new bonds – two carbon–oxygen double bonds, C=O, in a carbon dioxide molecule and four oxygen–hydrogen bonds in two water molecules.

Bond enthalpies always refer to breaking a bond in the gaseous compound. This means you can make fair comparisons between different bonds. You do not have to break all the old bonds before you can make new ones. New bonds start forming as soon as the first of the old bonds have broken.

In the combustion of methane, the energy taken in during the bond-breaking steps is less than the energy given out during the bond-making steps (Figure 4) so the overall reaction is exothermic. (If the reverse is true, the reaction is endothermic.)

As you need to break bonds before product molecules can begin to form, many reactions need heating to get them started. All reactions initially need energy to stretch and break bonds. Some reactions need only a little energy and there is enough energy available in the surroundings at room temperature. The reaction of acids with alkalis is an example.

Study tip

Bond-breaking is an endothermic process, so bond enthalpies are always positive.

Bond-making is exothermic.

gaseous atoms

breaking bonds
(takes in energy)

making new bonds
(gives out energy)

energy

reactants

products

▲ **Figure 4** *Breaking and making bonds in the reaction between methane and oxygen*

Other reactions need heating to get them started for example, applying a lit match to ignite methane on a gas cooker.

It isn't necessary for all the bonds to break before a reaction gets going. If it was, you would have to heat things to very high temperatures to make them react. Once a few bonds have broken, new bonds can start to form and this usually gives out enough energy to keep the reaction going. This is what happens when fuels burn. Some other reactions need continuous heating, for example, reactions that are only slightly exothermic.

Bonds and enthalpy cycles

You can represent bond-breaking and bond-making in an enthalpy cycle, such as the one given in Figure 5.

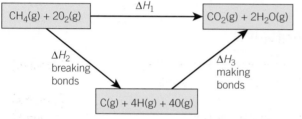

$CH_4(g) + 2O_2(g)$ — ΔH_1 → $CO_2(g) + 2H_2O(g)$

ΔH_2
breaking
bonds

ΔH_3
making
bonds

$C(g) + 4H(g) + 4O(g)$

▲ **Figure 5** *An enthalpy cycle to show bond-breaking and bond-making in the combustion of methane*

Worked example: Enthalpy changes from bond enthalpies

Use the enthalpy cycle shown above to work out a value for the enthalpy change of combustion of methane.

Step 1: Calculate the enthalpy change when bonds are broken, ΔH_2.

$$= 4 \times (C—H) + 2 \times (O=O)$$
$$= +2648 \, kJ \, mol^{-1}$$

Step 2: Calculate the enthalpy change when bonds are made, ΔH_3.

$$= -[2 \times (C=O) + 4 \times (O—H)]$$
$$= -3466 \, kJ \, mol^{-1}$$

The minus sign is there because energy is *released* when bonds are made.

Step 3: Write the expression for the enthalpy change of combustion, ΔH_1.

$$\Delta H_1 = \Delta H_2 + \Delta H_3$$

Step 4: Substitute in the values for ΔH_2 and ΔH_3.

$$= +2648 \, kJ \, mol^{-1} + (-3466 \, kJ \, mol^{-1})$$
$$= -818 \, kJ \, mol^{-1}$$

Study tip

Some people use the following shortcut here:

$$\text{enthalpy change} = \sum \text{bonds broken} - \sum \text{bonds made}$$

The value calculated in the Worked Example is different from the $-890\,\text{kJ}\,\text{mol}^{-1}$ given in data sheets for the standard enthalpy change of combustion of methane. There are two main reasons for this:

1 The value of ΔH_1 calculated here is not actually the standard value. In the equation the water product is gaseous, $H_2O(g)$, not liquid, $H_2O(l)$, as it would be under standard conditions. $H_2O(g)$ is used because when using bond enthalpies you have to work in the gaseous state.

2 The bond enthalpies given are often averages from several compounds. This makes them very useful as a 'toolkit' of values but it does mean that the results of such calculations are not always precise. However bond enthalpies are useful because they enable enthalpy changes to be measured when there is little specific data for a compound.

Study tip

The following are common errors made in these type of calculations:

● forgetting that bonds in O_2 are broken

● not noticing that there are two $C{=}O$ bonds per CO_2 and two $O{-}H$ bonds per H_2O.

Summary questions

You may need to look up values of bond enthalpies for these problems.

1 a Write an equation, with state symbols, for the combustion of one mole of propane, C_3H_8. *(2 marks)*
  b Write this equation again using full structural formulae, showing all the bonds within each molecule. *(2 marks)*
  c Make a list of the types and numbers of bonds that are broken in the reaction. *(1 mark)*
  d Make a list of the types and numbers of bonds that are made in the reaction. *(1 mark)*
  e Use bond enthalpies to calculate a value for the total enthalpy change involved in breaking all the bonds in the reaction of propane and oxygen. Decide whether bond breaking is an exothermic or endothermic process and put a negative or positive sign in front of the value calculated. *(2 marks)*
  f Use bond enthalpies to calculate a value for the total enthalpy change involved in making all the new bonds in the reaction of propane and oxygen. Decide whether bond making is an exothermic or endothermic process and put a negative or positive sign in front of the value calculated. *(2 marks)*
  g Add together the enthalpy changes involved to calculate the overall enthalpy change of the reaction. *(1 mark)*

2 Ethene reacts with bromine to form 1,2-dibromoethane.

$$H_2C{=}CH_2(g) + Br_2(g) \rightarrow BrH_2C{-}CH_2Br(g)$$

Use bond enthalpies to calculate a value for the enthalpy change of the reaction. Bond enthalpies: *(3 marks)*

$$Br{-}Br = +193\,\text{kJ}\,\text{mol}^{-1}; C{-}Br = +290\,\text{kJ}\,\text{mol}^{-1}.$$

3 Some apparently simple organic molecules do not exist because they are unstable and form another compound or compounds very easily. One such compound is ethenol, CH_2CHOH, which converts to ethanal, CH_3CHO.

Use bond enthalpies to calculate the enthalpy change of the reaction. *(3 marks)*

4 The difference between the enthalpy change of combustion of successive straight-chain alkanes is around $-650\,\text{kJ}\,\text{mol}^{-1}$. This is because each alkane differs from the previous one by a methyl fragment, $-CH_2$.
  a Write an equation for the combustion of the one carbon and two hydrogen atoms in this fragment. *(1 mark)*
  b Use bond enthalpies to calculate a value for the enthalpy change of the reaction in part a (include the breaking of the $C{-}C$ bond). *(2 marks)*
  c Suggest why this value is not $-650\,\text{kJ}\,\text{mol}^{-1}$. *(2 marks)*

DF 5 Getting the right sized molecules

Specification reference: DF(i), DF(h), DF(j), DF(l)

In Topic DF 3 you saw that gasoline is produced by fractional distillation of crude oil. This leaves two problems. The first is that the 'straight-run' gasoline from the primary distillation makes poor petrol. Some is used directly in petrol but most is treated further. The second is a problem of supply and demand. Crude oil contains a surplus of the high boiling fractions, such as the gas oil and the residue, and not enough of the lower boiling fractions, such as gasoline. Although demand for gas oil itself is comparatively low, it can be cracked and used in car petrol, therefore increasing demand. Figure 1 compares the supply and demand for different fractions of crude oil. The demand is greater than supply for both petrol and diesel.

The job of the refinery is to convert crude oil into useful components. In order to do this, the structure of the alkane molecules present must be altered to produce different alkanes. The alkanes are also converted into other types of hydrocarbon that are used in petrol. These include cycloalkanes, arenes (aromatic hydrocarbons), and alkenes. The products are blended to produce high-grade petrol.

Using the whole barrel

Cracking is one of the most important reactions in the petroleum industry. It starts with alkanes that have large molecules that are too big to use in petrol, for example, alkanes from the gas oil fraction. These large molecules are broken down to give alkanes with shorter chains that can be used in petrol. What's more, these shorter-chain alkanes tend to be highly branched so petrol made by cracking has a higher octane number. Another benefit of cracking is that it also helps to solve the supply and demand problem.

Cracking

The term **cracking** is used to describe any reaction in which a larger molecule is made into smaller molecules.

$$C_{11}H_{24} \rightarrow C_8H_{18} + C_3H_6$$

In this example a long-chain alkane from the kerosene fraction is made into octane (an alkane suitable for car petrol) and another compound, propene. Propene is **unsaturated** – it does not have as many hydrogen atoms as it could for the three carbon atoms. It has a carbon–carbon double bond (Figure 2) and is an **alkene**. Alkenes are described in more detail in DF 7.

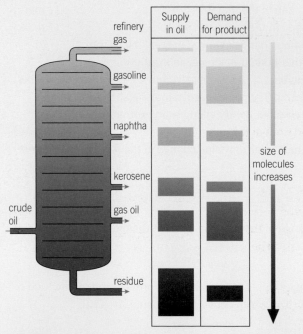

▲ **Figure 1** *Supply and demand for different fractions of crude oil*

▲ **Figure 2** *Propene*

Cracking: how is it done?

Much of the cracking carried out to produce petrol is done by heating heavy oils, such as gas oil, in the presence of a **catalyst**. You will be able to find out about catalysts later in this topic. This is catalytic cracking. The molecules in the feedstock can have 25–100 carbon atoms, although most will usually have 30–40 carbon atoms.

Cracking reactions are quite varied. Some of the types of reactions are:

- alkanes → branched alkanes + branched alkenes
- alkanes → smaller alkanes + cycloalkanes
- cycloalkanes → alkenes + branched alkenes
- alkenes → smaller alkenes.

The alkenes that are produced are important starting materials for other parts of the petrochemicals industry. Cracking always produces many different products, which need to be separated in a fractionating column.

In a modern catalytic cracker, the cracking takes place in a 60-metre high vertical tube about two metres in diameter. It is called a riser reactor because the hot vaporised hydrocarbons and zeolite catalyst are fed into the bottom of the tube and forced upwards by steam. The mixture is a moving fluidised bed where the solid particles flow like a liquid.

It takes the mixture about two seconds to flow from the bottom to the top of the tube – so the hydrocarbons are in contact with the catalyst for a very short period of time.

One of the problems with catalytic cracking is that, in addition to all the reactions you have already met, coke (carbon from the decomposition of hydrocarbon molecules) forms on the catalyst surface so that the catalyst eventually becomes inactive. The powdery catalyst needs to be regenerated to overcome this problem.

After the riser reactor, the mixture passes into a separator where steam carries away the cracked products leaving behind the solid catalyst. The catalyst goes into the regenerator, where it takes about 10 minutes for the coke to burn off in the hot air that is blown through the regenerator. The catalyst is then reintroduced into the base of the reactor ready to repeat the cycle.

The energy released from the burning coke heats up the catalyst. The catalyst transfers the energy to the feedstock so that cracking can occur without additional heating.

Catalytic crackers have been in operation since the late 1940s and have become very flexible and adaptable. They can handle a wide range of different feedstocks. The conditions and catalyst can be varied to give the maximum amount of the desired product – in this case branched alkanes for blending in petrol.

Cracking

Any reaction in which a larger molecule is made into smaller molecules.

Unsaturated

Any organic compound that has a double (or triple) bond between carbon atoms.

Synoptic link 🧪

You can find a general account of how to carry out cracking in Techniques and procedures.

Activities DF 5.1 and DF 5.2

You can find out more about cracking, and try cracking alkanes for yourself, in these activities.

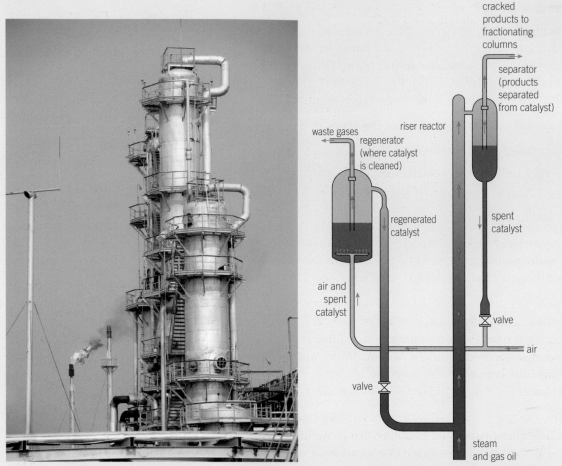

▲ **Figure 3** *A catalytic cracker (left) and how a catalytic cracker works (right). The feedstock is gas oil and the cracking reaction takes place in the riser reactor*

Chemical ideas: Rates of reactions 10.3

Catalysts

A **catalyst** is a substance that speeds up a reaction but can be recovered chemically unchanged at the end. The process of speeding up a chemical reaction using a catalyst is called **catalysis**.

Catalysts do not undergo any permanent *chemical* change, though sometimes they may be changed physically, for example, the surface of a solid catalyst may crumble or become roughened. This suggests that the catalyst is taking some part in the reaction, but is being regenerated.

Usually only small amounts of a catalyst are needed. The catalyst does not affect the amount of product formed, only the rate at which it is formed. A catalyst does not appear as a reactant in the overall equation for a reaction.

Homogenous catalysis

If the reactants and catalyst are in the same physical state (e.g., both are in aqueous solution) then the reaction is said to involve **homogeneous**

Catalyst

A substance which speeds up a reaction but can be recovered chemically unchanged at the end.

Catalysis

The process of speeding up a chemical reaction using a catalyst.

Synoptic link

Catalysts in living systems are called enzymes. You can find out more about enzymes in Chapter 7, Polymers and life.

catalysis. Enzyme-catalysed reactions in cells take place in aqueous solution and are examples of this type of catalysis.

Heterogeneous catalysis

Many important industrial processes involve **heterogeneous catalysis**, where the reactants and the catalyst are in different physical states. This usually involves a mixture of gases or liquids reacting in the presence of a solid catalyst.

When a solid catalyst is used to increase the rate of a reaction between gases or liquids, the reaction occurs on the surface of the solid (Figure 4). The reactants form bonds with atoms on the surface of the catalyst – they are **adsorbed** onto the surface. As a result, bonds in the reactant molecules are weakened and break. New bonds form between the reactants, held close together on the surface, to form the products. This in turn weakens the bonds to the catalyst surface and the product molecules are released.

It is important that the catalyst has a large surface area for contact with reactants. For this reason, solid catalysis are used in a finely divided form or as a fine wire mesh. Sometimes the catalyst is supported on a porous material to increase its surface area and prevent it from crumbling. This happens in the catalytic converters fitted to car exhaust systems.

Many of the heterogeneous catalysts used in industrial processes are transition metals (the metals in the central block of the periodic table) or transition metal compounds. You will meet a number of examples at different stages in the course.

Some examples of heterogeneous catalysis you will study more of are the use of platinum and rhodium in catalytic converters in cars (Topic DF 10) and the use of nickel powder in the hydrogenation of unsaturated oils to give saturated fats (Topic DF 6).

Catalyst poisoning

Catalysts can be poisoned so that they no longer function properly. Many substances which are poisonous to humans operate as a **catalyst poison**, blocking an enzyme-catalysed reaction.

In heterogeneous catalysis, the poison molecules are adsorbed more strongly to the catalyst surface than the reactant molecules. The catalyst cannot catalyse a reaction of the poison and so becomes inactive, with poison molecules blocking the active sites on its surface. This is the reason why leaded petrol cannot be used in cars fitted with a catalytic converter – lead is strongly adsorbed to the surface of the catalyst.

Catalyst poisoning is also the reason why it is not possible to replace the very costly metals (platinum and rhodium) in catalytic converters by cheaper metals (such as copper and nickel). These metals are vulnerable to poisoning by the trace amounts of sulfur dioxide present in car exhaust gases. Once the catalyst in a converter becomes inactive it cannot be regenerated. A new converter has to be fitted and this can be very costly.

Heterogeneous catalysis

Where the catalyst and the reactants are in different physical states.

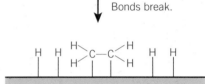

catalyst surface

↓ Reactants get adsorbed onto catalyst surface. Bonds are weakened.

↓ Bonds break.

↓ New bond forms.

Second bond forms, and product diffuses away from catalyst surface, leaving it free to absorb fresh reactants. ↓

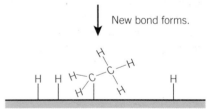

▲ **Figure 4** *An example of heterogeneous catalysis. The diagrams show a possible mechanism for nickel catalysing the reaction between ethene and hydrogen to form ethane*

Catalyst poison

A substance that stops a catalyst functioning properly.

Catalyst poisoning can be a problem in industrial processes. In the UK, nearly all the hydrogen for the Haber process is prepared by steam reforming of methane. Methane reacts with steam in the presence of a nickel catalyst.

$$CH_4(g) + H_2O(g) \xrightarrow{Ni(s)} CO(g) + 3H_2(g)$$

If the feedstock for the process contains sulfur compounds, these must be removed first to prevent severe catalyst poisoning.

Sometimes it is possible to clean or regenerate the surface of a catalyst. For example, in the catalytic cracking of long-chain hydrocarbons, carbon is produced and the surface of the catalyst becomes coated in a layer of soot. This blocks the adsorption of reactant molecules and the activity of the catalyst is reduced. The catalyst is constantly recycled through a separate container where hot air is blown through the powder. The oxygen in the air converts the carbon to carbon dioxide and cleans the catalyst surface.

Synoptic link

You will learn more about catalysts in Chapter 4, The ozone story.

Synoptic link

Catalysts are used in chemical industry and you will find out more about this in Chapter 6, The chemical industry.

Summary questions

1 Name a catalyst involved in each of the following industrial processes. In each case, state whether the process involves homogeneous or heterogeneous catalysis.
   a Catalytic cracking of long-chain hydrocarbons.   (1 mark)
   b Oxidation of carbon monoxide and unburnt petrol in a car exhaust.   (1 mark)

2 When carbon monoxide and nitrogen monoxide in car exhaust gases pass through a catalytic converter, carbon dioxide and nitrogen are formed.
   a Write a balanced chemical equation for this reaction.   (2 marks)
   b Explain why it is important to reduce the quantities of carbon monoxide and nitrogen monoxide released into the atmosphere.   (1 mark)
   c Explain the meaning of the term adsorbed.   (1 mark)
   d Suggest why catalytic converters do not work effectively until a car engine has warmed up.   (2 marks)

3 Figure 4 in this section shows a possible mechanism for the nickel-catalysed reaction between ethene and hydrogen. The reactants are adsorbed onto the surface of the nickel catalyst, where the reaction takes place. Using Ⓝ–Ⓞ to represent nitrogen monoxide and Ⓒ–Ⓞ to represent carbon monoxide, draw out a possible mechanism for the formation of carbon dioxide, Ⓞ=Ⓒ=Ⓞ as in the equation you have written in 2a.   (3 marks)

DF 6 Alkenes – versatile compounds

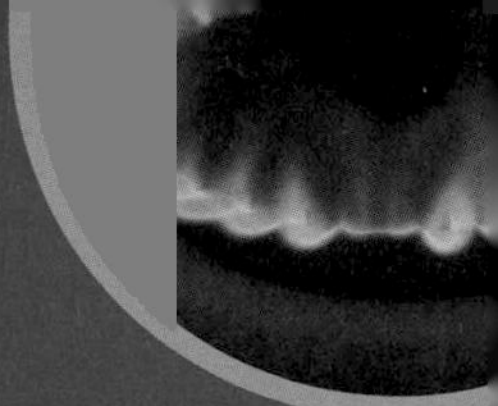

Specification reference: DF(b), DF(m), DF(o), DF(q)

A source of energy is essential for human beings to function. The source of our energy, our fuel, is food. If you look at the labels on many foods, you will see references to fats, and these described as saturated or unsaturated. Unsaturated fats contain carbon-carbon double bonds and are related to alkenes.

▲ **Figure 1** *Assorted fats and oils including olive oil, sunflower oil, butter, goose fat, duck fat, lard, and margarine*

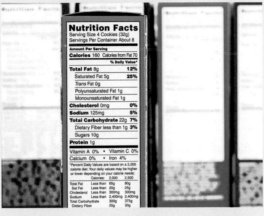

▲ **Figure 2** *The nutrition label from a cookie box emphasizing that the product is high in sugar and fat content. The label states that the food contains 0 g of trans fat*

Animal fats, for example, butter and lard, contain a much higher proportion of saturated fat than oils and fats derived from plants. (Oils have the same chemical make-up as fats, just lower melting points.) These saturated fats were for a long time associated with heart disease, as they were thought to cause a build-up of cholesterol in the arteries. However, the modern villains are trans fats. *cis* and *trans* refer to the arrangement of the groups across a carbon–carbon double bond. Naturally occurring unsaturated fats are *cis*. However, these often have melting points that are slightly too low for use in the food industry, so partial hydrogenation was carried out to react the fats with hydrogen in order to saturate some of the unsaturated bonds. In the process, some *cis* compounds were turned to *trans* and these compounds are found to be very likely to lead to a build-up of cholesterol. *Trans* fats are therefore being phased out in most food areas despite the fact they did add to the crispiness of things like biscuits.

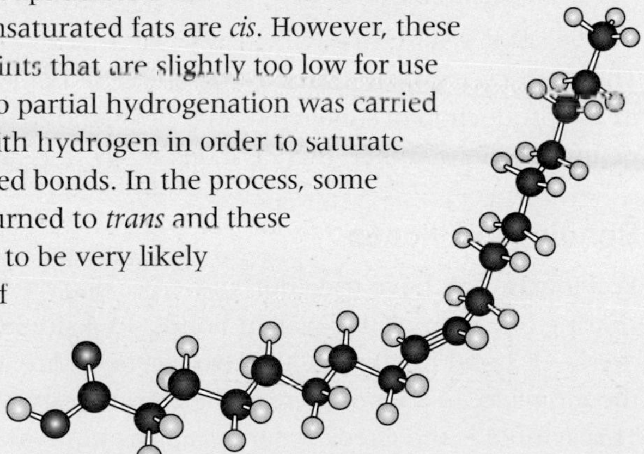

▲ **Figure 3** *Molecular model of elaidic acid. Elaidic acid is derived from a major trans fat found in hydrogenated vegetable oils. The trans group can be seen in the middle of the molecule*

Learning outcomes

Demonstrate and apply knowledge and understanding of:

→ the nomenclature, general formula, and structural formulae for alkenes

→ the bonding in organic compounds in terms of σ- and π-bonds

→ the terms addition, electrophile, carbocation; the mechanism of electrophilic addition to alkenes using curly arrows; how the products obtained when other anions are present can be used to confirm the model of the mechanism

→ the addition reactions of alkenes with the following, showing the greater reactivity of the C=C bond compared with C—C:

  • bromine to give a dibromo compound, including techniques and procedures for testing compounds for unsaturation using bromine water

  • hydrogen bromide to give a bromo compound

  • hydrogen in the presence of a catalyst to give an alkane (Ni with heat and pressure or Pt at room temperature and pressure)

  • water in the presence of a catalyst to give an alcohol (concentrated H_2SO_4, then add water; or steam/ H_3PO_4/ heat and pressure).

H   H
 \ /
  C = C
 / \
H   H

▲ **Figure 4** *Ethene*

▼ **Table 1** *Other members of the alkene family*

Name of alkene	Formula
propene	$CH_3-CH=CH_2$
but-1-ene	$CH_3-CH_2-CH=CH_2$
but-2-ene	$CH_3-CH=CH-CH_3$
pent-1-ene	$CH_3-CH_2-CH_2-CH=CH_2$

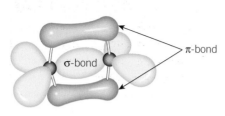

▲ **Figure 5** *σ-bonds and π-bonds*

Electrophile

A positive ion, or a molecule with a partial positive charge, that will be attracted to a negatively charged region and react by accepting a lone pair of electrons to form a covalent bond.

Chemical ideas: Organic chemistry: frameworks 12.2a

Alkenes

Naming alkenes and electrophilic addition

Ethene (Figure 4) is the simplest example of a class of hydrocarbons called **alkenes**.

Alkenes are distinguished from other hydrocarbons by the presence of the carbon–carbon double bond, C=C. The double bond implies that they are unsaturated hydrocarbons.

Examples of other members of the alkene family are listed in Table 1.

As with the alkanes, the boiling points of alkenes increase as the number of carbon atoms increases. Ethene, propene, and butene are gases. After that they are liquids and, eventually, solids.

All non-cyclic alkenes have the general formula, C_nH_{2n}, where n is the number of carbon atoms and can be any whole number. This is the same as the general formula of the cycloalkanes (e.g., cyclohexane) but the double bond makes alkenes react very differently from cycloalkanes.

Naming alkenes

Alkenes are named using similar rules to those used for alkanes.

1 Identify the longest chain and use the appropriate prefix to the number of carbons (Topic DF 3).

2 Follow the prefix with 'ene' to indicate the compound in an alkene.

3 For alkenes where the longest carbon chain is four carbon atoms or longer, insert a number before 'ene' to indicate the location of the carbon–carbon double bond.

4 For branched alkenes, you follow the same rules for branched alkanes (Topic DF 3).

For example, in but-1-ene the C=C is between C_1 and C_2 so the lowest number is used.

There are cycloalkenes, such as cyclohexene (an important intermediate in the production of some types of nylon) and there are dienes, such as penta-1,3-diene. When there is a diene, an 'a' is added to the prefix.

Bonding in alkenes

Previously, you have used dot-and-cross diagrams to illustrate the sharing of electrons in covalent bonds. In a single bond, for example, the C—C bond in alkanes, the two electrons are arranged between the atoms in an area of increased electron density, called a σ-bond. This symbol is the Greek letter *sigma*, the equivalent to s. It is so-called because the overlap of s-orbitals gives rise to a σ-bond, though there are other ways in which they can be formed. A double bond consists of one σ-bond and another type of bond, called a π-bond. *Pi* is the Greek letter for p. *One* π-bond consists of *two* areas of negative charge, one above and the other below the line of the atoms (Figure 5).

Chemical reactions of ethene

The four electrons in the double bond of ethene give the region between the two carbon atoms a higher than normal density of negative charge. Positive ions, or molecules with a partial positive charge on one of the atoms, will be attracted to this negatively charged region. They can react by accepting a pair of electrons from the C=C double bond. Substances that do this are called **electrophiles**.

Electrophilic addition reactions

Reaction with bromine

When ethene gas is bubbled through bromine, the red-brown bromine becomes decolorised – this is a good general test for unsaturation in an organic compound.

In order to interpret how a reaction occurs, chemists use reaction mechanisms. This involves logically deciding the movement of electrons using ideas such as bond polarity and charges. Look at the mechanism for the reaction of ethene with bromine. Chemists believe that the bromine molecule becomes **polarised** as it approaches the alkene. This means that the electrons in the bromine are repelled by the alkene electrons and are pushed back along the molecule. The bromine atom nearest the alkene becomes slightly positively charged and the bromine atom furthest from the alkene becomes slightly negatively charged. The positively charged bromine atom now behaves as an **electrophile** and reacts with the alkene double bond.

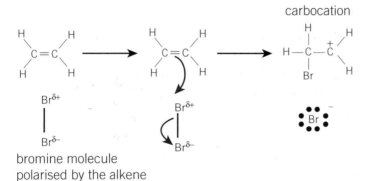

bromine molecule
polarised by the alkene

Remember, curly arrows like these represent the movement of a pair of electrons in chemical reactions. One of the carbon atoms now only has six outer electrons – it has become positively charged. It is a **carbocation**.

Carbocations react very rapidly with anything that has electrons to share – such as the bromide ion. A pair of electrons moves from the bromide ion to the positively charged carbon to form a new carbon–bromine covalent bond.

> ### Study tip
> An organic reaction mechanism represents the sequence of events in a reaction using the movement of electrons, represented by curly arrows.

> ### Activity DF 6.1
> This activity enables you to test for unsaturation used bromine.

> ### Study tip
> Make sure you clearly show precisely where electrons move to and where they have moved from using your curly arrows.

> ### Study tip
> The Br⁻ could attack from either side of the positively charged carbon atom – here it is shown attacking from below.

> ### Carbocation
> An ion with a positively-charged carbon atom.

Synoptic link 🧪

Adding bromine water to an alkene can be used as a test for unsaturation. You can find out more about this in Techniques and procedures.

Addition reaction

A reaction where two or more molecules react to form a single larger molecule.

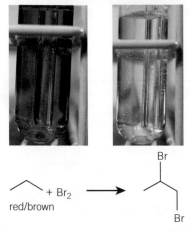

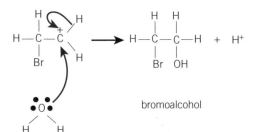

red/brown

▲ **Figure 6** *Red-brown bromine (left) and decoloured bromine after ethene has been passed through (right). The colourless 1,2-dibromoethane has been formed*

bromoalcohol

▲ **Figure 7** *Formation of a bromoalcohol*

Activity DF 6.2

This activity helps you to check your understanding of electrophilic addition mechanisms.

The overall reaction for the two steps in the mechanism is

This overall scheme is an example of an organic **reaction mechanism**.

An **addition reaction** is one where two or more molecules react to form a single, larger molecule. This equation represents an addition reaction and since the initial attack is by an electrophile the process is called an **electrophilic addition**.

How do you know the proposed mechanisms are correct? The mechanism for electrophilic addition via a carbocation is supported by experimental evidence. For example, if hex-1-ene reacts with bromine in the presence of chloride ions, two products are formed – 1,2-dibromohexane and 1-bromo-2-chlorohexane. However, 1,2-dichlorohexane is not formed as the electrophile must attack first, followed by an anion.

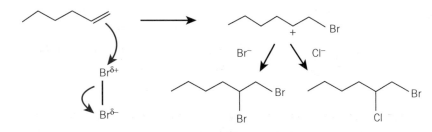

Often the test for an alkene involves shaking the alkene with bromine water rather than pure bromine. In this case, there is an alternative to the second stage in the reaction. Water molecules have lone pairs of electrons and can act as nucleophiles in competition with the bromide ions.

If the bromine water is dilute, there will be many more water molecules than bromide ions present and the bromoalcohol will be the main product of the reaction. This does not affect what you see – the bromine water is still decolorised (Figure 7).

Reaction with hydrogen bromide

The conditions under which a reaction is carried out can be very important in determining the mechanism. For example, ethene reacts readily at room temperature with a solution of hydrogen bromide, HBr, in a polar solvent. It is another example of electrophilic addition.

Alkenes also react with gaseous hydrogen bromide but ions are not involved and the mechanism involves a radical addition.

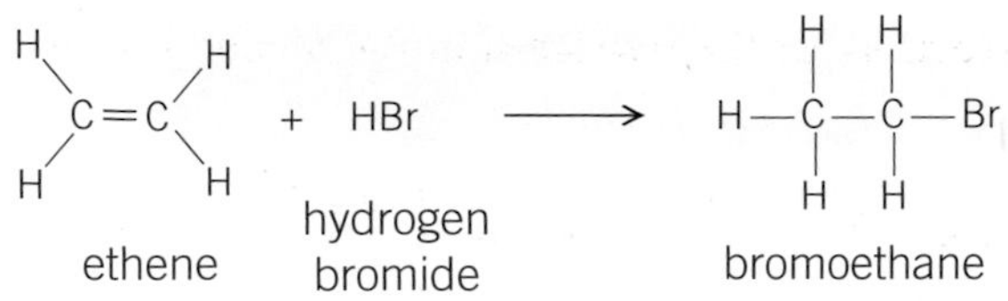

ethene + hydrogen bromide → bromoethane

Reaction with water

At high temperature, high pressure, and in the presence of a catalyst (phosphoric acid adsorbed onto solid silica), ethene and water (as steam) undergo an addition reaction. The process is used for the industrial manufacture of ethanol.

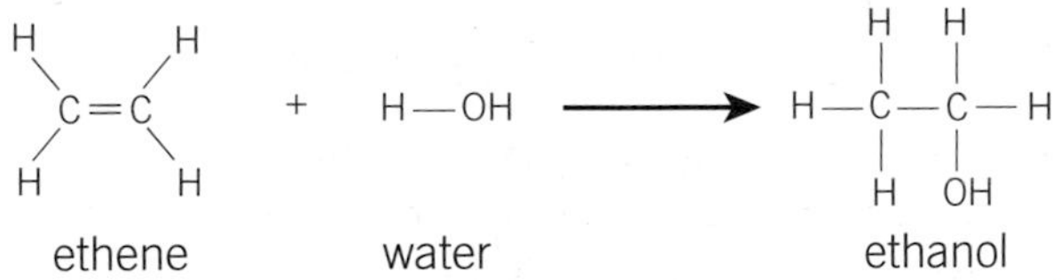

ethene + water → ethanol

In the laboratory, ethene can be converted to ethanol by first adding concentrated sulfuric acid, and then diluting with water.

Step 1

Step 2

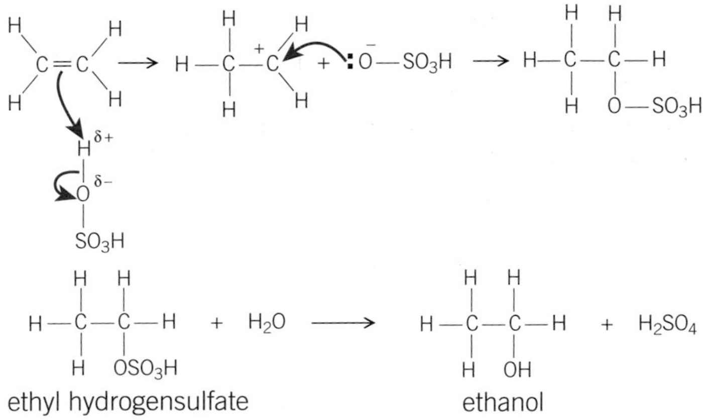

ethyl hydrogensulfate + H_2O → ethanol + H_2SO_4

The overall reaction is addition of water across the double bond. The addition of water to an alkene is an example of a **hydration reaction**.

Reaction with hydrogen

The reaction of ethene with hydrogen is another example of an addition reaction, but here the mechanism involves hydrogen atoms, and takes place on the surface of a catalyst. A catalyst is needed to help break the strong hydrogen–hydrogen bond and form hydrogen atoms that can react with the alkene. If a platinum catalyst is used the process takes place under standard laboratory conditions. Nickel is a cheaper but less efficient catalyst. It needs to be very finely powdered and the gases need heating to approximately 150 °C under a pressure of 5 atm for hydrogenation to occur.

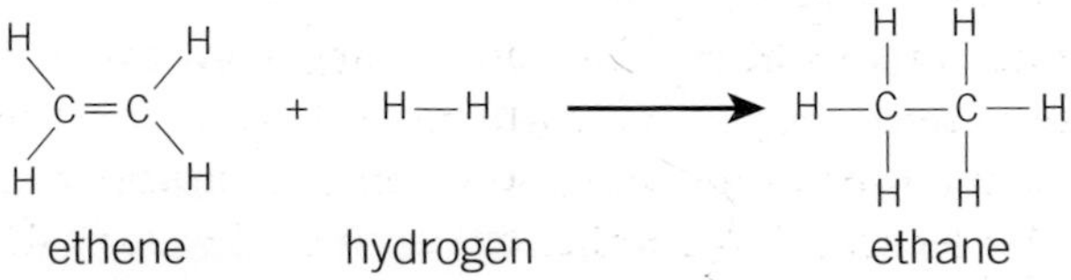

ethene + hydrogen → ethane

This hydrogenation reaction is the reaction used to make the unsaturated fats and oils more saturated. Ethene can also undergo addition reactions to form polymers, as you will see in Topic DF 7.

Summary questions

1 Complete the table of names, structural formulae, and skeletal formulae for the following alkenes. (12 marks)

Name	Structural formula	Skeletal formula
pent-2-ene		
		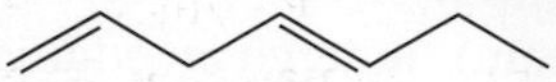
	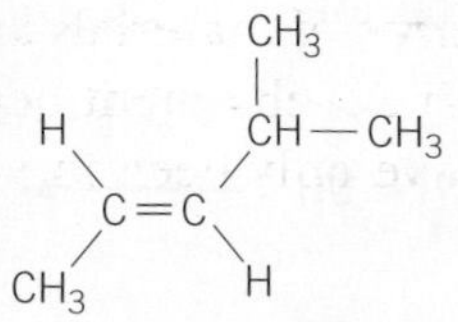	
3-ethylhept-1-ene		

cyclopenta-1,3-diene		

2 Use full structural formulae to write an overall equation for:
  a the reaction of propene with bromine (2 marks)
  b the reaction of propene with hydrogen. (2 marks)

3 The reaction of propene with hydrogen bromide can give two different products. Draw the full structural formulae of these products. (2 marks)

4 Give the reagents and conditions needed to carry out each of the following reactions.
  a bromination of alkenes (2 marks)
  b industrial production of ethanol from ethene (2 marks)
  c laboratory hydrogenation of alkenes to alkanes (2 marks)
  d industrial production of margarine from sunflower oil (2 marks)

5 Describe a test you could carry out in the laboratory to show the presence of unsaturation in an organic compound. State what you would do and what you would see. (2 marks)

6 a Draw the mechanism for the reaction of propene with hydrogen bromide, HBr, to form $CH_3CHBrCH_3$. (4 marks)
  b Draw the structure of another compound that might be formed when propene reacts with HBr. (1 mark)

7 Ethene reacts with dilute aqueous bromine containing some dissolved sodium chloride.
  a Draw the mechanism for the formation of CH_2BrCH_2Cl using these reagents. (4 marks)
  b Give the formula of another compound that would be formed using these reagents and draw the mechanism for its formation. (4 marks)
  c Suggest why $CHBr_2CH_3$ is not formed. (2 marks)

DF 7 Polymers and plastics
Specification reference: DF(p)

You have seen how fuels can be produced from oil. There are many other useful products, including plastics, that can be made from oil based chemicals. These plastics are types of **polymers**. A polymer is a long molecule made up from lots of small molecules called monomers. Polymers are produced in abundance by nature – in plants, animals, and in our bodies. Synthetic polymers are so much part of our lives, both in terms of materials and culture, that it is difficult to believe that their development began as recently as the 1940s. Indeed, polymers have only been in widespread use since the 1950s.

▲ **Figure 1** *Plastics are synthetic polymers that we use in many different ways such as plastic packaging, pipes, skis, and non-stick frying pans*

In the late nineteenth century, **plastics** were produced by modifying natural polymers. Celluloid, for example, was produced by reacting cellulose (from plants) with nitric acid. The first plastic to be made in significant quantities from manufactured chemicals was Bakelite, made from phenol and methanal (Figure 2). Bakelite is still used to make electrical fittings such as sockets and plugs. Although it was first made in 1872 by accident, it was not until 1910 that the process was patented and Bakelite was manufactured.

The polythene story

Imperial Chemical Industries (ICI) was formed in 1926 by the joining together of a number of smaller chemical companies. The prime aim of the merger was to form a strong competitor to the huge German chemical company IG Farben.

In 1930 Eric Fawcett, who was working for ICI, got the go-ahead to carry out research at high pressures and temperatures aimed at producing new dyestuffs. His results were disappointing and his project was eventually abandoned.

His team then moved into the field of high-pressure gas reactions and was joined by Reginald Gibson. On Friday 24th March 1933, Gibson and Fawcett carried out a reaction between ethene and benzaldehyde using a pressure of about 2000 atm. They were hoping to make the two chemicals add together to produce a ketone (Figure 3)

Their apparatus leaked and at one point they had to add extra ethene. They left the mixture to react over the weekend.

▲ **Figure 2** *A celluloid film reel (top) and a Bakelite radio (bottom). Both are examples of polymers*

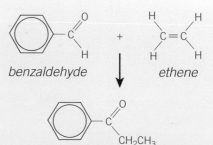

benzaldehyde + *ethene*

▲ **Figure 3** *The reaction Gibson and Fawcett were attempting*

They opened the vessel on the following Monday and found a white waxy solid. When they analysed it they found that it had the empirical formula CH_2. They were not always able to obtain the same results from their experiment – sometimes they got the white solid, on other occasions they had less success, and sometimes their mixture exploded leaving them with just soot!

The work was halted in July 1933 because of the varied results and dangerous nature of the reaction.

Learning to control the process

In December 1935 the work was restarted. Fawcett and Gibson found that they could control the heat given out during the reaction if they added cold ethene at the correct rate. This kept the mixture cooler and prevented an explosion. They also found that they could control the reaction rate and relative molecular mass of the solid formed by varying the pressure.

A month later they had made enough of the material to show that it could be melted, moulded, and used as an insulator.

Most crucial of all was the identification of the role of oxygen in the process – this was done by Michael Perrin, who took charge of the programme in 1935. When oxygen was not present, the polymerisation did not occur. Too much oxygen caused the reaction to run out of control.

▲ **Figure 4** *As well as more obvious uses, this tunnel uses poly(ethene) sheeting to protect the plants from the cold*

The trick was to add just enough oxygen. The leak in Fawcett and Gibson's original apparatus had accidentally let in a small amount of oxygen. If this had not happened then the discovery of poly(ethene) may not have been made. It was also Perrin who showed that even if benzaldehyde is left out of the reaction mixture, the polymer still forms.

Poly(ethene) is tough, durable, and has excellent electrical insulating properties. Unlike rubber, which had previously been used for insulating cables, poly(ethene) is not adversely affected by weather or water. It also has almost no tendency to absorb electrical signals. Its first important use was for insulating a telephone cable laid between the UK mainland and the Isle of Wight in 1939. Its unique electrical properties were again essential during the Second World War in the development of radar.

▲ **Figure 5** *Poly(ethene) was essential in the development of radar equipment in the Second World War*

The first poly(ethene) washing-up bowls appeared in the shops in 1948 and were soon followed by carrier bags, squeezy bottles, and sandwich bags. Sadly, poly(ethene) and some other early polymer materials were over-exploited. They were used for all manner of novelty items and as cheap but poor substitutes for many natural materials. This gave plastic a bad name – the word plastic is often used to describe something which looks cheap and does not last. The reputation still sticks despite the durability and wide range of uses of modern polymers.

Elastomers, plastics, and fibres

Polymer properties vary widely. Polymers that are soft and springy, which can be deformed and then go back to their original shape, are called **elastomers**. Rubber is an elastomer.

Poly(ethene) is not so springy and when it is deformed it tends to stay out of shape, undergoing permanent or plastic deformation. Polymers like this can be incorporated into plastics.

Stronger polymers, which do not deform easily, are just what is needed for making clothing materials. Some can be made into strong, thin threads which can then be woven together. These polymers, such as nylon, can be used as fibres.

Poly(propene) is on the edge of the plastic/fibre boundary. It can be used as a plastic, like poly(ethene), but it can also be made into fibres for use in carpets.

▲ **Figure 6** *This copy of the famous sculpture Venus Di Milo and was made from nylon using a 3D printer*

Chemical ideas: Organic chemistry: frameworks 12.2b

Addition polymerisation

What is a polymer?

A polymer molecule is a long molecule made up from lots of small molecules called monomers. If all the monomer molecules, A, are the same, an A–A polymer forms:

$$- - A + A + A + A - - \rightarrow - -A–A–A–A - -$$

Poly(ethene) and poly(chloroethene), PVC, are examples of A–A polymers. If two different monomers are used, an A–B polymer may be formed, in which A and B monomers alternate along the chain:

$$- - A + B + A + B - - \rightarrow - - A–B–A–B - -$$

Polyamides (nylons) and polyesters are examples of this type of A–B polymer.

Many polymers are formed in a reaction known as **addition polymerisation**. The monomers usually contain carbon–carbon double bonds, for example, in alkenes. The addition polymerisation of propene is typical example (Figure 7).

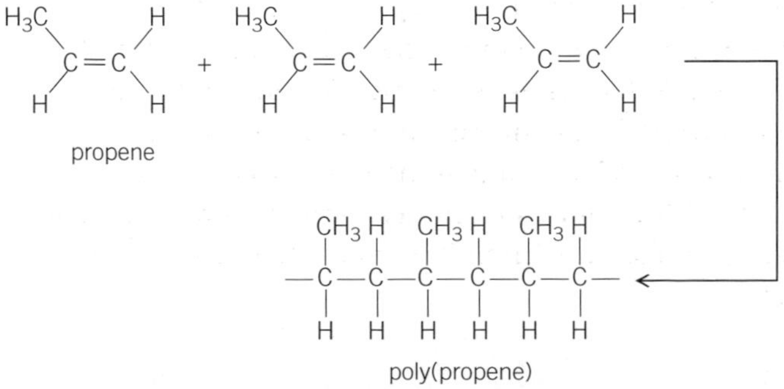

propene

poly(propene)

▲ **Figure 7** *The polymerisation of propene*

> ### Synoptic link
>
> You will find out about another type of polymerisation (condensation) in Chapter 7, Polymers of life.

> ### Polymerisation
>
> small molecules called monomers join together to produce long chain polymers.

> ### Activity DF 7.1
>
> This activity allows you to learn about the properties and uses of some addition polymers.

> ### Study tip
>
> In addition polymerisation, when the monomers join together there is no other product except the polymer.

Study tip

Many students are tempted to draw the poly(propene) formula incorrectly with three carbons in a row. The repeating unit is the two carbons with the methyl group 'sticking out'.

Activity DF 7.2

This activity helps you to appreciate the relationship between monomers and their associated addition polymer.

In the chain, the same basic unit is repeated over and over again, so the polymer structure can be shown simply as:

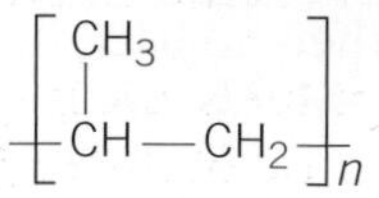

$$\left[\begin{array}{c} CH_3 \\ | \\ CH - CH_2 \end{array} \right]_n$$

The polymer is named by placing the name of the monomer in brackets and adding poly to the beginning, for example, poly(propene).

Sometimes more than one type of monomer is used in addition polymerisation. For example, if some ethene is added to the propene during the polymerisation process, both monomers become incorporated into the final polymer. This is called **copolymerisation**. A section of the copolymer chain could look like this:

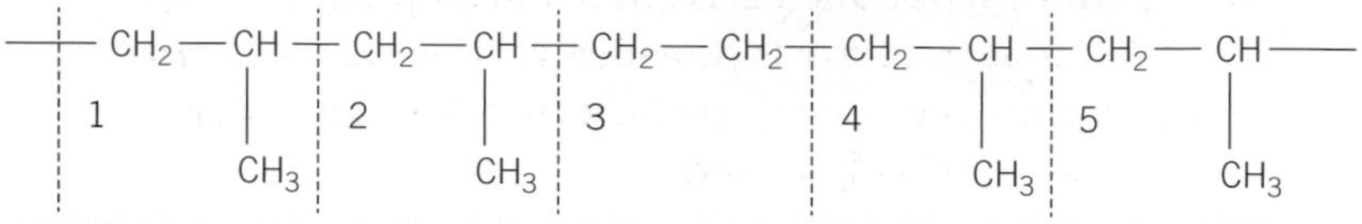

```
    CH₂—CH    CH₂—CH    CH₂—CH₂   CH₂—CH    CH₂—CH
    1         |         2         |         3              4         |         5        |
              CH₃                 CH₃                                 CH₃                 CH₃
```

Units 1, 2, 4, and 5 derive from propene, whilst unit 3 derives from ethene.

Summary questions

1 Write down the structure of a length of polymer formed from:
   a three monomer units of poly(ethene)   (*1 mark*)
   b two monomer units of poly(propene)   (*1 mark*)

2 The polymer PVC is made from a monomer with the structure $H_2C{=}CHCl$. Write down the structure of a length of polymer formed from 6 monomer units.   (*1 mark*)

3 Name the two alkenes used to make this copolymer.   (*2 marks*)

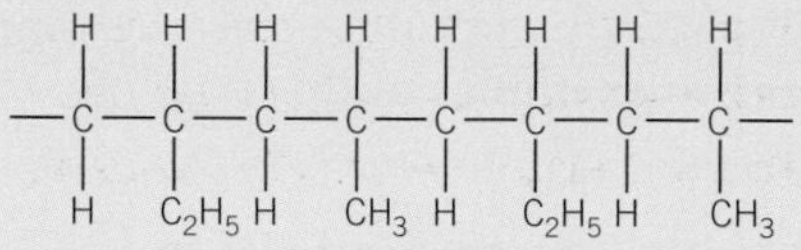

```
    H   H   H   H   H   H   H   H
    |   |   |   |   |   |   |   |
  —C — C — C — C — C — C — C — C—
    |   |   |   |   |   |   |   |
    H  C₂H₅ H  CH₃  H  C₂H₅ H  CH₃
```

4 Three addition polymers have the general structure shown below.

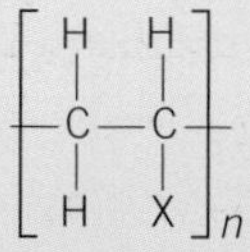

$$\left[\begin{array}{c} H \quad H \\ | \quad | \\ C - C \\ | \quad | \\ H \quad X \end{array} \right]_n \qquad X = H, CH_3, \text{ and } Cl.$$

   Name the three polymers.   (*3 marks*)

5 Draw a section of the polymer chain formed when chloroethene and ethenyl ethanoate copolymerise in an addition polymerisation reaction to form an A–B polymer.   (*2 marks*)

DF 8 Burning fuels

Specification reference: DF(a)

You saw in DF 1 and DF 2 that fuels release energy when they burn in oxygen and that this is because of the large bond enthalpies of the products, particularly C=O. This means relatively large amounts of energy are released when C=O bonds in carbon dioxide are formed.

Using the fuel

About 30–40% of each barrel of crude oil goes to make petrol, but it's not as simple as just distilling off the right bit at the refinery and sending it to the petrol stations. Petrol has to be blended to get the right properties. One important property is its volatility.

A mixture of petrol vapour and air is ignited in a cylinder in a car engine. The vapour-air mixture is provided via the fuel injection system. When the weather is very cold the petrol is difficult to vaporise, so the car is difficult to start.

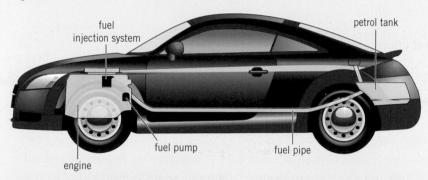

▲ **Figure 1** *The fuel supply system. When you start a car engine, fuel is pumped to the electronic fuel injection system where a fine spray of the fuel is mixed with air in the correct proportions prior to ignition*

To get over this problem, petrol companies make different blends for different times of the year. During winter they put more volatile components in the petrol so it vaporises more readily. This means putting in more of the hydrocarbons with small molecules, such as butane and pentane.

In hot weather too many of these more volatile components will mean the petrol will vaporise too easily. Petrol would vaporise from the tank by evaporation – a process which is costly and polluting. Also, if the fuel vaporises too readily then pockets of vapour form in the fuel supply system. The fuel pump then delivers a mixture of liquid and vapour to the carburettor instead of mainly liquid. This means that not enough fuel gets through to keep the engine running – it's called vapour lock.

All petrols are a blend of hydrocarbons of high, medium, and low volatility. As well as altering the petrol blend for the different seasons in a particular country, the blend will be different in different countries depending on the climate. The colder the climate, then more volatile components are added to the blend. Petrol companies change their blend throughout the year – and you don't even notice.

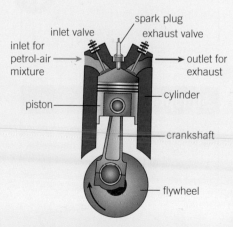

▲ **Figure 2** *How a four-stroke petrol engine works. The petrol air mixture is drawn into the cylindor. The piston compresses the mixture, then a spark makes the mixture explode, pushing the piston down and turning the crankshaft. In a diesel engine, air alone enters via the inlet valve. There is no spark plug and fuel is injected via a separate inlet during the compression stroke. The high pressure causes the fuel–air mixture to ignite*

In both petrol and diesel engines, the fuel–air mixture has to ignite at the right time, usually just before the piston reaches the top of the cylinder.

Look at Figure 2. As the fuel–air mixture is compressed it heats up, and the more it is compressed the hotter it gets. Modern cars achieve greater efficiency than in the past by using higher compression ratios, often compressing the gases in the cylinder by about a factor of 10. These combustion reactions involve reacting gases.

Chemical ideas: Measuring amounts of a substance 1.5a

Calculations involving gases

Volumes of gases

When carbon-based fuels burn, *complete* combustion forms carbon dioxide and water, though some carbon monoxide is also formed as a result of *incomplete* combustion.

For example:

$$C_7H_{16} + 11O_2 \rightarrow 7CO_2 + 8H_2O$$ complete combustion of heptane

$$C_7H_{16} + 10O_2 \rightarrow 7CO_2 + 2CO + 8H_2O$$ incomplete combustion of heptane

$$C_2H_5OH + 3O_2 \rightarrow 2CO_2 + 3H_2O$$ complete combustion of ethanol

$$C_6H_6 + 7.5O_2 \rightarrow 6CO_2 + 3H_2O$$ complete combustion of benzene

Chemists use chemical equations to work out the masses of reactants and products involved in a reaction. If one or more of these is a gas, it is sometimes more useful to know the volume of gas rather than its mass.

Measuring volumes

The units used to measure volume depend on how big the volume is. In chemistry, large volumes are usually measured in cubic decimetres (dm^3), with smaller volumes measured in cubic centimetres (cm^3). A decimetre is a tenth of a metre, 10 cm. A cubic decimetre is therefore 10 cm × 10 cm × 10 cm, or 1000 cm³.

$$1\,dm^3 = 1000\,cm^3$$

Molar volume

The number of molecules in one mole of any gas is always equal to the Avogadro constant N_A, 6.02×10^{23}. The molecules in a gas are very far apart compared to the actual size of each molecule so the molecule has a negligible effect on the total volume the gas occupies. Therefore, one mole of any gas always occupies the same volume, no matter which gas it is. Avogadro realised this in 1811, when he put forward his famous law (sometimes called Avogadro's hypothesis).

> Equal volumes of all gases at the same temperature and pressure contain an equal number of molecules.

The volume occupied by one mole of any gas at a particular temperature and pressure is called the **molar volume**. At standard

---

Synoptic link

You will look at measuring volumes of gas in Chapter 4, The ozone story.

Activity DF 8.1

In this activity you will calculate the volume of one mole of hydrogen gas.

Synoptic link

You first encountered the concept of Avogadro's constant in Chapter 1, Elements of Life.

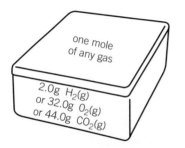

one mole of any gas

2.0g $H_2(g)$
or 32.0g $O_2(g)$
or 44.0g $CO_2(g)$

▲ **Figure 3** *A mole of any gas at room temperature and pressure occupies 24.0 dm³ (about the volume of a large biscuit tin)*

temperature and pressure (s.t.p.) the molar volume of a gas is 22.4 dm³. Standard temperature and pressure means a temperature of 0 °C (273 K) and a pressure of 1 atmosphere (101.3 kPa).

At room temperature, around 25 °C (298 K), and 1 atmosphere pressure, referred to as room temperature and pressure (RTP), the volume of a mole of any gas is about 24.0 dm³ (Figure 3).

The idea of molar volume allows you to calculate the amount in moles from the volume of a gas, and vice versa, provided you know the temperature and pressure of the gas. Here are two examples.

 Worked example: Calculating the volume of a gas from a given mass

Calculate the volume occupied by 4.4 g of carbon dioxide at room temperature and pressure. Assume that the molar volume of a gas at room temperature and pressure (RTP) is 24.0 dm³.

Step 1: Calculate the amount in moles from the mass. Relative formula mass of CO_2 = 44.0

$$\text{amount of } CO_2 = \frac{\text{mass}}{\text{molar mass}}$$

$$= \frac{4.4\,g}{44.0\,g\,mol^{-1}} = 0.1\,mol$$

Step 2: Calculate the volume.

1 mol CO_2 at r.t.p. has a volume of 24.0 dm³

$$24.0 \times 0.1 = 2.4$$

0.1 mol carbon dioxide at r.t.p. has a volume of 2.4 dm³

 Worked example: Calculating the mass of a gas from a given volume

Calculate the mass of 1.2 dm³ of methane gas, CH_4, at room temperature and pressure. Assume that the molar volume of a gas at room temperature and pressure (RTP) is 24.0 dm³.

Step 1: Calculate the amount in moles.

$$\text{amount of } CH_4 = \frac{\text{volume}}{\text{molar volume}}$$

$$= \frac{1.2\,dm^3}{24.0\,dm^3\,mol^{-1}} = 0.05\,mol$$

Step 2: Calculate the mass.

relative formula mass of CH_4 = 16.0

mass of CH_4 = amount in moles × molar mass
$$= 0.05\,mol \times 16.0\,g\,mol^{-1} = 0.8\,g$$

Reacting volumes of gases

Molar volumes can be used to work out the volumes of gases involved in a reaction. Consider the manufacture of ammonia from nitrogen and hydrogen:

$$N_2(g) + 3H_2(g) \rightarrow 2NH_3(g)$$

From the equation one mole of nitrogen reacts with three moles of hydrogen to produce two moles of ammonia. Using the idea that one mole of each gas occupies $24.0\,dm^3$ at RTP:

$$24.0\,dm^3\ N_2 + (3 \times 24.0\,dm^3)\ H_2 \rightarrow (2 \times 24.0\,dm^3)\ NH_3$$

or $\quad 1\,dm^3\ N_2 + 3\,dm^3\ H_2 \rightarrow 2\,dm^3\ NH_3$

or $\quad 1$ volume $N_2 + 3$ volumes $H_2 \rightarrow 2$ volumes NH_3

So, if you had $10\,cm^3$ of nitrogen it would react with $30\,cm^3$ of hydrogen to form $20\,cm^3$ of ammonia, provided all the measurements were taken at the same temperature and pressure.

For a reaction involving only gases, you can convert a statement about the numbers of moles of each substance involved to the same statement about volumes.

$N_2(g)$	+	$3H_2(g)$	$\rightarrow$	$2NH_3(g)$
1 mole	+	3 moles	$\rightarrow$	2 moles
1 volume	+	3 volumes	$\rightarrow$	2 volumes

🖩 Worked example: Calculations involving reacting volumes of gases

What volume of air is needed to completely burn $15\,cm^3$ of hexane vapour? (Assume all volumes are measured at the same temperature and pressure and that air contains 21% oxygen)

Step 1: Write the equation.

$$C_6H_{14} + 9\tfrac{1}{2}O_2 \rightarrow 6CO_2 + 7H_2O$$

Step 2: Work out the volume of oxygen.

$15\,cm^3$ of hexane combines with $142.5\,cm^3$ (15×9.5) oxygen from the ratio in the equation.

Step 3: Work out the volume of air required.

$142.5\,cm^3$ of oxygen is contained in $142.5 \times \dfrac{100}{21}\,cm^3$ air

$= 680\,cm^3$ of air (two significant figures, to match data)

Synoptic link ⚗

You can find a general description of how to measure the volumes of gases in Techniques and procedures.

Working with masses and volumes

You have seen how to calculate reacting masses using chemical equations and working with moles. Very similar calculations can be used to calculate the volumes of gases produced in reactions. This is because one mole of molecules of all gases occupies $24.0\,dm^3$ at room temperature and pressure.

Study tip

The molar gas volume at room temperature and pressure (RTP) is $24.0\,dm^3\,mol^{-1}$.

Worked example: Calculation involving reacting masses and volumes

What volume of carbon dioxide is produced when 15g of calcium carbonate completely decompose? Assume that one mole of gas occupies 24.0 dm³ at r.t.p.

Step 1: Write the equation.

$$CaCO_3 \rightarrow CaO + CO_2$$

Step 2: Work out the moles of the calcium carbonate and carbon dioxide from the equation.

one mole of $CaCO_3$ decomposes to produce one mole of CO_2

Step 3: Write out the mass and volume of one mole for calcium carbonate and carbon dioxide.

mass of 1 mole of $CaCO_3$ = 100.1 g   volume of one mole of CO_2 = 24.0 dm³

Step 4: Calculate the volume of carbon dioxide.

$$1\,g = \frac{24.0}{100.1}$$

$$15\,g = 15 \times \frac{24.0}{100.1}$$

$$= 3.6\,dm^3$$

The ideal Gas Equation

When doing calculations involving gas volumes, one mole of a gas occupies 24.0 dm³ at room temperature and pressure. However, when temperature or pressure are different from room temperature and pressure the ideal gas equation is used.

pressure P (Pa) × volume V (m³) =
amount of gas n (moles) × gas constant R (J K⁻¹ mol⁻¹) × temperature T (K)

$$PV = nRT$$

The gas constant R has the value 8.31 J K⁻¹ mol⁻¹.

 Worked Example: Calculating volume from the ideal gas equation

What is the volume of 1 mole of gas at 100 kPa pressure and 16 °C?

Step 1: Convert kPa to Pa.

1 kPa = 1000 Pa

So to convert from kPa to Pa, multiply the kPa value by 1000.

$100 \times 1000 = 1 \times 10^5\,Pa$

Step 2: Convert °C to K by adding 273 to the °C value.

$16 + 273 = 289\,K$

Study tip

It is vital to keep a very careful eye on the units and powers of ten when using the ideal gas equation.

For volumes:

- $1\,cm^3 = 1 \times 10^{-6}\,m^3$
- $1\,dm^3 = 1 \times 10^{-3}\,m^3$.

For pressure:

- 1 Pa is equal to $1\,N\,m^{-1}$ (Newton per square metre).

119

Step 3: Rearrange the ideal gas equation to make volume the
subject and substitute in the values.

$$V = \frac{nRT}{P} = \frac{1 \times 8.31 \times 289}{1 \times 10^5} = 0.0240 \, m^3 = 24.0 \, dm^3$$

16 °C and 1×10^5 Pa are the values for room temperature and
pressure.

 **Worked example: Calculating pressure from the ideal
gas equation**

What is the pressure if 5.0 g of nitrogen is present in a volume of
50 cm³ at 300 K?

Step 1: Calculate the moles of nitrogen. $M_r(N_2) = 28$

$$n = \frac{mass}{relative\ molecular\ weight} = \frac{5.0}{28} = 0.179 \, mol$$

Step 2: Convert cm³ to m³.

$$1 \, cm^3 = 1 \times 10^{-6} \, m^3$$

So to convert from cm³ to m³, multiply the cm³ value
by 1×10^{-6}.

$$50 \times 1 \times 10^{-6} = 5 \times 10^{-5} \, m^3$$

Step 3: Rearrange the ideal gas equation to make pressure the
subject and substitute in the values.

$$P = \frac{nRT}{V} = \frac{0.179 \times 8.31 \times 300}{5.0 \times 10^{-5}} = 8.9 \times 10^6 \, Pa$$

This is about nine times atmospheric pressure.

Summary questions

In the following problems, unless stated otherwise, you should assume
that the volumes of all gases are measured at the same temperature and
pressure.

1 The volumes of one mole of all gases are the same when measured at
the same temperature and pressure. The volumes of one mole of liquids
or solids are almost always different. Why do gases differ from liquids
and solids in this way? *(2 marks)*

2 10 cm³ of hydrogen are burned in oxygen to form water.
   a Write a balanced equation, including state symbols for
   this reaction. *(2 marks)*
   b What volume of oxygen is needed to burn the hydrogen
   completely? *(3 marks)*

3 $10 cm^3$ of a gaseous hydrocarbon reacts completely with $40 cm^3$ of oxygen to produce $30 cm^3$ of carbon dioxide.
   a How many moles of carbon dioxide must have been formed from one mole of the hydrocarbon? *(2 marks)*
   b How many carbon atoms must there be in the formula of the hydrocarbon? *(2 marks)*
   c How many moles of oxygen were used in burning one mole of the hydrocarbon? *(2 marks)*
   d How many moles of water must have been formed in burning one mole of the hydrocarbon? *(2 marks)*
   e What is the formula of the hydrocarbon? *(1 mark)*

4 1.2 g of magnesium react with excess sulfuric acid. Calculate the volume of hydrogen produced at room temperature and pressure. *(3 marks)*

5 18 g of pentane, C_5H_{12}, are completely burnt in a car engine to form carbon dioxide and water.
   a How many moles of pentane are burned? *(2 marks)*
   b How many moles of oxygen are needed to burn all the pentane? *(1 mark)*
   c What volume of oxygen is needed, assuming that one mole of gas occupies $24.0 dm^3$? *(2 marks)*
   d What volume of air is needed (assume that air contains 21% oxygen by volume)? *(2 marks)*
   e What volume of carbon dioxide is formed? *(1 mark)*

6 7.5 g of a gras occupies $4.11 dm^3$ at 100 kPa and 290 K. Calculate the M_r of the gas. *(2 marks)*

7 a Calculate the amount (in moles) of air in a $1.7 dm^3$ car tyre at 1.5 bar and 0 °C.
    (1 bar = 100 kPa) *(2 marks)*
   b Assuming the volume does not change, calculate the pressure (in bar) in the tyre at 20 °C. *(2 marks)*

8 $10 cm^3$ of nitrogen is reacted with $20 cm^3$ hydrogen. Calculate the total volume of gases formed. *(4 marks)*

9 Two bulbs of $50 cm^3$ capacity are connected by a tube of negligible volume. Initially both bulbs are in melting ice and the pressure is 1.2×10^5 Pa.

One bulb is then placed in boiling water whilst the other is left in melting ice. Calculate the new pressure in the bulbs. *(4 marks)*

DF 9 What do the molecules look like?

Specification reference: DF(c), DF(s), DF(t)

Learning outcomes

Demonstrate and apply knowledge and understanding of:

→ the relationship of molecular shape to structural formulae and the use of wedge and dotted lines to represent 3D-shape

→ structural isomerism and structural isomers

→ stereoisomerism in terms of lack of free rotation about C═C bonds when the groups on each carbon differ; description and naming as:

- *E/Z* for compounds that have an H on each carbon of C═C

- *cis/trans* for compounds in which one of the groups on each carbon of C═C is the same.

When considering fuels you have talked about the molecules. However, chemists cannot see molecules. Instead, chemists use models of molecular structures to explain how molecules behave in order to make predictions about further behaviour. When, in the past, these models led to correct predictions, chemists found they were *useful*. You cannot say they are *true*, however, and most models have their failings as well as successes.

For example, dot-and-cross diagrams are useful to describe bonding. However, try to draw such a diagram for nitrogen monoxide, NO, or diborane, B_2H_6, and you will see that they have failings.

You are going to look at some different ways of representing molecules, the concept of isomers, and naming conventions.

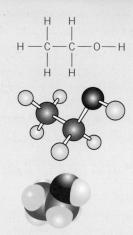

▲ **Figure 1** *Three ways of representing an ethanol molecule – full structural formula, ball and stick, and space filling*

▲ **Figure 2** *A representation of a more complicated molecule – the enzyme lysozyme*

Chemical ideas: Organic chemistry: frameworks 12.1b

Alkanes

Shapes of alkanes

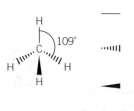

▲ **Figure 3** *A regular tetrahedron*

— represents a bond in the plane of the paper

⫶⫶⫶⫶ represents a bond in a direction behind the plane of the paper

◀ represents a bond in a direction in front of the plane of the paper

▲ **Figure 4** *Three-dimensional shape of methane*

Representing structures in a two-dimensional way on paper can give a misleading picture of what the molecule looks like. The pairs of electrons in the covalent bonds repel one another, and so arrange themselves round the carbon atom as far apart as possible. Therefore, the carbon–hydrogen bonds in methane are directed so they point towards the corners of a regular tetrahedron (Figure 3). The carbon atom is at the centre of the tetrahedron, and the H—C—H bond angles are 109° (109° 28′ to be precise).

The best way to show this is to use a molecular model. Figure 4 shows how you represent the three-dimensional shape of the methane molecule on paper.

The structure of ethane in three dimensions is shown in Figure 5. Each carbon atom is at the centre of a tetrahedral arrangement.

Figure 6 shows the three-dimensional structure of butane. You can see that hydrocarbon chains are not really straight, but a zig-zag of carbon atoms. All the bond angles are 109°. The shape of a hydrocarbon chain is often represented by a **skeletal formula**, also shown in Figure 6.

Skeletal formulae show only the carbon skeleton and associated functional groups of the molecule (Figure 7), so do not assist in visualising shapes of molecules).

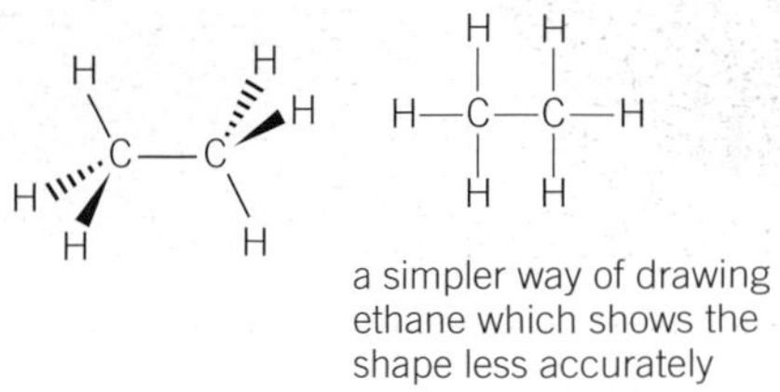

a simpler way of drawing ethane which shows the shape less accurately

▲ **Figure 5** *The three-dimensional shape of ethane*

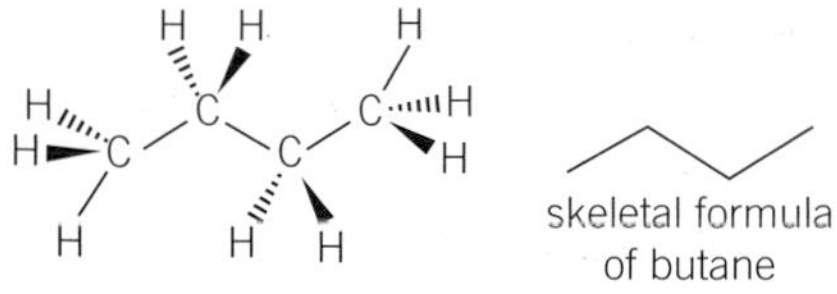

skeletal formula of butane

▲ **Figure 6** *The three-dimensional shape of butane and its skeletal formula*

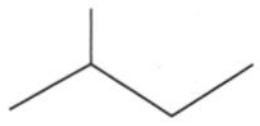

▲ **Figure 7** *The skeletal formula of 2-methylbutane*

Chemical ideas: Bonding, shapes, and sizes 3.3

Structural and *E/Z* isomerism

Isomerism

Two molecules that have the same molecular formula but differ in the way their atoms are arranged are called **isomers**. Isomers are distinct compounds with different physical properties, and often different chemical properties too. The occurrence of isomers (isomerism) is very common in carbon compounds because of the great variety of ways in which carbon atoms can form chains and rings. However, you will also meet examples of isomerism in inorganic chemistry.

There are two ways in which atoms can be arranged differently in isomers.

1   The atoms are bonded together in a different order in each isomer – these are called structural isomers.

2   The order of bonding in the isomers is the same, but the arrangement of the atoms in space is different in each isomer – these are called stereoisomers.

Structural isomerism

Structural isomers have the same molecular formula but the atoms are bonded together in a different order. They have different *structural formulae*. There are various ways in which structural isomerism can arise.

Chain isomerism

There is only one alkane corresponding to each of the molecular formulae CH_4, C_2H_6, and C_3H_8. With four or more carbon atoms in a chain, different arrangements are possible (Figure 8). These are called **chain isomers**.

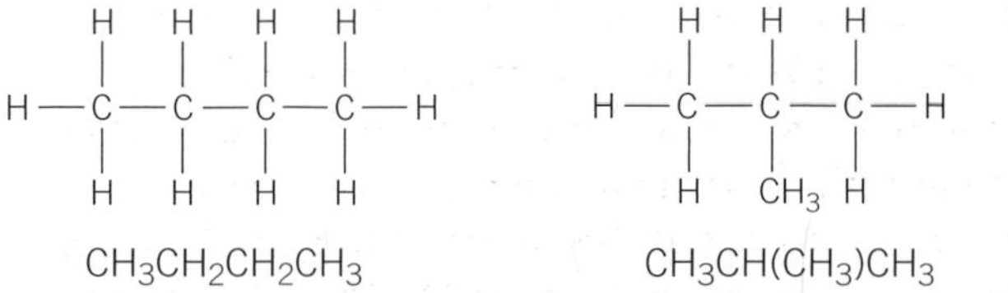

$CH_3CH_2CH_2CH_3$ $CH_3CH(CH_3)CH_3$

▲ **Figure 8** *Two structures with the molecular formula of C_4H_{10}*

Both butane and methylpropane have the same molecular formula C_4H_{10}. Their different structures lead to different properties, for example, the boiling point of methylpropane is 12 K lower than that of butane.

Activity DF 9.1

This activity checks your understanding of structural isomers and structural isomerism.

As the number of carbon atoms in an alkane increases, the number of possible isomers increases. There are over four thousand million isomers with the molecular formula $C_{30}H_{62}$.

Position isomerism

Position isomerism can occur where there is an atom, or group of atoms, substituted in a carbon chain or ring. These are called functional groups. You first encountered functional groups in Topic DF 3. Position isomerism occurs when the functional group is situated in different positions in the molecules.

The isomers of C_3H_7OH are shown in Figure 9. The –OH functional group is situated at two different places on the hydrocarbon chains.

Study tip

In alcohols the functional group is –OH and the general formula is $C_nH_{2n+1}OH$.

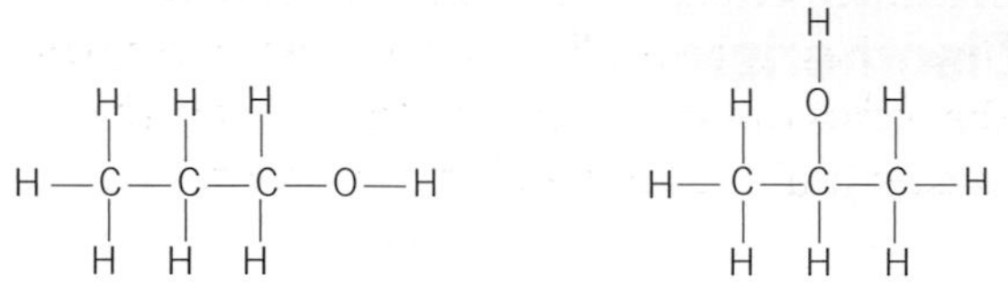

propan-1-ol   propan-2-ol

▲ **Figure 9** *Two position isomers for the molecular formula C_3H_7OH*

Functional group isomerism

It is sometimes possible for compounds with the same molecular formula to have different functional groups, and because they have different functional groups they will belong to a different homologous series. These are called **functional group isomers**. The isomers of molecular formula C_3H_8O are shown in Figure 10.

Synoptic link

You will learn more about reactions of alcohols in Chapter 5, What's in a medicine?

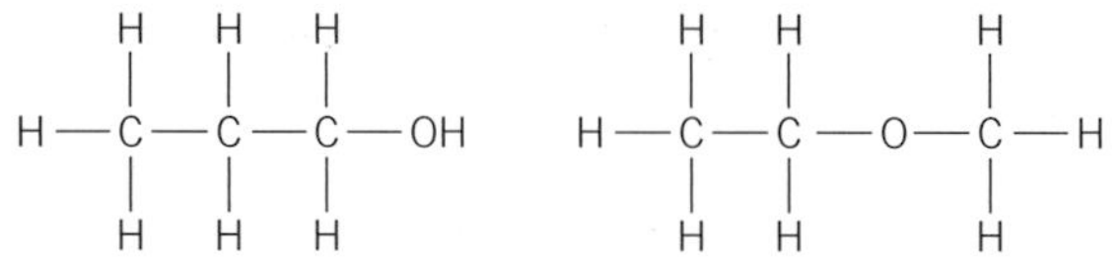

propan-1-ol
(an alcohol)
functional group
–OH

methoxyethane
(an ether)
functional group
–O–

▲ **Figure 10** *Two functional group isomers for the molecular formula C_3H_8O*

Study tip

As part of the course, you do not need to be able to name ethers.

The isomer on the left is called propan-1-ol. Its functional group is –OH and it is a member of the homologous series known as the alcohols. The isomer on the right is methoxyethane. Its functional group, –O–, is quite different from that of propan-1-ol and it belongs to another family of organic compounds, known as the ethers.
As well as having different physical properties, propan-1-ol and methoxyethane have very different chemical properties due to the different functional groups.

Synoptic link

You will study optical isomerism, along with other types of isomerism, in Chapter 8, Polymers and life.

Stereoisomerism

Stereoisomerism is when molecules have the same structural formula but differ in how their atoms are arranged in space. There are two types of stereoisomerism – *E*/*Z* isomerism and optical isomerism.

E/Z isomerism

E/Z **isomerism**, or *cis/trans* isomerism is one type of stereoisomerism. Stereoisomers have identical molecular formulae and the atoms are bonded together in the same order, but the arrangement of atoms in space is different in each isomer.

But-2-ene has two *E/Z* isomers (Figure 11).

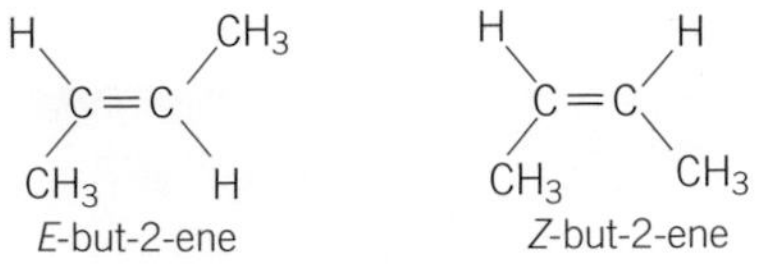

▲ **Figure 11** *E/Z isomers of but-2-ene*

To turn the second form into the first form you would have to spin one end of the molecule round in relation to the other end. This can only be done by first breaking the π-bond in the double bond. You can easily prove this by using molecular models.

The average bond enthalpy for the bond that has to be broken here is about $+270\,kJ\,mol^{-1}$. This much energy is not available at room temperature. A covalent bond has to be broken and another reformed in order to interconvert the forms of but-2-ene. In other words, the process is an example of a chemical reaction and the two forms are different chemicals.

> **Activity DF 9.2**
>
> This activity helps you check your understanding of *E/Z* isomerism.

Naming stereoisomers of alkenes

The two different but-2-enes need different names. The older way of naming them uses *cis/trans* nomenclature. The form of but-2-ene with the same groups (in this case both are methyl groups) on the same side of the double bond is called *cis*-but-2-ene (Figure 12). When the substituents are on opposite sides of the double bond the molecule is said to be the *trans* form (Figure 13).

> **Synoptic link**
>
> You will be looking at *cis/trans* isomers later in Chapter 10, Colours by design.

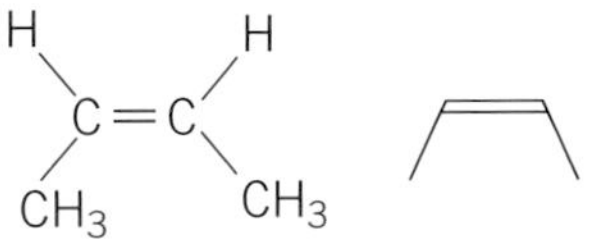

▲ **Figure 12** *cis-but-2-ene*

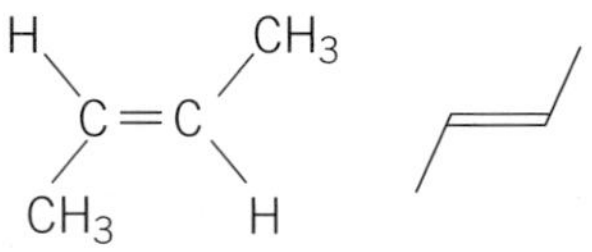

▲ **Figure 13** *trans-but-2-ene*

There is a second nomenclature system for naming alkenes and this is the *E/Z* system. In the *E/Z* notation. *E* means opposite and corresponds generally to the term *trans* whilst *Z* means together and corresponds to the term *cis*. This means that *cis*-but-2-ene is also called (*Z*)-but-2-ene and *trans*-but-2-ene is also called (*E*)-but-2-ene.

This rule works when either the two groups of either end of the double bond are the same or there is a hydrogen on both carbons.

The *cis/trans* system has its limitations, for example, the two isomers in Figure 14 cannot be named using the *cis/trans* system. The *E/Z* system has its own set of rules to help inform the naming.

You will only be expected to name isomers with the same groups on either end of a double bond and name them as *E* or *Z* isomers.

Z and E isomers are different compounds, so they have different properties. Table 1 gives some information on some isomers which are clearly different substances.

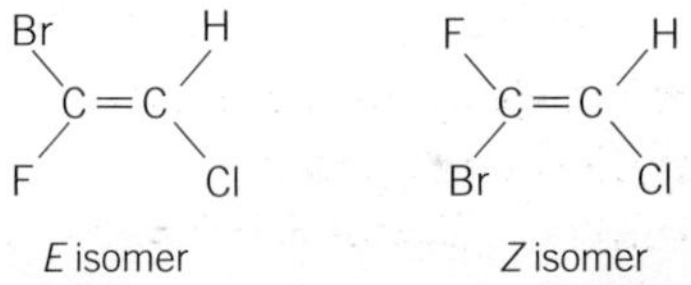

> **Study tip**
>
> If you have difficulty remembering which isomer is Z and which is E, the phrase 'Z-zame-zide' may help!

▲ **Figure 14** *E/Z isomers of a haloalkane*

▼ **Table 1** *Physical properties of some E/Z isomers*

E or *trans* isomer	Melting point / K	Density / g cm⁻³	*Z* or *cis* isomer	Melting point / K	Density / g cm⁻³
H, CH₃ C=C CH₃, H	168	0.604	H, H C=C CH₃, CH₃	134	0.621
Br, H C=C H, Br	267	2.23	Br, Br C=C H, H	220	2.25
H, COOH C=C HOOC, H	573	1.64	H, H C=C HOOC, COOH	412	1.59

Chemical ideas: Organic compounds: modifiers 13.1

Naming organic compounds

You have already learnt how to name alkanes and alkenes. Alcohols and haloalkanes follow similar rules for naming.

Naming alcohols

1 Name the alcohol using the root prefix of the longest chain.

2 Determine the position of the alcohol group, –OH, by counting the carbons to give the lowest number.

3 Add 'ol' onto the end of the name.

H—C—C—OH (with H atoms)	CH_3CH_2OH	ethanol
H—C—C—C—C—C—H (with H, OH atoms)	$CH_3CH_2CH_2(CHOH)CH_3$	pental-2-ol

When an alcohol is branched, just follow the rules for a branched alkane, but replace the 'e' at the end of the name with 'ol' and include the number the –OH group is attached to.

$$H_3C \overset{1}{-} \overset{2}{\underset{\underset{OH}{|}}{C}} \overset{3}{-} CH_3 \quad (CH_3)$$

2-methylpropan-2-ol

Activity DF 9.3, DF 9.4, and DF 9.5

These activities all help you check your understanding of the nomenclature and formulae of alkanes, alkenes, and alcohols.

General rules for naming alkanes, haloalkanes, and alcohols

A systematic naming system makes it possible to name compounds unambiguously, using a few simple rules. The following rules summarise what you have learnt so far about naming organic compounds.

1 Count the carbon atoms in the longest chain. This gives the root name, for example, an alkane with *five* carbons is called *pent*ane. Beware – occasionally the longest carbon chain goes 'round corners'.

2 Use alkyl group prefixes to indicate chain branches, for example, a $-CH_3$ group attached to a carbon chain is called a methyl group – methylpropane, $CH_3CH(CH_3)CH_3$.

3 Use chloro-, bromo-, iodo-, and so on to indicate a halogen atom substituted for a hydrogen atom, for example, bromoethane, CH_3CH_2Br. (These are called haloalkanes and you will meet them later in the course.)

4 Add suffixes to indicate other groups that have been substituted for a hydrogen atom, for example 'ol' for alcohols – ethanol, CH_3CH_2OH.

5 If necessary indicate the number(s) of the carbon atom(s) on which the substitution has occurred, keeping the numbers as low as possible. In none of the above examples is a number necessary, as they are unambiguous, but longer carbon chains need numbers to be unambiguous.
For example, 2-methylpentane $CH_3CH(CH_3)CH_2CH_2CH_3$ and not 4-methylpentane.
If there are two groups attached to the chain, the name would be 2,2-dimethylpentane $CH_3C(CH_3)_2CH_2CH_2CH_3$ not 4,4-dimethylpentane.

6 Prefixes are listed in alphabetical order, for example, 2-bromo-1-chloropropane, $CH_3CHBrCH_2Cl$.

There are other rules that occur in more complicated situations, and these will be introduced as you encounter them.

Study tip

The numbers are separated from each other by commas and from words by dashes. There are no gaps between prefixes or suffixes and the root name.

Also, if there are two of the same group, the number must be repeated (e.g., '2,2') and 'di' must be placed before the group.

Activity DF 9.1 and DF 9.2

You can practise naming organic compounds in these activities.

Summary questions

1 Write the molecular formula for each of the compounds in the following pairs. Use the molecular formulae to decide whether the compounds in the pair are isomers.

a
$CH_3-CH_2-CH_2-CH_2-CH_2-CH_3$ and

(2 marks)

b
$CH_3-CH_2-CH_2-CH_2-Cl$ and

(2 marks)

c
$CH_3-CH_2-CH_2-OH$ and

(2 marks)

d

and

(2 marks)

e

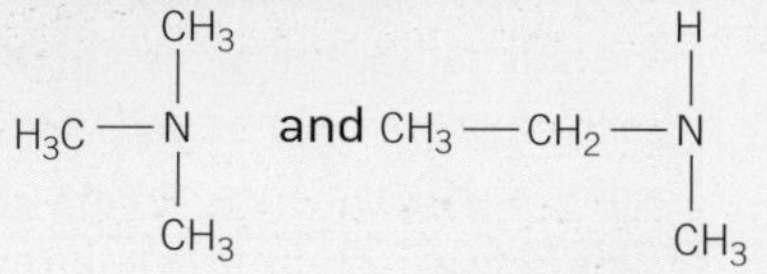

(2 marks)

2 There are two different compounds with the molecular formula C_4H_{10}. For each of them:
   a Draw full structural formulae.   (2 marks)
   b Draw three-dimensional shapes using wedges and dotted lines.   (2 marks)
   c Draw skeletal formulae.   (2 marks)
   d Name each compound.   (2 marks)

3 Draw the full structural formula of ethanol. Mark the values of a H—C—H
   bond angle and the C—O—H bond angle. Explain the values you have given.   (3 marks)

4 There are several structural isomers with the molecular formula C_4H_9Br.
   Draw their skeletal formulae and name them.   (4 marks)

5 There are four structural isomers with the molecular formula C_8H_{10} in which each
   isomer contains a benzene ring and at least one side chain. Draw their structures.   (4 marks)

6 Draw the skeletal formulae of all the structural isomers which have the molecular
   formula $C_4H_{10}O$. Name all those that are alcohols.   (4 marks)

7 a Draw and label the *E/Z* isomers of 1,2-dichloroethene.   (2 marks)
   b Suggest which of the isomers has the higher boiling point. Explain your answer.   (2 marks)

8 Nerol, which occurs in bergamot oil, geraniol (in roses), linalool (in lavender),
   and cilronellol (in geraniums) are four compounds from the lerpene family.
   Their skeletal structures are shown below.

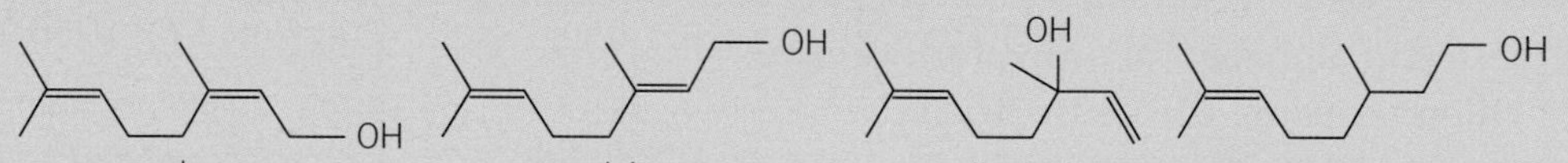

nerol   geraniol   linalool   citronellol

   a How are the structures of nerol and geraniol related?   (1 mark)
   b How many moles of hydrogen (H_2) would be required to saturate one mole of geraniol?   (1 mark)
   c How are nerol and geraniol related to citronellol?   (1 mark)
   d How are the structures of nerol and geraniol related to linalool?   (1 mark)

DF 10 The trouble with emissions
Specification reference: DF(k), DF(n)

Figure 1 shows what goes into – and what comes out of – a car engine. Some of these exhausts are causing issues worldwide. There are slight differences between the emissions of petrol and diesel engines but many of the pollutants are similar. The oxides of nitrogen – nitrogen oxide and nitrogen dioxide – are grouped together as NO_x. Similarly, SO_x represents the oxides of sulfur – sulfur dioxide and sulfur trioxide – and C_xH_y represents the various hydrocarbons present in the exhaust fumes.

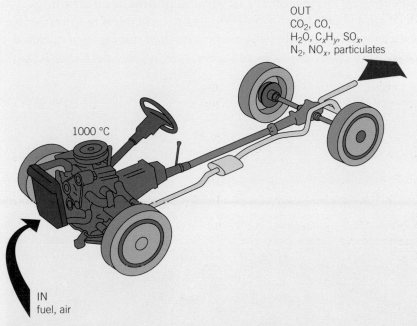

OUT
CO_2, CO,
H_2O, C_xH_y, SO_x,
N_2, NO_x, particulates

1000 °C

IN
fuel, air

▲ **Figure 1** *What goes into – and comes out of – a car engine*

Learning outcomes

Demonstrate and apply knowledge and understanding of:

→ the origin of atmospheric pollutants from a variety of sources: particulates, unburnt hydrocarbons, CO, CO_2, NO_x, SO_x; the environmental implications and methods of reducing these pollutants

→ environmental implications of these pollutants

→ balanced equations for the complete and incomplete combustion (oxidation) of alkanes, cycloalkanes, alkenes, and alchols.

Furthermore, because petrol is so volatile, on a warm day a parked car gives off hydrocarbon fumes from the petrol tank and carburettor. This is **evaporative emission** and accounts for about 10% of emissions of volatile organic compounds from petrol vehicles.

The oxides of sulfur in vehicle exhausts come from sulfur compounds in the fuel. These combine with the oxygen from the air in the heat of the engine. Oxides of nitrogen are formed mainly from the components of the air itself. At the high temperatures in vehicle engines, nitrogen and oxygen in the air react to form nitrogen oxide. Some of this reacts with more oxygen to form nitrogen dioxide. Oxides of sulfur and nitrogen are acidic and give rise to acid rain in the atmosphere. This can cause health problems (particularly for asthmatics), corrode limestone buildings, and damage forests and lakes.

Carbon monoxide is formed by the incomplete combustion of hydrocarbon fuels. It is very toxic to humans and is oxidised to carbon dioxide in the atmosphere.

Particulates are very small carbon particles, not visible to the naked eye, that can get into our lungs and cause irritation and disease. Particulates are also produced by incomplete combustion of the hydrocarbon fuels in diesel.

Photochemical smog

The substances in Figure 1 are not the only pollutants caused by vehicles. Ozone is a *secondary pollutant* – it is not released directly into the atmosphere. Ozone is formed from chemical reactions that occur when sunlight shines on a mixture of *primary pollutants* (nitrogen oxides and hydrocarbons), oxygen, and water vapour. Other irritating compounds are formed by the breakdown and further reaction of the hydrocarbons. These reactions all occur in **photochemical smogs**, which are a great cause for concern. (A photochemical reaction occurs when a molecule absorbs light energy and then undergoes a chemical reaction.)

▲ **Figure 2** *Photochemical smog in Beijing*

Ozone has a vital role in the troposphere, however it can also be an irritating, toxic gas and high concentrations near ground level are damaging to human health. It weakens the body's immune system and attacks lung tissue. Furthermore, ozone in the troposphere acts as a greenhouse gas, contributing to global warming.

Photochemical smogs contain a mixture of primary and secondary pollutants. The exact composition varies and depends on the nature of the primary pollutants, the local geography, weather conditions, the time of day, and the length of the smog episode. Photochemical smogs normally occur in the summer during high pressure (anticyclonic) conditions.

Synoptic link

You will find out more about the role of ozone in Chapter 4, The ozone story.

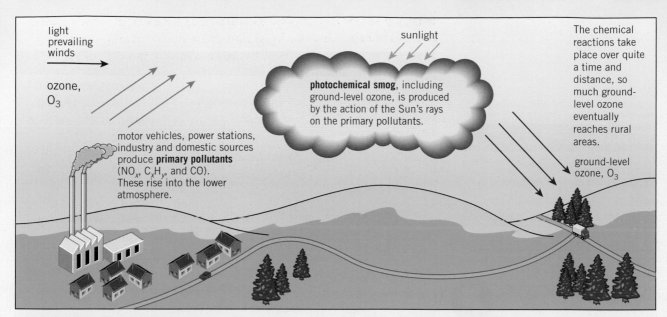

▲ **Figure 3** *Formation of a photochemical smog*

What are the effects of photochemical smog?

Photochemical smogs cause haziness and reduced visibility in the air close to the ground. For many people they can cause eye and nose irritation and some difficulty in breathing. In vulnerable groups, such as asthmatics who already have respiratory problems, very young children, and many older people, this effect can be enhanced.

High ozone concentrations affect animals and plants too. Ozone is a highly reactive substance that attacks most organic matter. Compounds with carbon–carbon double bonds are particularly vulnerable so many materials such as plastics, rubbers, textiles, and paints can be damaged.

Chemical ideas: Human impacts 16.1

Atmospheric pollutants

The formation of pollutants

Table 1 summarises the sources and polluting effects of the primary pollutants formed by vehicle engines.

Complete and incomplete combustion

When a hydrocarbon (alkane, cycloalkane, or alkene) undergoes combustion the products of combustion will vary, depending on the amount of oxygen available. In a plentiful supply of air:

hydrocarbon + oxygen → carbon dioxide + water

▼ **Table 1** *Sources and polluting effects of primary pollutants from vehicle engines and other sources*

Pollutant	Major sources	Major polluting effects
particulates – particularly small carbon particles below 2.5×10^{-12} m	volcanoes, burning fuels, burning coal	penetrate deep into the human body causing heart attacks and lung cancer
volatile organic compounds, VOC	plants, unburnt fuel from petrol engines	photochemical smog
carbon monoxide, CO	Incomplete combustion of hydrocarbons in fossil fuels, burning biomass	toxic gas, photochemical smog
carbon dioxide, CO_2	combustion of fossil fuels	greenhouse effect
nitrogen oxides, NO_x	combustion of fuels in power stations and vehicles	acid rain, photochemical smog
sulfur oxides, SO_x	volcanoes, burning of fuels containing sulfur	toxic gas, acid rain

Study tip

Complete and incomplete combustion of hydrocarbons and alcohols always lead to the production of water. It is the carbon compounds that differ.

Activity DF 10.1

In this activity you can develop your research skills to learn about aspects of pollutants.

If the air (oxygen) supply is limited, combustion of hydrocarbons will lead to the production of water and either carbon monoxide or carbon (soot).

$$C_7H_{16} + 11O_2 \rightarrow 7CO_2 + 8H_2O \qquad \text{complete combustion of heptane}$$

$$C_7H_{14} + 10\tfrac{1}{2}O_2 \rightarrow 7CO_2 + 5H_2O \qquad \text{complete combustion of hept-1-ene}$$

$$C_8H_{18} + 7\tfrac{1}{2}O_2 \rightarrow 6CO + 2C + 9H_2O \quad \text{incomplete combustion of octane}$$

$$C_6H_{12} + 6O_2 \rightarrow 6CO + 6H_2O \qquad \text{incomplete combustion of cyclohexene}$$

Alcohols can also be used as fuels. The same rules apply. For example:

$$C_2H_5OH + 3O_2 \rightarrow 2CO_2 + 3H_2O \qquad \text{complete combustion of ethanol}$$

$$C_2H_5OH + 2O_2 \rightarrow 2CO + 3H_2O \qquad \text{incomplete combustion of ethanol}$$

The burning of sulfur compounds in fuels produces sulfur dioxide, SO_2.

$$S + O_2 \rightarrow SO_2$$

Some nitrogen oxides, NO_x, are produced by burning nitrogen compounds in fuels, but these are present in very low proportions, especially in vehicle fuels. However, nitrogen and oxygen in the air react in the high temperatures of a vehicle engine.

$$N_2 + O_2 \rightarrow 2NO$$

Production of acid rain

Sulfur dioxide reacts with water in the atmosphere to form sulfuric(IV) acid, a weak acid.

$$SO_2 + H_2O \rightarrow H_2SO_3$$

However sulfur dioxide is oxidised to sulfur trioxide, SO_3, in the stratosphere, which then reacts with water in the atmosphere to form sulfuric(VI) acid, a strong acid.

$$SO_3 + H_2O \rightarrow H_2SO_4$$

▲ **Figure 4** *A stone sculpture eroded by acid rain*

Nitrogen oxide, NO, and nitrogen dioxide, NO_2, react with water and oxygen in the atmosphere to form nitric(V) acid, a strong acid.

$$2NO + H_2O + 1\frac{1}{2}O_2 \rightarrow 2HNO_3$$

$$2NO_2 + H_2O + \frac{1}{2}O_2 \rightarrow 2HNO_3$$

Acid rain causes breathing difficulties, corrodes limestone buildings, and kills forests and life in lakes.

Photochemical smog

This is formed when primary pollutants are acted upon by sunlight to produce secondary pollutants, for example, ozone, nitrogen dioxide, and nitric acid. Photochemical smogs cause haziness and reduced visibility in the air (Figure 2) as well as causing respiratory problems in humans.

Tackling the emissions problem

Many countries are now bringing in legislation to limit exhaust emissions.

▼ **Table 2** *European emission limits in mg km^{-1}. Diesel engines produce less carbon dioxide per km than petrol engines*

Pollutant	Petrol		Diesel	
	Emissions limit in year 2000	Emissions limit in year 2015	Emissions limit in year 2000	Emissions limit in year 2015
carbon monoxide	2300	1000	600	500
NO_x	500	60	200	80
particulates	–	5	50	5
hydrocarbons	200	100	–	–

There are important indirect methods for tackling the emissions problem, such as limiting the traffic entering towns and encouraging car-sharing schemes for people travelling to and from work. However, there are two ways of tackling the emissions problem directly. One involves changing the design of cars, and the other involves changing the fuel used by the car (Topic DF 11).

Using catalysts

Catalysts speed up reactions that involve pollutants in car exhausts. In the following reactions, pollutants are being converted to carbon dioxide, water, and nitrogen, all naturally present in the air.

1 using oxygen to turn carbon monoxide to carbon dioxide
2 using oxygen to turn hydrocarbons to carbon dioxide and water
3 reacting nitrogen oxide with carbon monoxide to form carbon dioxide and nitrogen

These reactions occur naturally, but under the conditions inside an exhaust system they go too slowly to get rid of the pollutants.

Catalytic converters in petrol cars contain catalysts of platinum or rhodium on a honeycomb structure. They are called three-way catalysts because they speed up the three reactions above.

This kind of catalyst system works *only* if the air-petrol mixture is carefully controlled so that it is exactly the stoichiometric mixture for the fuel. This means it has the exact calculated ratio of hydrocarbon to oxygen for complete combustion. If the mixture is too rich (too much fuel) then there is not enough oxygen in the exhaust fumes to remove carbon monoxide and the hydrocarbons. Cars fitted with three-way catalyst systems need to have oxygen sensors in the exhaust gases, linked back to electronically controlled fuel injection systems.

Catalytic converters work only when they are hot. A platinum catalyst starts working around 240 °C, but you can get the catalyst to start working at about 150 °C by alloying the platinum with rhodium. These catalysts are poisoned by lead, so the converters can only be used with lead-free fuel.

▲ **Figure 5** *A catalytic converter*

The catalyst is used in the form of a fine powder spread over a ceramic support with a surface that has a network of tiny holes (Figure 4). The surface area of the catalyst exposed to the exhaust gases is about the same as two or three football fields.

Diesel engines have a higher concentration of oxygen, so any attempt to reduce NO_x to nitrogen would fail as the reducing agent would simply be oxidised by the oxygen. Diesel engines do have oxidation catalysts that turn carbon monoxide to carbon dioxide, and hydrocarbons to carbon dioxide and water. There have been attempts to reduce the amount of particulates in diesel exhaust gases by burning them at high temperatures. Unfortunately this reduces the fuel efficiency.

Catalytic converters

Petrol engines

The main pollutants in the exhaust are carbon monoxide, hydrocarbons, and nitrogen monoxide. These are removed by the following reactions:

- carbon monoxide $2CO + O_2 \rightarrow 2CO_2$
- hydrocarbons $C_7H_{16} + 11O_2 \rightarrow 7CO_2 + 8H_2O$
- nitrogen monoxide $2NO + CO \rightarrow N_2 + 2CO_2$

All three of these reactions take place in a three-way catalytic converter consisting of platinum or rhodium on a porous support.

Sulfur oxide pollutants are best prevented, by removing sulfur impurities from the fuels, before they are made available to the motorist.

Diesel engines

The main pollutants present in the exhaust are carbon monoxide, hydrocarbons, particulates, and nitrogen oxide compounds. Carbon monoxide and hydrocarbons are removed in the same way as in petrol engines. Particulates are removed by diesel particulate filters that contain a variety of materials, the most common being a ceramic. Regeneration (burning off the carbon particles) is accomplished by increasing the temperature at times decided by the vehicle's computer. This increases fuel consumption. Nitrogen oxides can by reduced by recycling some of the exhaust gases through the cylinder, lowering the temperature and thus the amount of NO_x formed. Alternatively, a reagent such as ammonia is used in the presence of a catalyst.

$$4NO + 4NH_3 + O_2 \rightarrow 4N_2 + 6H_2O$$

Again sulfur dioxide is best avoided in the first place by using ultra low-sulfur fuels.

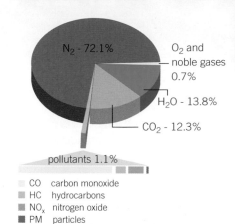

▲ **Figure 6** *Whilst the amount of harmful pollutants in a diesel engine exhaust is small, it is important they are removed*

Summary questions

1 a Explain the difference between primary and secondary pollutants. *(2 marks)*
b Choose one secondary pollutant and say how it is made from a primary pollutant. *(2 marks)*
c What are the main disadvantages of tropospheric ozone? *(1 mark)*

2 a Why is it important that catalytic converters start working at as low a temperature as possible? *(1 mark)*
b What is meant by a catalyst poison? *(1 mark)*
c Why are these catalysts used in the form of a fine powder? *(1 mark)*
d Suggest a reason why a catalytic converter has to be replaced eventually. *(1 mark)*
e Catalytic converters convert the pollutants carbon monoxide, CO, hydrocarbons, C_xH_y, and nitrogen oxides, NO_x, into less harmful gases. This is still only a partial solution to the emissions problem. Why? *(2 marks)*

3 a Explain how oxides of nitrogen are removed from the exhaust of a petrol engine. *(2 marks)*
b Explain why this method does not work for a diesel engine. *(2 marks)*
c Describe a method that is used to reduce the oxides of nitrogen in a diesel exhaust. *(2 marks)*

4 Explain, with equations, how each of the following primary pollutants is made in a car engine. *(4 marks)*
a nitrogen oxide, NO
b carbon monoxide, CO
c hydrocarbons
d sulfur dioxide, SO_2

5 a Use the data in Table 2 to decide how *you* would choose between a petrol and a diesel car. *(2 marks)*
b Write equations for the reactions involving a diesel oxidation catalyst which remove the following pollutants.
i carbon monoxide *(1 mark)*
ii cetane, $C_{16}H_{34}$ *(1 mark)*

DF 11 Other fuels

Specification reference: DF(k), DF(u)

In addition to changing the design of the car (Topic DF 10), the other way to tackle the emissions problem is to change the fuel used by cars.

Aromatic hydrocarbons make up as much as 40% of petrol. Aromatic hydrocarbons may cause higher carbon monoxide, CO, hydrocarbon, C_xH_y, and nitrogen oxide, NO_x, emissions, and some may cause cancer. Benzene for example is strictly controlled, but others may be controlled in the future.

Butane content is also likely to be lowered in the future. Butane is volatile and responsible for evaporative emissions leading to ozone formation and photochemical smogs. However, both butane and aromatic compounds help petrol to perform well in modern engines. If they are removed then they must be replaced by something else. This is why the petrol companies looked to oxygenates (compounds such as ethanol that contain oxygen) as a possible solution.

There are a range of options of different fuel sources for the future, each of which has advantages and disadvantages.

Other hydrocarbon fuels

Liquified petroleum gas (LPG) comes from the distillation of crude oil and is often called autogas when it is used in cars. It is a mixture of propane and butane in varying quantities – often around 60% propane. It has to be kept under pressure to store the hydrocarbons as liquids. Petrol vehicles can be converted fairly easily to run on both fuels – one of the main changes needed is a larger fuel tank.

Autogas works in high performance engines and produces about 20% less carbon dioxide per mile than petrol. Because of the higher ratio of carbon to hydrogen it releases less carbon monoxide. It also produces fewer unburnt hydrocarbons and nitrogen oxides than petrol. For vehicle owners, road taxes and fuel taxes are lower than for petrol. The main disadvantage is that, although numbers are increasing, LPG filling stations are still relatively rare.

▲ **Figure 1** *A bus powered by natural gas, pictured in Bristol*

Liquid natural gas (LNG) is mainly methane and comes from oil and natural gas fields. Methane cannot be liquefied by pressure alone and it must be cooled to below −160 °C. LNG is therefore most suitable for large vehicles, especially as it works in modified diesel engines. Once again, there is a high carbon to hydrogen ratio so less carbon monoxide is produced, and much fewer NO_x than diesel.

Replacements for fossil fuels

You have seen how fossil fuels are useful to humans. However, even though energy is extracted from fossil fuels more economically and with modern developments such as fracking (which produces shale gas), fossil fuels will not last for ever.

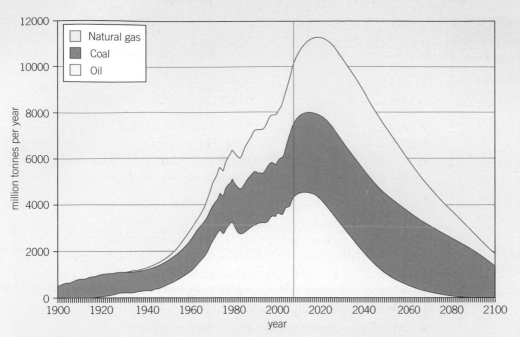

▲ **Figure 2** *Supplies of conventional fossil fuels, according to a 2008 study. Their use will decline in the future as supplies run out*

Because of this potential supply problem, coupled with concerns about emissions (Topic DF 10), alternative fuels are being considered, such as biofuels.

Chemical ideas: Human impacts 16.2

Alternatives to fossil fuels

Biofuels

Biofuels are alternative fuels derived from renewable plant and animal materials. Examples of biofuels include ethanol and biodiesel.

Ethanol

Ethanol can be made by fermentation of carbohydrate crops such as sugar cane. Cars cannot easily run on ethanol alone as it is too volatile, but mixtures of petrol containing up to 15% ethanol are used.

Ethanol is sometimes said to be carbon neutral as the carbon dioxide produced in fermentation and burning the ethanol matches the carbon dioxide absorbed in the growing plant. However, energy is used to produce and distribute the ethanol, producing carbon dioxide if the energy comes from fossil fuels. Another issue, especially in developing countries, is that the land used to grow crops for fermentation could be used to produce food instead.

Biodiesel

Biodiesel is typically made by chemically reacting fats and oils, such as vegetable oil or animal fat, with an alcohol producing fatty acid esters. This process is called trans-esterification.

▲ **Figure 3** *A biofuel pump in Belgium. The fuel, gasohol, is a mixture of petrol (85%) and ethanol*

Synoptic link

You will learn about esters and carboxylic acids in Chapter 5, What's in a medicine?

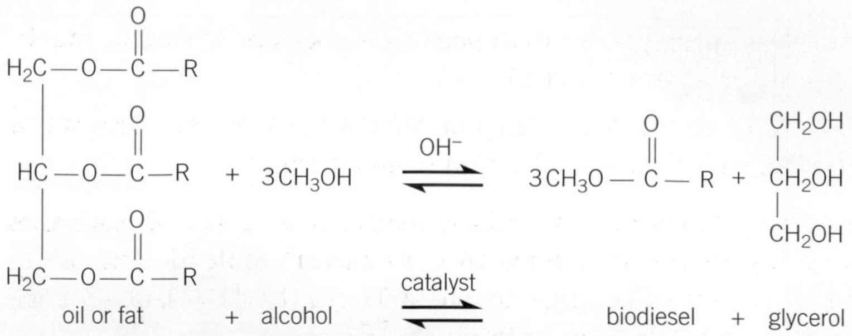

▲ **Figure 4** *The trans-esterification of an oil or fat*

The advantages of biodiesel over diesel are:

- it can be made from waste oil rather than using fossil-fuel based oil
- it is carbon-neutral (except for the energy required to produce and distribute it)
- some diesel vehicles can run on pure biodiesel, but most on mixtures with regular diesel
- it is biodegradable if spilled
- it contains virtually no sulfur, so reduces oxides of sulfur in emissions
- it produces less particulates, carbon monoxide, and hydrocarbons than petrol and diesel (though these reductions are limited by the ways in which nitrogen oxides are removed).

The disadvantage is that it produces more nitrogen oxides than conventional fossil fuels.

Green diesel (often derived from algae), and biogas (methane derived from animal manure and other digested organic material) are other examples of biofuels.

Hydrogen

Hydrogen is another potential fuel for the future. On combustion, hydrogen produces just water.

$$2H_2(g) + O_2(g) \rightarrow 2H_2O(g)$$

Its advantages are:

- it is renewable and can be made by electrolysis of water
- it can be stored and sent down pipelines, in much the same way that methane currently is
- it can be used in internal combustion engines or in a fuel cell to generate electricity
- it produces no carbon dioxide, carbon monoxide, or hydrocarbons when burnt.

Its disadvantages are:

- its production from water often depends on the use of electricity from fossil fuel power stations (though alternative energy sources for electricity are being developed)

- it is less energy dense than petrol – it does not release as much energy per gram as petrol

- oxides of nitrogen are still produced at the high temperatures a hydrogen internal combustion engine runs at.

The hydrogen economy would use hydrogen as a way of storing and distributing energy. If systems are costed over whole lifetime use in terms of money and energy, then distributing hydrogen by pipeline may be cheaper than transmitting electricity.

Fuel cells are being used to generate electricity on a small scale in cars. These fuel cells convert the chemical energy from a fuel into electricity through a chemical reaction with oxygen or another oxidizing agent in an electrochemical cell. In the case of fuel cells for cars the fuel is usually oxygen. This means the main product is water.

The main problem in the design of cars run on hydrogen is that a large volume of gaseous hydrogen is required to get the mileage equivalent to a fuel tank of petrol. Some way of storing hydrogen more compactly is needed. One solution is to store it as a liquid in a high-pressure fuel tank.

Schemes for developing alternative fuel for use on a large scale depend on long-term and large-scale investments in new infrastructures, so success will depend on political as well as economic factors. The reward could be cleaner renewable fuels.

▲ **Figure 5** *A hydrogen fuel station in Iceland*

▲ **Figure 6** *Refilling a car with hydrogen*

Synoptic link

You will find out about redox reactions and electrolysis in Chapter 3, Elements from the sea.

Summary questions

1   a   Write equations for the complete combustion of one mole of octane, cetane, and ethanol. (*3 marks*)
    b   Look up the enthalpy changes of combustion of these compounds. Work out the volume of carbon dioxide produced per kJ of energy transferred by the burning of each of the three fuels. (Assume one mole of CO_2 occupies $24.0\,dm^3$.) (*6 marks*)
    c   Comment on the values you obtain. (*1 mark*)

2   The petrol tank of a typical car holds about 45 litres of petrol (approximately 10 gallons). Calculate the amount of energy released by burning 45 litres of petrol. Use the following information to help you.
    - assume that petrol is octane, C_8H_{18}
    - the standard enthalpy change of combustion of C_8H_{18} is $-5500\,kJ\,mol^{-1}$
    - the density of octane is $0.70\,g\,cm^{-3}$
    - 1 litre is $1000\,cm^3$. (*2 marks*)

3   a   Calculate the mass and volume (at 20 °C and 1 atmosphere pressure) of hydrogen needed to provide the same amount of energy as 45 litres of octane in **2**. (*2 marks*)
    - The standard enthalpy change of combustion of H_2 is $-286\,kJ\,mol^{-1}$.
    - One mole of a gas at 20 °C and atmospheric pressure has a volume of about 24 litres.

    Hydrogen engines are efficient. In motorway driving conditions, a hydrogen engine can be over 20% more efficient than a petrol engine. In city 'stop-go' driving conditions, the hydrogen engine is about 50% more efficient than a petrol engine.
    b   Taking efficiencies into account, what mass and volume of hydrogen are needed to give the same mileage as 45 litres of petrol in
        i   motorway driving conditions   (*2 marks*)   ii   city driving conditions.   (*2 marks*)

Practice questions

1 The amount of energy needed to raise 1 g of a substance 1 °C, is called its:

  A heat capacity

  B thermal capacity

  C molar heat capacity

  D specific heat capacity   *(1 mark)*

2 Which row in the table is correct?

A	bond-making	endothermic	$+\Delta H$
B	bond-breaking	exothermic	$-\Delta H$
C	bond-making	exothermic	$+\Delta H$
D	bond-breaking	endothermic	$+\Delta H$

(1 mark)

3 $10\,cm^3$ of $2.0\,mol\,dm^{-3}$ NaOH is mixed with $10\,cm^3$ $2.0\,mol\,dm^{-3}$ H_2SO_4. The temperature rises by y °C.

The enthalpy change of neutralisation of the reaction of NaOH with H_2SO_4 (in kJ) is given by:

  A $4.18\,y$ C $20 \times 4.18\,y$

  B $4.18\dfrac{y}{2}$ D $15 \times 4.18\dfrac{y}{2}$ *(1 mark)*

4 Chloroethene has the formula of $CH_2{=}CHCl$. Poly(chloroethene) can be represented as:

  A $-(CH_3CH_2Cl)-$ C $-(CHCHCl)-$

  B $-(CHClCHCl)-$ D $-(CH_2CHCl)-$

(1 mark)

5 Which of the following *cannot* be formed from ethene in a one-step reaction?

  A CH_2BrCH_2Br C CH_3CH_2OH

  B CH_3CHBr_2 D CH_3CH_2Br *(1 mark)*

6 Which of the following might be found in the exhaust when hydrogen burns in an internal combustion engine?

  A NO_x C hydrocarbons

  B CO   D SO_2 *(1 mark)*

7 A fuel is burnt in a small lamp and heats up a copper calorimeter containing water. Look at these three statements.

  1 The specific heat capacity of the water is not accurately known.

  2 Fuel may evaporate from the wick.

  3 The fuel may not burn completely.

Which statements are limitations to accuracy in measuring the energy transferred?

  A 1, 2, and 3 correct   C 2 and 3 correct

  B 1 and 2 correct   D 1 correct   *(1 mark)*

8 Which of the following are isomers of propan-1-ol?

  1 $CH_3CH(OH)CH_3$

  2 $CH_3CH_2OCH_3$

  3 CH_3CH_2CHO

  A 1, 2, and 3 correct   C 2 and 3 correct

  B 1 and 2 correct   D 1 correct   *(1 mark)*

9 Disadvantages of fossil fuels compared with biofuels include:

  1 they give off CO_2 when burned

  2 they have not (recently) absorbed CO_2 from the atmosphere

  3 supplies are running out.

  A 1, 2, and 3 correct   C 2 and 3 correct

  B 1 and 2 correct   D 1 correct   *(1 mark)*

10 Poly(propene) is a polymer which is now being used to make 'polymer banknotes'. One useful property is that it is unreactive to reagents such as acids.

  a Draw **full** structural formulae for propene and poly(propene)   *(2 marks)*

  b Why is poly(propene) less reactive than propene?   *(1 mark)*

  c Propene reacts with bromine.

    (i) Describe a test for propene based on this reaction.   *(1 mark)*

    (ii) Give the mechanism for the reaction of propene with bromine, using curly arrows and showing lone pairs and partial charges. Give the formula of the product. *(4 marks)*

    (iii) Say what you understand by the term *electrophile* and identify an electrophile from your mechanism.   *(2 marks)*

    (iv) Some chloride ions are added to the reaction in (ii).

       Give the formula of a product containing chlorine that would be formed.   *(1 mark)*

d Butene is the next alkene in the homologous series after propene.

  (i) What do you understand by the term *homologous series?* (*1 mark*)

  (ii) What term is used to describe compounds like alkenes that have C=C bonds? (*1 mark*)

  (iii) One structural isomer of butene has two stereoisomers.

  Draw the structures for these two stereoisomers and name them.
  (*2 marks*)

11 Sherbet sweets get their fizz from the reaction between sodium bicarbonate and citric acid.

$$C_6H_8O_7(aq) + 3NaHCO_3(s) \rightarrow$$
$$Na_3C_6H_5O_7 + 3CO_2(g) + 3H_2O$$
Equation 12.1

a (i) Some students decide to investigate this **endothermic** reaction and measure the enthalpy change of reaction.

  They have available 10.0 g portions of sodium bicarbonate and 25.0 cm^3 portions of a solution of citric acid. The citric acid portions represent an excess over the sodium bicarbonate portions in the reaction in **Equation 12.1**.

  Describe how they would carry out their experiment and how they would work out ΔH per mole of sodium bicarbonate from their results.
  (*6 marks*)

  (ii) A student says that the temperature change they measure would be inaccurate because of heat losses. Comment on this statement.
  (*2 marks*)

b Calculate the volume of carbon dioxide (measured at room temperature and pressure) that the students would collect if they reacted 10.0 g of sodium bicarbonate with excess citric acid. Assume none of the carbon dioxide dissolves.

  Give your answer to an appropriate number significant figures. (*3 marks*)

12 Petrol cars produce less NO$_x$ and particulates than diesel cars but more CO and hydrocarbons.

a (i) Suggest why diesel cars produce fewer hydrocarbons. (*1 mark*)

  (ii) Write an equation to show how CO is formed from the combustion of hexane in a petrol engine. (*1 mark*)

  (iii) Give a reason why particulates are a pollutant. (*1 mark*)

b NO and CO can be removed from the exhaust of a petrol engine by reacting them together over a catalytic converter. This uses a heterogeneous catalyst.

  (i) Give the equation for this reaction.
  (*1 mark*)

  (ii) Explain the term *heterogeneous* in the context of catalysis and describe the first stage in the mechanism of this type of catalysis. (*2 marks*)

c Heterogeneous catalysts are also used for cracking hydrocarbons. Write the equation for a reaction in which nonane is cracked to produce ethene and one other product.
(*1 mark*)

d Ethanol is one example of a biofuel.

  (i) Write the equation for the complete combustion of ethanol.

  Show state symbols under standard conditions. (*1 mark*)

  (ii) Use the bond enthalpy values in the table to calculate a value for the enthalpy change of combustion of ethanol. (*3 marks*)

Bond	Average bond enthalpy / kJ mol^{-1}
C—H	413
C—C	347
C—O	358
O—H	464
C=O	805
O=O	498

  (iii) The enthalpy change of combustion of ethanol in a Data Book is different from your answer to (ii). Suggest **two** reasons for this. (*2 marks*)

  (iv) Biofuels are said to be sustainable. Explain the word *sustainable* in this context. (*1 mark*)

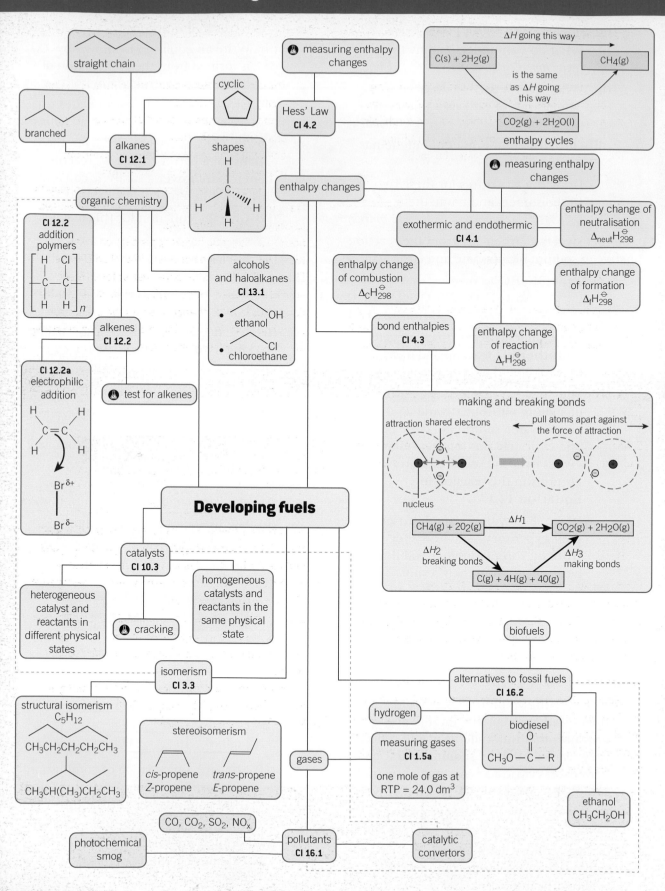

straight chain

branched

cyclic

alkanes
Cl 12.1

shapes

organic chemistry

Cl 12.2
addition
polymers

alkenes
Cl 12.2

Cl 12.2a
electrophilic
addition

test for alkenes

alcohols
and haloalkanes
Cl 13.1
- ethanol
- chloroethane

measuring enthalpy
changes

Hess' Law
Cl 4.2

enthalpy changes

$C(s) + 2H_2(g)$

ΔH going this way

is the same
as ΔH going
this way

$CO_2(g) + 2H_2O(l)$

$CH_4(g)$

enthalpy cycles

measuring enthalpy
changes

exothermic and endothermic
Cl 4.1

enthalpy change of
neutralisation
$\Delta_{neut}H_{298}^{\ominus}$

enthalpy change
of combustion
$\Delta_cH_{298}^{\ominus}$

enthalpy change
of formation
$\Delta_fH_{298}^{\ominus}$

bond enthalpies
Cl 4.3

enthalpy change
of reaction
$\Delta_rH_{298}^{\ominus}$

making and breaking bonds

attraction shared electrons

pull atoms apart against
the force of attraction

nucleus

$CH_4(g) + 2O_2(g)$ $\xrightarrow{\Delta H_1}$ $CO_2(g) + 2H_2O(g)$

ΔH_2
breaking bonds

ΔH_3
making bonds

$C(g) + 4H(g) + 4O(g)$

Developing fuels

catalysts
Cl 10.3

heterogeneous
catalyst and
reactants in
different physical
states

cracking

homogeneous
catalysts and
reactants in the
same physical
state

biofuels

alternatives to fossil fuels
Cl 16.2

hydrogen

biodiesel
$CH_3O - C - R$
(with O double bond)

isomerism
Cl 3.3

structural isomerism
C_5H_{12}

$CH_3CH_2CH_2CH_2CH_3$

$CH_3CH(CH_3)CH_2CH_3$

stereoisomerism

cis-propene
Z-propene

trans-propene
E-propene

gases

measuring gases
Cl 1.5a

one mole of gas at
RTP = 24.0 dm³

ethanol
CH_3CH_2OH

CO, CO_2, SO_2, NO_x

photochemical
smog

pollutants
Cl 16.1

catalytic
convertors

Carbon dioxide, enthalpy cycles, and alternative fuels

1 Below are two diagrams representing the bonding in carbon dioxide. The first is a simple dot-and-cross diagram whilst the second shows σ- and π-orbitals. Evaluate the two models of bonding, identifying phenomena that can be explained by each.

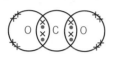

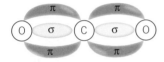

2 In this module you have learnt about different enthalpy cycles. Produce a summary of Hess's Law to include how enthalpy cycles can be used to calculate:

a enthalpy changes of combustion from enthalpy changes of formation;

b enthalpy changes of formation from enthalpy changes of combustion;

c enthalpy changes of reaction from enthalpy changes of formation or combustion;

d enthalpy changes of reaction from bond enthalpies.

Carry out research to include a wide range of worked examples in your summary.

3 Society is currently highly dependent on fossil fuels, yet the Intergovernmental Panel on Climate Change (IPCC) has stated that use of fossil fuels should be phased out by 2100 if the world is to avoid dangerous climate change. Discuss the challenges and opportunities of moving to a low-carbon economy and the role that chemists can play.

Extension

Silanes and silenes are the silicon analogues of alkanes and alkenes. They contain silicon and hydrogen atoms only.

Silanes contain single bonds only. At standard temperature and pressure, the two simplest silanes are gases, whereas the next two (Si_3H_8 and Si_4H_{10}) are liquids. Silanes are highly reactive when mixed with air, spontaneously catching fire.

Silenes contain silicon–silicon double bonds. The simplest contain just two silicon atoms and are known as disilenes. They are extremely reactive and polymerise readily. The first stable disilene, tetramesityldisilene, was synthesised in 1972. It contains very bulky side groups to stabilise the molecule.

Bond	Average bond enthalpy / kJ mol^{-1}
Si—Si	222
Si=Si	113

1 What is the general formula for silanes and silenes?

2 Draw dot-and-cross diagrams for SiH_4 and Si_2H_4 and deduce the bond angles around each silicon atom.

3 Write a balanced chemical reaction for the oxidation of SiH_4 when it mixes with air.

4 Look up the structure of tetramesityldisilene. Suggest why silenes readily polymerise, and why bulky side groups in tetramesityldisilene help to stabilise the molecule.

5 What volume of hydrogen would react with 0.9 g of Si_2H_4? (one mole of H_2 occupies 24.0 dm^3 at r.t.p.)

6 Would SiH_4 make a good fuel?

CHAPTER 3
Elements from the sea

Topics in this chapter

Why a chapter on Elements from the sea?

The first chapter in the course – Elements of life – told the story of how the elements were formed. The theme is taken further in this chapter, which tells how extract some elements, such as bromine, from the natural resources that contain them and turn them into useful substances. In order to understand the extraction of bromine from sea water, redox chemistry is used. This leads on to looking at the manufacture of another useful halogen – chlorine and the chemistry of the halogens, needing an understanding of electrolysis. The risks and benefits of using chlorine are discussed. Chlorine is used to make bleach which leads to a study of dynamic equilibria, including a quantitative treatment of this subject. The industrial processes you look at in this chapter introduce the concept atom economy.

Knowledge and understanding checklist

From your Key Stage 4 study you will have studied the following. Work through each point, using your Key Stage 4 notes and the support available on Kerboodle.

☐ The halogens.

☐ Oxidation and reduction.

☐ Electrolysis.

☐ Dynamic equilibria.

☐ Atom economy.

You will learn more about some ideas introduced in earlier modules:

☐ amounts in moles (**Elements of life**)

☐ electronic structure of atoms (**Elements of life**)

☐ covalent bonding (**Elements of life** and **Developing fuels**)

☐ the periodic table (**Elements of life**).

Maths skills checklist

In this chapter, you will need to use the following maths skills. You can find support for these skills on Kerboodle and through MyMaths.

☐ Recognise and make use of appropriate units in calculation.

☐ Recognise and use expressions in decimal and ordinary form.

☐ Use ratios, fractions, and percentages.

☐ Understand and use the symbols =, <, <<, >>, >, ∝, ~, ⇌.

☐ Interpret data in tables and graphs and make estimations.

MyMaths.co.uk
Bringing Maths Alive

Learning outcomes

Demonstrate and apply knowledge and understanding of:

→ a description of the following physical properties of the halogens – appearance and physical state at room temperature, volatility, solubility in water and organic solvents

→ the relative reactivity of the halogens in terms of their ability to gain electrons

→ the details of the redox changes which take place when chlorine, bromine, and iodine react with other halide ions, including observations, equations, and half-equations

→ the reactions between halide ions, Cl^-, Br^-, and I^-, and silver ions, Ag^+, and ionic equations to represent these precipitation reactions, the colours of the precipitates, and the solubility of silver halides in ammonia.

In Figure 1, the person is floating on water in the Dead Sea. They float because of the high **density** of the water there. The high density is due to the water containing about $350\,g\,dm^{-3}$ of salts compared with $40\,g\,dm^{-3}$ in water from the oceans. A coffee mug could hold about $350\,g$ of salts and $1\,dm^3$ is about four coffee mugs – the solution is very concentrated.

The Dead Sea is the lowest point on Earth, almost $400\,m$ below sea level in the Rift Valley that runs from East Africa to Syria (Figure 2). It is like a vast evaporating basin – water flows in at the north end from the River Jordan, but there is no outflow. The countryside around it is desert and in the scorching heat so much water evaporates that the air is thick with haze, making it hard to see across to the mountains a few kilometres away on the other side. This steady evaporation of water for thousands of years has resulted in huge accumulations of **salts**.

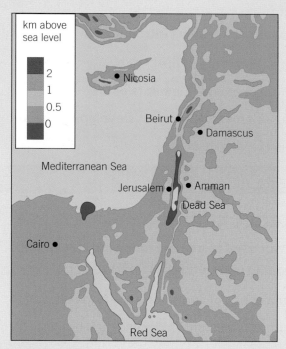

▲ **Figure 1** *The Dead Sea – humans float easily due to the high density of the water*

◄ **Figure 2** *The region around the Dead Sea in relief*

Salts found in the Dead Sea

Surveys of the salt concentration of the Dead Sea were conducted as early as the seventeenth century. Estimates suggest that there are about 43 billion tonnes of salts in the Dead Sea, and a particular feature is the relatively high proportion of bromides.

The sea is the major source of minerals in the region. A chemical industry has grown up around the Dead Sea in Israel and it has become one of the largest exporters of bromine compounds in the world. The annual production of bromine compounds in Israel exceeds 230 000 tonnes. Bromide, chloride, and iodide ions are colourless in solution and the crystalline salts from the Dead Sea look white. All of these ions are from Group 7. The elements in Group 7 look very different to their salts.

19.0
F
9
35.5
Cl
17
79.9
Br
35
126.9
I
53
210
As
85

▲ **Figure 3** *The elements of Group 7*

The p-block – Group 7

The **halogens** are the elements in Group 7 of the periodic table (Figure 3).

All halogen atoms have seven electrons in the outer shell. The halogens are the most reactive group of non-metals and none of them are found naturally in the elemental form. They are all found naturally in compounds such as calcium fluoride and sodium chloride. These two compounds contain the halide ions F^- and Cl^-, respectively.

Fluorine and chlorine are the most abundant halogens, bromine occurs in smaller quantities, iodine is quite scarce, and astatine is artificially produced, short-lived, and radioactive.

All the halogen elements occur as **diatomic molecules**, for example, fluorine, F_2, and bromine, Br_2. The two atoms are linked by a single covalent bond (Figure 4).

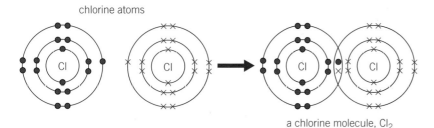

chlorine atoms

a chlorine molecule, Cl_2

▲ **Figure 4** *Covalent bonding in the chlorine, Cl_2, molecule*

In compounds, a halogen atom achieves stability by:

- gaining an electron from a metal atom, forming a halide ion in an ionic compound (Figure 5)

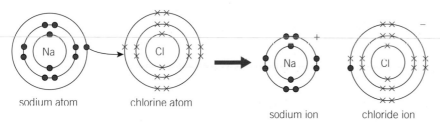

sodium atom   chlorine atom

sodium ion   chloride ion

▲ **Figure 5** *Halogens can achieve eight outer electrons by forming an ion, for example, Cl^-, via ionic bonding with a metal*

- sharing an electron from another non-metal atom in a covalently bonded compound (Figure 6).

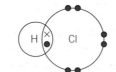

▲ **Figure 6** *Halogens can achieve eight outer electrons by covalent bonding with a non-metal*

Physical properties of the halogens

Table 1 shows some properties of the halogens.

▼ **Table 1** *Some physical properties of the halogens*

Element	fluorine	chlorine	bromine	iodine
Formula of the element	F_2	Cl_2	Br_2	I_2
Appearance at room temperature	pale yellow gas	green gas	dark red volatile liquid	shiny grey/black solid – sublimes to give purple vapour on warming
Melting point / K	53	172.0	266.0	387.00
Boiling point / K	85	239.0	332.0	457.00
Solubility at 298 K / g per 100 g of water	reacts with water	0.6	3.5	0.03

Synoptic link

Chapter 4, The ozone story, explains how intermolecular bonds cause the trends in boiling points for the halogens.

▲ **Figure 7** *Halogens dissolved in water (the lower layer) – chlorine is pale green (left), bromine is orange/yellow (middle), and iodine is brown (right)*

▲ **Figure 8** *Halogens dissolved in cyclohexane – chlorine is pale green (left), bromine is orange/brown/red (middle), and iodine is violet (right)*

This demonstrates some trends of properties going down the group:

● become darker in colour

● melting point and boiling point increase

● change from gases to liquids to solids at room temperature

● become less volatile.

There is no trend for the solubility of the halogens in water as you go down the group, but the halogens are more soluble in organic solvents (such as hexane and cyclohexane) than they are in water. The solutions of halogens in organic solvents have a much more distinct colour and this makes it easier to say which halogen is present without ambiguity (Figure 7 and Figure 8).

Chemical properties of the halogens

Which is most reactive and why?

The halogens are a group of reactive elements. They tend to remove electrons from other elements so that they can complete their outer shell. For example:

$$2Na(s) + Cl_2(g) \rightarrow 2NaCl(s)$$

● Sodium, Na, has lost an electron to become Na^+. It has been **oxidised**.

● Each chlorine atom, Cl, has gained an electron to become Cl^-. It is an **oxidising agent**.

● The elements at the top of the group are the most reactive and are the strongest oxidising agents.

The core of an atom is made up of the nucleus and inner shell electrons (Figure 9). Both fluorine and chlorine have atoms that consist of a core with a charge of +7, surrounded by an outer electron shell containing seven electrons. Both react by gaining one electron to complete their outer shell. Fluorine atoms are very small, so the attraction between the core and the electron that completes its outer shell is very strong. In chlorine, the outer shell is further from the core and the attraction for the extra electron is weaker. This means that fluorine gains an extra electron more readily to become a negative ion than chlorine does. This trend continues down the halogen group, as each successive member of the group has one more complete electron shell than the previous one.

Activity ES 1.1

In this activity you will find out more about the physical properties of the halogens.

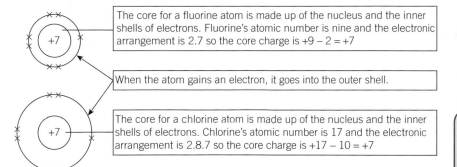

The core for a fluorine atom is made up of the nucleus and the inner shells of electrons. Fluorine's atomic number is nine and the electronic arrangement is 2.7 so the core charge is $+9 - 2 = +7$

When the atom gains an electron, it goes into the outer shell.

The core for a chlorine atom is made up of the nucleus and the inner shells of electrons. Chlorine's atomic number is 17 and the electronic arrangement is 2.8.7 so the core charge is $+17 - 10 = +7$

▲ **Figure 9** *Core charges and outer shells of electrons in fluorine (top) and chlorine (bottom)*

Activity ES 1.2

This activity helps you learn about the reactions of halogens and halides.

Overall, this means that fluorine is the most reactive member of the halogen group – it is the strongest oxidising agent. The reactivity decreases as you go down Group 7, for example, in the reaction with metals such as iron and sodium, chlorine reacts more vigorously than bromine.

Reactions of the halogens with halide ions

If you add a solution of chlorine (pale green) to a solution of potassium iodide ions (colourless) there is a chemical reaction. The solution turns brown and iodine is produced.

$$Cl_2 \text{ (aq)} + 2KI \text{ (aq)} \rightarrow 2KCl \text{ (aq)} + I_2 \text{(aq)}$$
pale green   brown

This equation can be simplified by writing an ionic equation – the K^+ ions can be left out because they are unchanged:

$$Cl_2\text{(aq)} + 2I^-\text{(aq)} \rightarrow 2Cl^-\text{(aq)} + I_2\text{(aq)}$$

Ions that are left out of ionic equations because they are unchanged are called **spectator ions**.

Here each iodide ion loses an electron and is oxidised. Each chlorine atom gains an electron and you call the chlorine an oxidising agent. In a reaction, the oxidising agent is said to have been **reduced**. You will learn more about reduction in Topic ES 2.

The **half-equations** make this clearer:

$$Cl_2\text{(aq)} + 2e^-\text{(aq)} \rightarrow 2Cl^-\text{(aq)}$$
$$2I^-\text{(aq)} \rightarrow I_2\text{(aq)} + 2e^-\text{(aq)}$$

Half-equations show more clearly what is happening in a redox reaction. In one of the half-equations a species gains electron(s) and is reduced (the Cl_2). In the other half-equation a species loses electron(s) and is oxidised (the I^-).

A similar thing happens if bromine solution is added to iodide ions:

$$Br_2(aq) + 2I^-(aq) \rightarrow 2Br^-(aq) + I_2(aq)$$

orange/yellow   brown

These are examples of **displacement reactions** because a halogen displaces or pushes out a less reactive halogen from a compound. They are also called **redox reactions** because both oxidation and reduction have occurred in the same reaction. In the example above, the bromine gains electrons and is reduced whilst the iodine loses electrons and is oxidised.

Table 2 summarises the reactions of halogens and halide ions. You won't be surprised by the entries in the table if you remember that:

- reactivity decreases down the group
- the strongest oxidising agent is fluorine
- a halogen displaces a less reactive halide.

▼ **Table 2** *Does a halogen displace a halide ion?*

		Halide			
		F$^-$	Cl$^-$	Br$^-$	I$^-$
Halogen	F$_2$	no reaction	yes	yes	yes
	Cl$_2$	no reaction	no reaction	yes	yes
	Br$_2$	no reaction	no reaction	no reaction	yes
	I$_2$	no reaction	no reaction	no reaction	no reaction

When you carry out these reactions, the colour changes in the solutions can be difficult to detect. Adding an organic solvent can make the colour changes more distinct. The upper layer will be the organic layer.

Reactions of halide ions

At GCSE you may have met the reactions of silver ions with halide ions when studying **qualitative analysis**.

Reactions of halide ions with silver ions

When two solutions mix and a solid is formed, the solid is called a **precipitate**. Silver halides are precipitated when a solution of silver ions is added to a solution containing chloride, bromide, or iodide ions. The general equation is:

$$Ag^+(aq) + X^-(aq) \rightarrow AgX(s)$$

In the equation, X represents a halide. These are examples of precipitation reactions. Figure 10 shows the colours of these precipitates.

It can be difficult to distinguish between the colours of these precipitates so ammonia solution is sometimes added (Figure 11). The solubility of silver chloride is greater than that of silver bromide and silver iodide is insoluble in ammonia solution.

▲ **Figure 10** *The colours of the silver halides (from left to right) – white silver chloride, AgCl, cream silver bromide, AgBr, and pale yellow silver iodide, AgI*

▲ **Figure 11** *Distinguishing between the silver halides using ammonia solution – silver chloride (left), silver bromide (middle), and silver iodide (right) is insoluble*

Study tip

When you write an ionic equation for precipitation, first put the solid on the right.

$$\rightarrow AgCl(s)$$

Then add the aqueous ions on the left.

$$Ag^+(aq) + Cl^-(aq) \rightarrow AgCl(s)$$

Always include state symbols. In this example, no balancing is needed.

Summary questions

1   Write a stoichiometric equation with state symbols for the following reactions. Remember that Group 7 ions have a charge of -1, Group 1 ions have a charge of $+1$, and the molecules of Group 7 elements are diatomic.
    a   Chlorine water is mixed with aqueous sodium iodide.   (2 marks)
    b   Bromine water is mixed with aqueous potassium iodide.   (2 marks)
    c   Bromine water is mixed with aqueous sodium chloride.   (1 mark)

2   Write the two half-equations with state symbols for the following ionic equation.

$$Br_2(aq) + 2I^-(aq) \rightarrow 2Br^-(aq) + I_2(aq) \qquad \text{(4 marks)}$$

3   Write balanced ionic equations with state symbols for the following precipitation reactions when silver nitrate solution is added to the following.
    a   potassium iodide solution   (2 marks)
    b   sodium bromide solution   (2 marks)
    c   copper(II) chloride solution   (2 marks)

4   Given that water in the Dead Sea has a bromide ion concentration of $5.2\,g\,dm^{-3}$ and a chloride ion concentration of $208\,g\,dm^{-3}$, calculate the following:
    a   Concentration in $mol\,dm^{-3}$ for bromide ions in Dead Sea water.   (1 mark)
    b   Concentration in $mol\,dm^{-3}$ for chloride ions in Dead sea water.   (1 mark)
    c   The simplest ratio of bromide ions to chloride ions in the Dead Sea. One of the numbers in $Br^-:Cl^-$ should be 1.   (1 mark)

5   Predict what you would observe after these chemicals have been added together and shaken.
    a   Silver chloride solution and excess dilute ammonia solution.   (1 mark)
    b   Chlorine water, potassium iodide solution, and cyclohexane. (2 marks)
    c   Sodium chloride solution, iodine solution, and cyclohexane.   (2 marks)

Synoptic link

You carried out calculations where you converted concentration in $g\,dm^{-3}$ to concentration in $mol\,dm^{-3}$ in Chapter 1, Elements of life. You may decide to revisit this before attempting the summary questions.

ES 2 Bromine from sea water

Learning outcomes

Demonstrate and apply knowledge and understanding of:

→ the explanation (given the necessary information) of the chemical processes occurring during the extraction of the halogens from minerals in the sea

→ redox reactions of s-, p- and d-block elements and their compounds in terms of electron transfer:

- use of half-equations to represent simple oxidation and reduction reactions
- the definition of oxidation and reduction as loss and gain of electrons
- identification of oxidising and reducing agents

→ the oxidation states assigned to and calculated for specified atoms in formulae (including ions) and explanation of which species have been oxidised and which reduced in a redox reaction

→ use of systematic nomenclature to name and interpret the names of inorganic compounds

→ use of oxidation states to balance redox equations that do not also involve acid–base reactions.

The high levels of bromide ions in the Dead Sea make it an ideal source of the element bromine. The Dead Sea Bromine Group Ltd at S'Dom in Israel opened on the shore of the Dead Sea in the 1930s and still produces bromine. Bromine compounds have a wide range of uses in the pharmaceutical industry. Drugs that use bromine in their manufacture are undergoing trials for treatment of Alzheimer's disease. Most of the bromine produced is used in the manufacture of flame-retardants, which have been instrumental in saving many thousands of lives.

Getting bromine from dissolved bromide ions involves simple chemistry – in a laboratory a chlorine solution is added to a solution containing bromide ions. Industrially it is more complicated, and involves the use of some ingenious engineering (Figure 1).

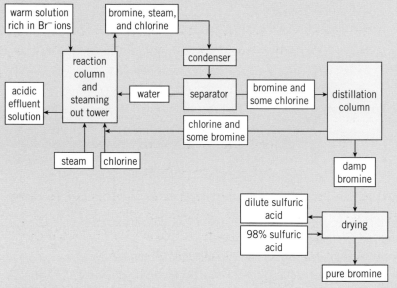

▲ **Figure 1** *A scheme for the industrial manufacture of bromine*

The industrial production of bromine

Chlorine is added to warm, partially evaporated, acidified sea water to displace bromine from the bromide ions. Bromine is volatile (boiling point 331 K) and bromine vapour and water is given off when steam is blown through the solution. The vapours are then condensed and two layers form because liquid bromine is not very soluble in water. The dense bromine layer is run off from the water layer that floats on top. The impure bromine is then distilled and dried.

The chlorine required for the process is produced by electrolysis on site. The reaction of chlorine with bromide ions is an example of a very important type of chemical reaction in which both oxidation and reduction take place – redox reactions.

Chemical ideas: Redox 9.1

Oxidation and reduction

Definitions of oxidation and reduction

At GCSE, you learnt that oxidation is gain of oxygen, and reduction is loss of oxygen. When magnesium and copper(II) oxide are heated together the following reaction happens:

$$Mg(s) + CuO(s) \rightarrow MgO(s) + Cu(s)$$

- Magnesium is oxidised because it gains oxygen – copper(II) oxide causes this to happen and so it is the oxidising agent.
- Copper(II) oxide is reduced because it loses oxygen – magnesium causes this to happen and so it is the **reducing agent**.

This is a redox reaction because it involves both oxidation and reduction.

Redox defined by electrons

Not all redox reactions involve oxygen, for example, the reaction between chlorine and bromide ions used to make bromine at the Dead Sea chemical plant:

$$Cl_2(aq) + 2Br^-(aq) \rightarrow 2Cl^-(aq) + Br_2(aq)$$

The half-equations are:

$$Cl_2(aq) + 2e^- \rightarrow 2Cl^-(aq)$$

$$2Br^-(aq) \rightarrow Br_2(aq) + 2e^-(aq)$$

- Chlorine is reduced because it gains electrons – bromide ions cause this to happen and so they are the reducing agent.
- Bromide ions are oxidised because they lose electrons – chlorine causes this to happen and so it is the oxidising agent.

This is also a redox reaction.

This is a much more useful definition. Half-equations can be written to show what happens to magnesium atoms and copper(II) ions.

$$Mg(s) \rightarrow Mg^{2+}(s) + 2e^-$$

$$Cu^{2+}(s) + 2e^- \rightarrow Cu(s)$$

- Magnesium is oxidised because it loses electrons.
- Copper(II) ions are reduced because they gain electrons.

Redox defined by oxidation state

A third definition is:

- when something is oxidised, its **oxidation state** (also called its oxidation number) increases
- when something is reduced, its oxidation state decreases.

> **Study tip**
>
> Remember, oxidising agents are reduced in chemical reactions and reducing agents are oxidised in chemical reactions.

> **Study tip**
>
> Some students find OIL RIG useful for remembering the definition involving electrons:
>
> **O**xidation **I**s **L**oss, **R**eduction **I**s **G**ain.

> **Synoptic link**
>
> You will develop your understanding of redox reactions in Chapter 9, Developing metals.

> **Study tip**
>
> *Oxidation* is:
> - gain of oxygen
> - loss of electrons
> - an increase in oxidation number.
>
> *Reduction* is:
> - loss of oxygen
> - gain of electrons
> - a decrease in oxidation number.

Oxidation states

The oxidation state assigned to an element in a chemical tells you how many electrons have been lost or gained compared to the unreacted element. It is useful for:

- naming inorganic compounds
- deciding what has been oxidised and what has been reduced in a reaction
- identifying the oxidising agent and the reducing agent in a redox reaction
- balancing redox equations.

Working out the oxidation state for an element

The atoms in elements always have oxidation number 0. For example, oxygen, O_2, magnesium, Mg, and chlorine, Cl_2, all have the oxidation number 0.

Working out the oxidation state for a simple ion

In simple ions, the oxidation state is the same as the charge on the ion. For example, chlorine in Cl^- has the oxidation state −1, oxygen in O^{2-} has the oxidation state −2, and potassium in K^+ has the oxidation state +1.

Working out oxidation states in compounds

When compounds have no overall charge, the oxidation states of all the constituent elements must add up to zero. Some elements have oxidation states that never or rarely change. Table 1 shows the oxidation state for some elements in compounds.

▼ **Table 1** *Oxidation states of some elements in compounds*

Element	Oxidation state
fluorine	−1
oxygen	−2 (except when combined with F or in the peroxide ion $O_2{}^{2-}$)
chlorine	−1 (except when combined with O or F)
bromine	−1 (except when combined with O, F, or Cl)
iodine	−1 (except when combined with O, F, Cl, or Br)
hydrogen	+1 (except when in a metal hydride e.g., NaH)
Group 1	+1
Group 2	+2
aluminium	+3

 Worked example: Oxidation states in compounds

What is the oxidation state of bromine in BrF_3?

Step 1: Work out the oxidation state of fluorine in BrF_3.

Using the rules in Table 1, fluorine has an oxidation number of −1. There are three fluorine atoms in BrF_3.

$$(-1) \times 3 = -3$$

Step 2: Work out the oxidation state of bromine in BrF_3.

The overall charge of BrF_3 is 0. The oxidation state of bromine and fluorine must equal 0.

$$0 - (-3) = +3$$

The oxidation state of bromine is +3.

Working out oxidation state in more complicated ions

The same rules that are used above can also be used for the elements that make up complicated ions containing more than one element, such as $PO_4{}^{3-}$. This time the oxidation state of the constituent elements add up to the overall charge on the ion.

Worked example: Oxidation states in complex ions

What is the oxidation state of phosphorus in PO_4^{3-}?

Step 1: Work out the contribution of oxygen to the overall PO_4^{3-} ion in terms of oxidation state.

Using the rules in Table 1, oxygen has an oxidation state of -2. There are four oxygen atoms in PO_4^{3-}.

$$(-2) \times 4 = -8$$

Step 2: Work out the oxidation state of phosphorus in PO_4^{3-}.

The overall charge of PO_4^{3-} is -3. The oxidation states of phosphorus and oxygen when added together must equal -3.

$$\underset{\substack{\text{overall charge} \\ \text{on the ion}}}{(-3)} \quad - \quad \underset{\substack{\text{total oxidation numbers of} \\ \text{oxygen in the ion}}}{(-8)} \quad = +5$$

The oxidation state of phosphorus is $+5$.

Activity ES 2.1

This activity helps you learn the rules for assigning oxidation states.

Systematic names

From a given formula you can work out the **systematic name**. A systematic name has a Roman numeral in brackets and the Roman numeral is the oxidation state of one of the elements in a compound.

At GCSE you may have met iron compounds in qualitative analysis. Iron can have a $+2$ or $+3$ oxidation state. The iron(II) ion is Fe^{2+} and the iron(III) ion is Fe^{3+}. As such, FeO is called iron(II) oxide and Fe_2O_3 is called iron(III) oxide. Copper also has two oxidation states, $+1$ and $+2$. Figure 2 shows the two oxides of copper – copper(II) oxide and copper(I) oxide.

The following rules are used for systematic names.

- Oxidation states are only used in systematic names for elements which have variable oxidation states, for example, transition metals, tin, lead, sulfur, nitrogen, and the halogens.

- The number shows the oxidation state of a preceding element. In iron(II) oxide the iron has oxidation state $+2$ but in potassium nitrate(V), KNO_3, it is the nitrogen that has the oxidation state $+5$.

- The number is placed close up to the element it refers to – there is no space between the name and the number.

▲ **Figure 2** *Two oxides of copper copper(II) oxide (left) and copper(I) oxide (right)*

 Worked example: Writing systematic names

What is the systematic name for $KMnO_4$?

Step 1: Identify the name of the compound.

$KMnO_4$ is potassium manganate.

▼ **Table 2** *Roman numerals*

Number	Roman numeral
1	I
2	II
3	III
4	IV
5	V
6	VI
7	VII
8	VIII

Step 2: Identify the element that has a variable oxidation state.

Manganese, like iron, is a transition metal. Transition metals have variable oxidation states, so the Roman numeral sits next to the manganate.

Step 3: Work out the oxidation state of manganese.

Using the rules from Table 1, the oxidation state of oxygen is −2.

$$(-2) \times 4 = -8$$

The oxidation state of potassium, a Group 1 metal, is +1.

The overall charge of $KMnO_4$ is 0.

$$0 - (-8) - (+1) = +7$$

Step 4: Write out the systematic name for $KMnO_4$.

The Roman numeral for seven is VII (Table 2), so the systematic name must be potassium manganate(VII).

Naming oxyanions

An **oxyanion** is a negative ion with oxygen in. The name always ends in –ate to show that oxygen is present. The nitrate(V) ion has the formula NO_3^- and the nitrate(III) ion has the formula NO_2^-. Oxyanions need to have some indication of the oxidation state of the element other than oxygen.

Chlorine, bromine, and iodine usually have the oxidation state −1 but not when they are in oxyanions. Table 3 shows the oxidation state of chlorine in different chlorate ions.

▼ **Table 3** *Oxidation state of chlorine in different chlorate ions*

Ion	Oxidation state of clorine	Name of ion
ClO^-	+1	chlorate(I)
ClO_2^-	+3	chlorate(III)
ClO_3^-	+5	chlorate(V)
ClO_4^-	+7	chlorate(VII)

Study tip

Be careful when adding oxidation states. Although there are two I⁻ ions in the example, it is the oxidation state on an individual iodine that is required, so it is just −1.

Using oxidation states to decide what has been oxidised and what has been reduced

Oxidation states can be used to find what has been oxidised and what has been reduced in a redox reaction. Look at the below reaction and allocate oxidation states for each element.

$$Cl_2(aq) + 2I^-(aq) \rightarrow 2Cl^-(aq) + I_2(aq)$$

Cl	0		−1	decrease in oxidation state, Cl is reduced
I		−1	0	increase in oxidation state, I is oxidised

Here the Cl_2 is the oxidising agent and the I⁻ is the oxidising agent.

Using oxidation states to help balance equations

Sometimes **stoichiometric equations** (balanced equations) are easy to balance but sometimes they are very difficult. If you are trying to balance a redox reaction then using oxidation states can speed things up. In a redox reaction, the number of electrons lost must equal the number of electrons gained.

 Worked examples: Balancing equations using oxidation states

Example 1

What is the balanced equation for the following reaction?

$$Na + Mg^{2+} \rightarrow Na^+ + Mg$$

Step 1: Identify how the oxidation state for each element changes in the reaction.

$$Na + Mg^{2+} \rightarrow Na^+ + Mg$$

Na	0		+1	loses 1e⁻ so oxidised
Mg		+2	0	gains 2e⁻ so reduced

Step 2: Balance the equation so the number of electrons lost is equal to the electrons gained.

For every Mg^{2+} that gains two electrons, **two** Na must lose one electron.

$$2Na + Mg^{2+} \rightarrow 2Na^+ + Mg$$

Example 2

What is the balanced equation for the following reaction?

$$H^+ + I^- + H_2SO_4 \rightarrow H_2S + H_2O + I_2$$

Step 1: Identify how the oxidation state for each element changes in the reaction.

$$H^+ + I^- + H_2SO_4 \rightarrow H_2S + H_2O + I_2$$

H	+1	+1	+1	+1	does not change
I	−1			0	loses one electron so oxidised
S		+6	−2		gains eight electrons so reduced
O		−2	−2		does not change

Step 2: Balance the equation so the number of electrons lost is equal to the electrons gained.

One H_2SO_4 molecule gains eight electrons from **eight** I^- ions, forming **four** I_2 molecules and one H_2S

$$H^+ + 8I^- + H_2SO_4 \rightarrow H_2S + H_2O + 4I_2$$

Activity ES 2.2

In this activity you can carry out a number of test-tube redox reactions and check your understanding of them.

Step 3: Balance the number of hydrogens and oxygens by adjusting the number of H^+ ions and H_2O molecules.

The elements in these chemicals have been neither oxidised nor reduced. You cannot change the number of I^-, H_2SO_4, H_2S, or I_2 because they were involved in the loss or gain of electrons.

There are four oxygen atoms in H_2SO_4, so there must be four H_2O molecules.

$$H^+ + 8I^- + H_2SO_4 \rightarrow H_2S + \mathbf{4}H_2O + \mathbf{4}I_2$$

Therefore, there must be **eight** H^+ ions to balance the equation.

$$\mathbf{8}H^+ + 8I^- + H_2SO_4 \rightarrow H_2S + 4H_2O + 4I_2$$

Summary questions

1 **a** Insert electrons, e^-, on the appropriate side of the following half-equations in order to balance and complete them, so that the electrical charges on both sides are equal.

  i $K \rightarrow K^+$
  ii $H_2 \rightarrow 2H^+$
  iii $O \rightarrow O^{2-}$
  iv $Cr^{3+} \rightarrow Cr^{2+}$ *(4 marks)*

  b Identify whether each process is oxidation or reduction. *(4 marks)*

2 Write down the oxidation states of the elements in the following examples.

  a Ag^+ *(1 mark)*
  b Al_2O_3 *(1 mark)*
  c SO_4^{2-} *(1 mark)*
  d P_4 *(1 mark)*
  e SF_6 *(1 mark)*
  f PO_4^{3-} *(1 mark)*

3 Work out the oxidation state of the chlorine in each of these species.

  a ClO_2 *(1 mark)*
  b $HClO_4$ *(1 mark)*
  c $MgCl_2$ *(1 mark)*
  d Cl_2O_7 *(1 mark)*
  e HCl *(1 mark)*
  f Cl_2O *(1 mark)*

4 **a** In the process described for the manufacture of bromine from Dead Sea water, bromine is separated from other materials involved in the process. Which properties of bromine make it possible to separate it from:

  i water
  ii chlorine. *(4 marks)*

b Write an ionic equation with state symbols for the reaction of chlorine gas with aqueous bromide ions to produce aqueous chloride ions and bromine liquid. *(2 marks)*

c In the production of 1.0 tonne of bromine, what mass of chlorine is required in tonnes? Give your answer to 1 d.p. *(2 marks)*

d In the production of 5.0 g of bromine, what volume of chlorine is required at RTP? Give your answer to 2 s.f. and in dm^3. The volume of one mole of gas at room temperature and pressure is $24.0\ dm^3$. *(3 marks)*

5 a Some reactions of the halogens are shown below – they are all examples of redox reactions. In each case state which element is oxidised and which is reduced, and give the oxidation states of *each* atom or ion before and after the reaction.

  i $H_2 + Cl_2 \longrightarrow 2HCl$

  ii $2FeCl_2 + Cl_2 \longrightarrow 2FeCl_3$

  iii $2H_2O + 2F_2 \longrightarrow 4HF + O_2$ *(6 marks)*

b For each of the redox reactions in part **a** identify by formula the:

  i oxidising agent

  ii reducing agent. *(6 marks)*

6 Use oxidation states to help you balance the following redox reactions.

a $Br^- + H^+ + H_2SO_4 \longrightarrow Br_2 + SO_2 + H_2O$ *(2 marks)*

b $I^- + H^+ + H_2SO_4 \longrightarrow I_2 + H_2S + H_2O$ *(2 marks)*

7 Use oxidation states to name the following ions and compounds.

a SnO_2 *(1 mark)*

b $FeCl_2$ *(1 mark)*

c NO_3^- *(1 mark)*

d $PbCl_4$ *(1 mark)*

e $Mn(OH)_2$ *(1 mark)*

f CrO_4^{2-} *(1 mark)*

g VO_3^- *(1 mark)*

h SO_3^{2-} *(1 mark)*

8 Write formulae for the following compounds. In each case, the negative ion has a charge of -1.

a potassium chlorate(III) *(1 mark)*

b sodium chlorate(V) *(1 mark)*

c iron(III) hydroxide *(1 mark)*

d copper(II) nitrate(V) *(1 mark)*

Chlorine is needed for the extraction of bromine from Dead Sea water. The human demand for chlorine is high and it is manufactured worldwide on an enormous scale. Chlorine can be made by **electrolysis** (decomposing a compound using an electric current) of a concentrated solution of sodium chloride (brine).

The electrolysis of brine

Although the sea contains high concentrations of sodium chloride, rock salt is an even better source of sodium chloride. Rock salt can be recovered either by underground mining or by pumping water into the salt and collecting the salt solution at the surface. As well as chlorine, the electrolysis of brine generates hydrogen and sodium hydroxide (an alkali). As chlorine manufacture and sodium hydroxide manufacture are directly linked, the production of these chemicals is often referred to as the chlor–alkali industry.

Often in chemical reactions, additional products are made as well as the product you want. These are called **co-products**. The co-products, sodium hydroxide and hydrogen, can be sold and this helps to reduce the waste, as well as increasing profitability. The membrane cell is the most modern of the electrolysis methods for producing chlorine and has been used since the 1980s (Figure 1).

The half-equations involved in the electrolysis of sodium chloride solution are:

- at the positive electrode $\quad 2Cl^-(aq) \rightarrow Cl_2(aq) + 2e^-$
- at the negative electrode $\quad 2H_2O(l) + 2e^- \rightarrow 2OH^-(aq) + H_2(g)$

The equation representing the overall reaction occurring in the cell is:

$$2Cl^-(aq) + 2H_2O(l) \rightarrow Cl_2(aq) + 2OH^-(aq) + H_2(g)$$

▲ **Figure 2** *A mechanical seaweed harvester as used by Hebridean Seaweed Ltd off the north coast of Scotland*

Extracting iodine from seaweed

Just below the surface of the seas off the coast of Scotland lie hidden forests of a seaweed – kelp. This has historically been a valuable source of organic matter that local people used to fertilise their soil. In the early part of the nineteenth century, Napoleon Bonaparte was looking for a source of nitrate for making explosives. In 1811, the French entrepreneur Bernard Courtois responded to Napoleon's demands by attempting to make nitrate from rotting seaweed. He noticed some curious purple fumes and, by chance, discovered another of the halogens – iodine. This led to a whole industry developing in the coastal areas of Scotland.

The extraction of iodine from kelp can be carried out in the laboratory.

1. Heat the seaweed strongly on a tin lid using a blue Bunsen burner flame until only a small quantity of ash remains.
2. Boil the ash with distilled water. Filter whilst still hot and allow the filtrate to cool.
3. Add dilute sulfuric acid and then hydrogen peroxide solution to the filtrate. A deep brown coloured solution of iodine will be seen.
4. Transfer the mixture to a separating funnel and add cyclohexane. Shake for 30 seconds then clamp the separating funnel and allow the two layers to separate.
5. Discard the lower aqueous layer then run the upper cyclohexane layer into an evaporating dish. Leave this in a fume cupboard so that the cyclohexane evaporates leaving iodine crystals behind.

1. Why does the seaweed have to be heated strongly before adding the acid and hydrogen peroxide?
2. The filtrate contains chloride and bromide ions as well as iodide ions. Why are the iodide ions preferentially oxidised by the hydrogen peroxide?
3. Write a balanced equation with state symbols for the redox reaction which occurs between the iodide ions and the acidified hydrogen peroxide.
4. What colour will the upper cyclohexane layer be?
5. What does the final step tell you about the relative volatilities of iodine and cyclohexane?

Chemical ideas: Redox reactions 9.2

Electrolysis as redox reactions

When electricity is passed through a molten or aqueous ionic compound, the compound is broken down and the process is called electrolysis. The charged ions are free to move to the oppositely charged electrode and create a complete circuit.

Electrolysis of molten compounds

Solid ionic compounds do not conduct electricity since ions in a solid are not free to move. If the ionic compound is melted then the charged ions are free to move and carry a current. Electrons are lost or gained by ions at the **electrodes**.

Figure 3 shows the electrolysis of molten lead bromide to produce lead and bromine. The positive lead ions **migrate** to the negative electrode, called the **cathode**. The Pb^{2+} ions gain two electrons to form lead atoms. Since the lead bromide is hot, molten lead metal collects at the bottom of the container. Bromide ions have a negative charge and migrate to the positive electrode or **anode**. At the temperature needed to melt the lead bromide, bubbles of bromine gas will be seen. The molten lead bromide has been broken down into its elements.

cathode reaction: $Pb^{2+}(l) + 2e^- \rightarrow Pb(l)$ The lead ion gains electrons and is reduced.

anode reaction: $2Br^-(l) \rightarrow Br_2(g) + 2e^-$ The bromide ions lose electrons and are oxidised.

Synoptic link

You found out about the structure of ionic lattices in Chapter 1, Elements of life.

Study tip

Cations migrate to the **cat**hode.

Anions migrate to the **an**ode.

Study tip

Reduction occurs at the cathode.

Oxidation occurs at the anode.

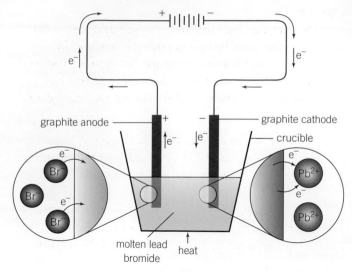

▲ **Figure 3** *Electrolysis of molten lead bromide*

It is easy to predict the products of the electrolysis of molten salt:

- product at cathode – metal
- product at anode – non-metal, apart from hydrogen.

Electrolysis of solutions

The electrolysis of solutions is much easier to carry out than electrolysis of molten compounds. When an ionic compound dissolves, the ions become free to move and carry the current. Figure 4 shows the apparatus that can be used to electrolyse a solution and collect any gases made so they can be identified.

Water can take part

When you electrolyse salt solutions, predicting the products at the electrodes is more difficult. In the electrolysis of molten salts there was no competition at the electrodes. In the electrolysis of solutions, water competes with the ions from the salt.

Water can be reduced at the cathode:

$$2H_2O(l) + 2e^- \rightarrow 2OH^-(aq) + H_2(g)$$ Gain of electrons so reduction

Water can also be oxidised at the anode:

$$2H_2O(l) \rightarrow O_2(g) + 4H^+(aq) + 4e^-$$ Loss of electrons so oxidation.

Reduction at the cathode

At the cathode there will be both metal ions from the salt and the water the salt is dissolved in. More reactive metals such as sodium or potassium remain as ions and so hydrogen gas is produced by reduction of the water. Less reactive metals such as copper and zinc are plated (deposited) on the cathode.

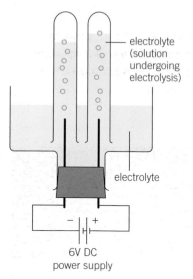

▲ **Figure 4** *Apparatus used for electrolysis of solutions*

Activity ES 3.1

In this activity you can carry out and check your understanding of electrolysis of aqueous solutions.

Oxidation at the anode

At the anode there will be negative ions from the salt and the water the salt is dissolved in. Halide ions have a greater tendency to be oxidised (or lose electrons) than water and so the halogen is produced at the anode. Other negative ions such as sulfates and nitrates have less tendency to be oxidised than water and so oxygen gas is produced at the anode.

There is another possibility at the anode. Usually in electrolysis the electrodes are themselves unreactive. Graphite and platinum are commonly used. A special case is when the anode and the metal ions in solution are the same. In the electrolysis of a copper compound with a copper anode, the metal anode loses mass because copper atoms change to copper ions and go into solution. An example would be the electrolysis of copper sulfate solution using copper electrodes.

▲ **Figure 5** *A copper plated oak leaf*

At the anode, oxidation occurs since electrons are lost by the copper electrode:

$$Cu(s) \rightarrow Cu^{2+}(aq) + 2e^-$$

At the cathode, copper is deposited and the concentration of the solution remains constant:

$$Cu^{2+}(aq) + 2e^- \rightarrow Cu(s)$$

Summary

You need to be able to predict the products of the electrolysis of solutions and be able to write half-equations. The information needed is summarised in Table 1.

> **Synoptic link** 🧪
>
> A summary of how to carry out electrolysis of aqueous solutions can be found in Techniques and procedures.

▼ **Table 1** *Products at electrodes from electrolysis of solutions*

Product at cathode (negative electrode)	Product at anode (positive electrode)	Lost at anode (positive electrode)
All electrodes	Unreactive electrode, e.g., graphite or platinum	Reactive anode, e.g., copper in $CuSO_4(aq)$
Reduction	Oxidation	Oxidation
Hydrogen if the metal in the salt comes from Group 1 or 2 or is aluminium $2H_2O(l) + 2e^- \rightarrow 2OH^-(aq) + H_2(g)$	Halogen if the salt is a halide $2Cl^-(aq) \rightarrow Cl_2(g) + 2e^-$	$Cu(s) \rightarrow Cu^{2+}(aq) + 2e^-$
Metal for all other salts $Cu^{2+}(aq) + 2e^- \rightarrow Cu(s)$	Oxygen if the salt is a sulfate or nitrate $2H_2O(l) \rightarrow O_2(g) + 4H^+(aq) + 4e^-$	
Hydrogen also made on electrolysis of acids, e.g., sulphuric acid $2H^+(aq) + 2e^- \rightarrow H_2(g)$	Oxygen also made on electrolysis of hydroxides, e.g., sodium hydroxide $4OH^-(aq) \rightarrow O_2(g) + 2H_2O(l) + 2e^-$	

Worked example: Half-equations for the electrolysis of zinc chloride

Write the half-equations for the reactions at the cathode and anode in the electrolysis of zinc chloride solution with graphite electrodes. Decide what is oxidised and what is reduced.

Step 1: Write the formula for zinc chloride – $ZnCl_2$

Step 2: Identify the cation – zinc, Zn^{2+}

Step 3: Identify the products of the reaction at the cathode – zinc is less reactive so hydrogen won't be produced.

Step 4: Write the half-equation for the reaction at the cathode.

Oxidation state of zinc in zinc chloride is +2. Zinc ions will *gain* two electrons at the cathode. Zinc is reduced.

$$Zn^{2+} (aq) + 2e^- \rightarrow Zn(s)$$

Step 5: Identify the anion – chloride, Cl^-

Step 6: Identify the products of the reaction at the anode – chloride is a halide so chlorine gas will be produced.

Step 7: Write the half-equation for the reaction at the anode.

Each chloride ion *loses* an electron to become an atom that pairs up with another atom to become a diatomic molecule. Chloride is oxidised.

$$2Cl^-(aq) \rightarrow Cl_2(g)$$

 Worked example: Half-equations for the electrolysis of zinc nitrate

Write the half-equations for the reactions at the cathode and anode in the electrolysis of zinc nitrate solution with zinc electrodes. Decide what is oxidised and what is reduced.

Step 1: Identify the products of the reaction at the cathode – Zinc is produced rather than hydrogen since the metal less reactive.

Step 2: Write the half-equation for the reaction at the cathode.

Each zinc ion gains two electrons to become an atom. The Zn^{2+} is reduced because it gains electrons.

$$Zn^{2+}(aq) + 2e^- \rightarrow Zn(s)$$

Step 3: Identify the products of the reaction at the anode – zinc electrode in a solution of a zinc compound so zinc atoms change into ions.

Step 4: Write the half-equation for the reaction at the anode.

Zinc atoms lose electrons to become zinc ions that go into solution. Zinc is oxidised because electrons are lost.

$$Zn(s) \rightarrow Zn^{2+}(aq) + 2e^-$$

Summary questions

1 Predict the products at the cathode and anode in the electrolysis of these molten compounds.
   a lead bromide   (*1 mark*)
   b sodium chloride   (*1 mark*)
   c zinc iodide.   (*1 mark*)

2 Look at the equation for the electrolysis of sodium chloride:

$$2Cl^-(aq) + 2H_2O(l) \rightarrow Cl_2(aq) + 2OH^-(aq) + H_2(g)$$

   a Calculate the amount (in moles) of sodium hydroxide, NaOH, in 1 tonne of solid sodium hydroxide.
      1 tonne is 1 000 000 g.   (*3 marks*)
   b What amount (in moles) of chlorine, Cl_2, is produced for each mole of NaOH?   (*2 marks*)
   c Calculate the mass of chlorine produced at the same time as 1 tonne of sodium hydroxide.   (*2 marks*)

3 Predict the products at the anode and cathode if the following solutions were electrolysed using the named electrodes.
   a sodium bromide with graphite electrodes   (*2 marks*)
   b aluminium nitrate with graphite electrodes   (*2 marks*)
   c zinc bromide with graphite electrodes.   (*2 marks*)

4 Write the half-equations for the cathode and anode in the electrolysis below. Say if they are reduction or oxidation.
   a zinc bromide solution with graphite electrodes   (*2 marks*)
   b sodium bromide solution with graphite electrodes   (*2 marks*)
   c sodium hydroxide solution with graphite electrodes   (*2 marks*)
   d nitric acid with platinum electrodes   (*2 marks*)
   e copper nitrate solution with copper electrodes.   (*2 marks*)

As well as its use in extracting bromine from sea water, chlorine is used to make bleaches. Chlorine gas is passed through a cold solution of sodium hydroxide. The sodium hydroxide solution reacts with the chlorine to form sodium chlorate(I).

$$Cl_2(aq) + 2NaOH(aq) \rightleftharpoons NaCl(aq) + NaOCl(aq) + H_2O(l)$$

In order to understand what the symbol $\rightleftharpoons$ means, you need to learn about equilibrium.

Chemical ideas: Equilibrium in chemistry 7.1

Chemical equilibrium

The general meaning of the term equilibrium is a state of balance where nothing changes. For example, a see-saw with two people of equal mass, sitting one on each side, is in a state of equilibrium. In chemistry, a state of equilibrium is also a state of balance, but it has a special feature – chemical equilibrium is **dynamic equilibrium**.

Why is equilibrium dynamic?

Figure 1 shows a sealed bottle of bromine. The bottle and its contents make up a **closed system**. Nothing can enter or leave the bottle. In this system, bromine is present in two states – as a liquid, $Br_2(l)$, and as a gas above the liquid, $Br_2(g)$.

When the bottle has been standing at a steady temperature for some time the depth of the orange colour above the liquid remains constant. The mass of bromine that is a gas and the mass of bromine that is a liquid in the closed system is constant.

▲ **Figure 1** Bromine liquid and gas at equilibrium

The system is at equilibrium and nothing appears to change on the macroscopic scale – the scale that you can see.

If you were able to see the individual molecules – the microscopic scale – the picture would be rather different. In the gas, the molecules are constantly moving rapidly in random directions, and inevitably collide with the molecules on the surface of the liquid (Figure 2). Some bounce back into the gas phase, but some enter the liquid phase.

At the same time, the molecules of bromine in the liquid are also constantly moving around, colliding with each other. Near the surface of the liquid, some of these molecules escape into the gaseous phase (Figure 2).

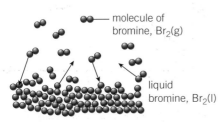

molecule of bromine, $Br_2(g)$

liquid bromine, $Br_2(l)$

▲ **Figure 2** The bromine equilibrium on a microscopic scale

There are molecules entering the liquid phase and molecules leaving the liquid phase – it is a **reversible change**. When the system is at equilibrium, the molecules enter and leave at the same rate. On the macroscopic scale it seems as though nothing is changing, but on the molecular scale molecules are constantly moving from one phase to the other. That is why the situation is described as dynamic equilibrium. It can be represented by the equation:

$$Br_2(g) \rightleftharpoons Br_2(l)$$

The $\rightleftharpoons$ sign represents dynamic equilibrium.

The road to equilibrium

Many chemical reactions are reversible reactions, for example, the reaction between hydrogen and iodine to produce hydrogen iodide.

$$H_2(g) \quad + \quad I_2(g) \quad \rightleftharpoons \quad 2HI(g)$$

colourless   purple   colourless

When reactions are represented using the $\rightleftharpoons$ symbol the reaction going from left to right is known as the **forward reaction** and the reaction going in the opposite direction is known as the **reverse reaction**. Even though the reaction is reversible, the substances on the left of the equilibrium sign are called the reactants and those on the right are called the products.

If the colourless hydrogen and the purple iodine are mixed in a closed container at 731 K, the purple colour becomes paler as the iodine reacts. The purple colour does not disappear and after a while the depth of the purple colour will stay constant (Figure 3).

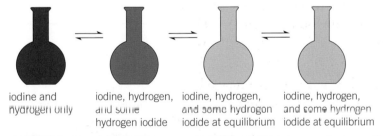

iodine and hydrogen only   iodine, hydrogen, and some hydrogen iodide   iodine, hydrogen, and some hydrogen iodide at equilibrium   iodine, hydrogen, and some hydrogen iodide at equilibrium

▲ **Figure 3** *Changes in appearance of an iodine and hydrogen mixture until equilibrium is reached*

The rate of the forward reaction will decrease as hydrogen and iodine react to make hydrogen iodide and their concentrations decrease (Figure 4). The rate of the reverse reaction will be zero on first mixing the hydrogen and iodine because there will be no hydrogen iodide. As more hydrogen iodide is made, the rate of the reverse reaction increases because the hydrogen iodide concentration increases.

The system has reached equilibrium when the rates of the forward and reverse reactions are the same. Hydrogen, iodine, and hydrogen iodide are all present and their concentrations remain constant once equilibrium has been reached. Reactants are turning into products at the same rate as products are turning into reactants.

Activity ES 4.1

In this activity you can model a dynamic equilibrium.

Study tip

When talking about an equilibrium, it is important to state which reaction you are referring to. The easiest way to do this is to write an equation.

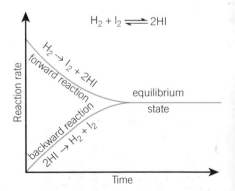

▲ **Figure 4** *Changes in the rate of reaction for the forward and reverse reaction*

So the full definition of dynamic equilibrium is:

- concentrations of reactants and products stay constant
- forward and reverse reactions are both happening (so dynamic)
- the rate of the forward and reverse reactions are equal to each other.

Although the concentrations of the reactants and products are constant, it is not true to say that the concentrations are the same at equilibrium. In Figure 5a, hydrogen and iodine were placed in a closed flask and allowed to reach equilibrium. The concentration of hydrogen iodide is greater than the concentration of the hydrogen and iodine at equilibrium. All the concentrations become constant at equilibrium as shown by the horizontal lines but the concentrations of reactants and products are not the same. In Figure 5b only hydrogen iodide was placed in the closed flask. At equilibrium the concentrations of reactants and products stay constant but are not the same.

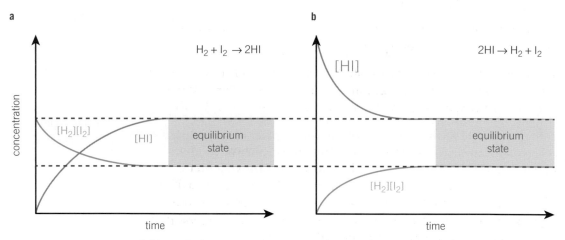

▲ **Figure 5** *Changes in concentration of a hydrogen and iodine mixture until equilibrium has been reached*

Figure 5 shows that under the same reaction conditions, the same equilibrium position is reached when the reaction starts with the products – hydrogen iodide – or the reactants – hydrogen and iodine. You cannot tell them apart.

Chemical ideas: Equilibrium in chemistry 7.2

The equilibrium constant K_c

Table 1 shows results obtained for the reaction between hydrogen and iodine to form hydrogen iodide. In the first three experiments, mixtures of hydrogen and iodine were put into sealed reaction vessels. In the final two experiments hydrogen iodide alone was sealed into the vessel. The mixtures were held at a constant temperature of 731 K until equilibrium was reached. The equilibrium concentrations of all three substances were then recorded.

▼ **Table 1** *Initial and equilibrium concentrations for the reaction* $H_2(g) + I_2(g) \rightleftharpoons 2HI(g)$

Experiment	Initial concentrations / mol dm^{-3}			Equilibrium concentrations / mol dm^{-3}			$K_c = \dfrac{[HI][HI]}{[H_2][I_2]}$
	[$H_2(g)$]	[$I_2(g)$]	[$HI(g)$]	[$H_2(g)$]	[$I_2(g)$]	[$HI(g)$]	
1	2.40×10^{-2}	1.38×10^{-2}	0.00	1.140×10^{-2}	0.120×10^{-2}	2.52×10^{-2}	46.4
2	2.44×10^{-2}	1.98×10^{-2}	0.00	0.770×10^{-2}	0.310×10^{-2}	3.34×10^{-2}	46.7
3	2.46×10^{-2}	1.76×10^{-2}	0.00	0.920×10^{-2}	0.220×10^{-2}	3.08×10^{-2}	46.9
4	0.00	0.00	3.04×10^{-2}	0.345×10^{-2}	0.345×10^{-2}	2.35×10^{-2}	46.9
5	0.00	0.00	7.58×10^{-2}	0.860×10^{-2}	0.860×10^{-2}	5.86×10^{-2}	46.4

For this reaction (looking at the last column) it was found that the following was constant:

$$\frac{[HI][HI]}{[H_2][I_2]}$$

This is the same as $\dfrac{[HI]^2}{[H_2][I_2]}$

This constant is called the **equilibrium constant** K_c. The square brackets [ ] mean concentration of.

The values of K_c are shown in Table 1. The equilibrium constant is the same whether the starting reaction mixture is $H_2 + I_2$ (Experiments 1 to 3) or HI (Experiments 4 and 5). The mean value of K_c for this reaction at 731 K is about 46.7.

K_c is greater than 1. This tells us that there must be more products than reactants – the top line must have a larger numerical value than the bottom line to calculate a value of 46.7 for K_c. The position of equilibrium lies to the products or to the right-hand side of the reaction. Since $K_c > 1$ (greater than 1), this means that, at equilibrium, most of the H_2 and I_2 has been converted to HI, but not all.

Writing the expression for K_c

The rules for writing K_c expressions have been discovered by using the results of many experiments. This makes it possible to write an expression for K_c for any reaction, without having to examine data. In the expression for K_c, the products of the forward reaction appear on the top line and the reactants on the bottom line. The power to which you raise the concentration of a substance is the same as the number which appears in front of it in the balanced equation.

In general, if an equilibrium mixture contains substances A, B, C, and D that react according to the equation:

$$aA + bB \rightleftharpoons cC + dD$$

then the expression for K_c is $K_c = \dfrac{[C]^c [D]^d}{[A]^a [B]^b}$

Once you have written the expression for K_c using the stoichiometric equation, use the concentrations of the substances to calculate a value for K_c.

Synoptic link

Acid–base equilibria are looked at in detail in Chapter 4, The ozone story.

Study tip

Always write a balanced stochiometric equation for a reaction before writing the equation for K_c. You will then be able to easily see the superscript values for K_c.

Worked example: Calculating K_c

Calculate the value for K_c (to 3 s.f.) for the reaction between nitrogen and hydrogen at 1000 K given the following equilibrium concentrations:

$$[N_2] = 0.142 \, mol \, dm^{-3}$$

$$[H_2] = 1.84 \, mol \, dm^{-3}$$

$$[NH_3] = 1.36 \, mol \, dm^{-3}$$

Step 1: Write the stoichiometric equation for the reaction.

$$N_2 + 3H_2 \rightleftharpoons 2NH_3$$

Step 2: Use the stoichiometric equation to write an expression for K_c.

$$K_c = \frac{[NH_3]^2}{[N_2][H_2]^3}$$

Step 3: Use the concentration values to calculate K_c.

$$K_c = \frac{(1.36)^2}{(0.142) \times (1.84)^3} = 2.09 \, mol^{-2} \, dm^6 \text{ at } 1000 \, k$$

Synoptic link

You will learn about units of K_c in Chapter 6, The chemical industry.

The size of K_c

Values of K_c vary enormously. Table 2 shows some values for two reactions at the same temperature. K_c is temperature-dependent so temperature is quoted alongside the equilibrium constant.

▼ Table 2 *Some values of K_c*

Reaction	K_c
$H_2(g) + Br_2(g) \rightleftharpoons 2HBr(g)$	1010 at 550 K
$2H_2(g) + S_2(g) \rightleftharpoons 2H_2S(g)$	0.000 094 at 1020 K

For the reaction of hydrogen with bromine, the top line in the K_c relationship is greater than the bottom line. At equilibrium there are more products than reactants.

$$K_c = \frac{[HBr]^2}{[H_2][Br_2]}$$

For the reaction of hydrogen with sulfur, the top line in the K_c relationship is much smaller than the bottom line. At equilibrium there are more reactants than products.

$$K_c = \frac{[H_2S]^2}{[H_2]^2 \, [S_2]}$$

All reactions are equilibrium reactions and even reactions that seem to go to completion actually have a little bit of reactant left in equilibrium with the product.

Study tip

$K_c > 1$ means there are more products than reactants at equilibrium.

$K_c \gg 1$ (greater than 10^{10}) means that the reaction appears to have gone to completion.

$K_c < 1$ means that there are more reactants than products.

$K_c \ll 1$ (less than 10^{-10}) means that the reaction appears not to have happened.

Chemical ideas: Equilibrium in chemistry 7.3

K_c and changed in concentration

Suppose a system is at equilibrium and you suddenly disturb it by adding more of a reagent. The composition of the system will change until equilibrium is reached again. The composition of the mixture will always adjust to keep the value of K_c constant, provided the *temperature stays constant.*

For example, in an experiment involving the formation of ethyl ethanoate the system was allowed to reach equilibrium. The equilibrium concentrations are shown in Table 3.

$$CH_3COOH(l) + C_2H_5OH(l) \rightleftharpoons CH_3COOC_2H_5(1) + H_2O(l)$$

▼ **Table 3** *Equilibrium is set up in Experiment 1, starting with equal concentrations of ethanoic acid and ethanol. In Experiment 2, the equilibrium is disturbed by adding extra ethanol. Both experiments are carried out at 298 K*

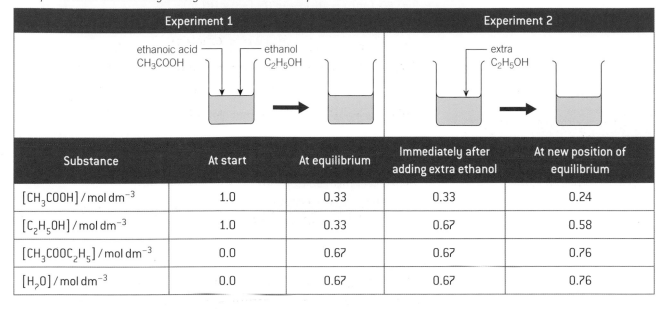

Substance	Experiment 1		Experiment 2	
	At start	At equilibrium	Immediately after adding extra ethanol	At new position of equilibrium
$[CH_3COOH]$ / mol dm^{-3}	1.0	0.33	0.33	0.24
$[C_2H_5OH]$ / mol dm^{-3}	1.0	0.33	0.67	0.58
$[CH_3COOC_2H_5]$ / mol dm^{-3}	0.0	0.67	0.67	0.76
$[H_2O]$ / mol dm^{-3}	0.0	0.67	0.67	0.76

Using the equilibrium concentrations from Experiment 1:

$$K_c = \frac{[CH_3COOC_2H_5][H_2O]}{[CH_3COOH][C_2H_5OH]}$$

$$= \frac{(0.67\,\text{mol dm}^{-3})(0.67\,\text{mol dm}^{-3})}{(0.33\,\text{mol dm}^{-3})(0.33\,\text{mol cm}^{-3})} = 4.1 \text{ at } 298\,\text{K}$$

In Experiment 2, one of the concentrations was deliberately changed by adding more C_2H_5OH to give a new concentration of $0.67\,\text{mol dm}^{-3}$. Immediately after adding the extra C_2H_5OH, before any changes occur, the new concentration ratio is:

$$\frac{(0.67\,\text{mol dm}^{-3})(0.67\,\text{mol dm}^{-3})}{(0.33\,\text{mol dm}^{-3})(0.67\,\text{mol dm}^{-3})} = 2.0$$

This value is smaller than K_c. In order to restore the value of K_c to 4.1 some C_2H_5OH and CH_3COOH must react (making the bottom line smaller) to produce $CH_3COOC_2H_5$ and H_2O (making the top line numerically larger). The system was left to reach equilibrium again, and the new equilibrium concentrations were measured.

$$K_c = \frac{(0.76\,\text{mol dm}^{-3})(0.76\,\text{mol dm}^{-3})}{(0.24\,\text{mol dm}^{-3})(0.58\,\text{mol cm}^{-3})} = 4.1 \text{ at } 298\,\text{K}$$

When the equilibrium was disturbed it moved in such a way that K_c remained constant. Some C_2H_5OH and CH_3COOH had to react to keep the K_c value at 4.1. More $CH_3COOC_2H_5$ and H_2O were produced. The equilibrium position had moved to the right, or moved to the product side to re-establish K_c.

Activity ES 4.2

In this activity you will be able to observe the effect of changing concentration on the position of equilibrium.

Study tip

Changing concentrations do not alter K_c once equilibrium as been reached, assuming the temperature remains constant.

Summary questions

1 Write expressions for K_c for the following reactions.
   a $2NO(g) + O_2(g) \rightleftharpoons 2NO_2(g)$ (1 mark)
   b $C_2H_6(g) \rightleftharpoons C_2H_4(g) + H_2(g)$ (1 mark)
   c $2HI(g) \rightleftharpoons H_2(g) + I_2(g)$ (1 mark)
   d $CO_2(aq) + H_2O(l) \rightleftharpoons HCO_3^-(aq) + H^+(aq)$ (1 mark)
   e $CH_3COOH(l) + C_3H_7OH(l) \rightleftharpoons CH_3COOC_3H_7(l) + H_2O(l)$ (1 mark)

2 The equilibrium constant K_c for a reaction is given by the expression:

$$K_c = \frac{[SO_3(g)]^2}{[SO_2(g)]^2 [O_2(g)]}$$

Write the balanced chemical equation for the reaction.   (1 mark)

3 A mixture of nitrogen and hydrogen was sealed in a steel vessel and held at 1000 K until equilibrium was reached. The contents were then analysed. The results are given in the following table.

Substance	Equilibrium concentration / mol dm^{-3}
$N_2(g)$	0.142
$H_2(g)$	1.840
$NH_3(g)$	1.360

   a Write an expression for K_c for the reaction:
$$N_2(g) + 3H_2(g) \rightleftharpoons 2NH_3(g)$$
                                                (1 mark)
   b Calculate a value for K_c.   (2 marks)

4 When PCl_5 is heated in a sealed container and maintained at a constant temperature, an equilibrium is established. At 523 K, the following equilibrium concentrations were determined.

Substance	Equilibrium concentration / mol dm^{-3}
PCl_5	0.077
PCl_3	0.123
Cl_2	0.123

   a Write an expression for K_c for the reaction:
$$PCl_5(g) \rightleftharpoons PCl_3(g) + Cl_2(g)$$
                                                (1 mark)
   b Calculate a value for K_c.   (2 marks)

5 For the reaction of aqueous chloromethane with alkali the equilibrium constant has a value of 1×10^{16} at room temperature.

$$OH^-(aq) + CH_3Cl(aq) \rightleftharpoons CH_3OH(aq) + Cl^-(aq)$$

What does this tell you about the concentration of chloromethane at equilibrium?   (2 marks)

6 Use K_c to explain how the position of equilibrium would change if acid was added to the reaction between carbon dioxide and water.
$$CO_2(aq) + H_2O(l) \rightleftharpoons HCO_3^-(aq) + H^+(aq)$$
                                                (2 marks)

ES 5 The risks and benefits of using chlorine

Specification reference: ES(f), ES(n)

There are a variety of risks associated with the production, storage, and transportation of chlorine, yet it makes a valuable contribution to improving our lives. The risks and benefits associated with toxic chlorine gas need to be weighed up before it is used.

Chlorine has a poor public image. It is associated with pollution – pollution of the land through pesticides that contain organochlorine compounds and pollution of the upper atmosphere through CFCs (chlorofluorocarbons). Chlorine has also been used as a poisonous gas. Both chlorine and phosgene, $COCl_2$, were used with deadly effect in the trenches in the First World War. They are also thought to have been used against civilian populations in recent years. ·

Learning outcomes

Demonstrate and apply knowledge and understanding of:

→ the risks associated with the storage and transport of chlorine; uses of chlorine which must be weighed against these risks, including: sterilising water by killing bacteria, bleaching

→ techniques and procedures in iodine–thiosulfate titrations.

Chemical ideas: Chemistry in industry 15.1

Risks and benefits of chlorine

Chlorine is a toxic gas detectable by smell at 1 part per million (ppm). Even in these small doses, chlorine can irritate the eyes, skin, and respiratory system. If inhaled at concentrations above 40 ppm, chlorine reacts in the lungs to form hydrochloric acid, HCl, which affects lung tissue and essentially causes drowning as liquid floods the lungs. Any leaks of chlorine during transport or storage cause danger for both workers and the general public.

Transporting chlorine

Although chlorine is usually prepared on site at the chemical plant requiring chlorine, this is not always the case and some chlorine is transported. Chlorine can be transported by road (Figure 1) or rail in specially designed pressurised tank containers. In some countries including the UK, a Hazchem warning plate (Figure 2) is attached to the tank during transport. In the event of an accident, this gives information for the fire brigade regarding what action is needed.

The chlorine is transported as a liquid as more chlorine can be stored in a fixed volume as a liquid under pressure than as a gas. If the temperature or pressure becomes too high, the tanks have pressure release devices designed to vent the tank and release some chlorine as gas. It is better to vent a small amount of chlorine to the atmosphere than to have a catastrophic explosion in case of a tank failure. Generally, the tanks are made and lined with steel. It is essential that the inside of the tank is dry as chlorine reacts with water to produce corrosive acids. Tanks have a cylindrical, protective housing at the top. This means that all loading and unloading is done through the protective housing at the top of the tank. Another safety feature

▲ **Figure 1** *A tank truck for transporting chlorine*

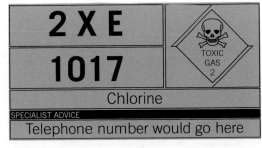

▲ **Figure 2** *The Hazchem code for chlorine*

Synoptic link

Aspects of the impact of the chemical industry on society are explored in Chapter 6, The chemical industry.

on large chlorine tanks is an excess flow valve, which is designed to close automatically if the angle valve which regulates the discharge of chlorine is broken or sheared off in the case of an accident in transport. It is activated if the discharge of liquid chlorine at the exit port exceeds some predetermined value.

Unloading chlorine

When chlorine is transferred on delivery from the rail tanker to a bulk trailer on site, a scrubber unit ensures that air being displaced from the bulk trailer has any chlorine removed from it (Figure 3). The scrubber has sodium hydroxide solution that reacts with the chlorine to produce sodium chlorate(I) – bleach. The bleach can be sold on.

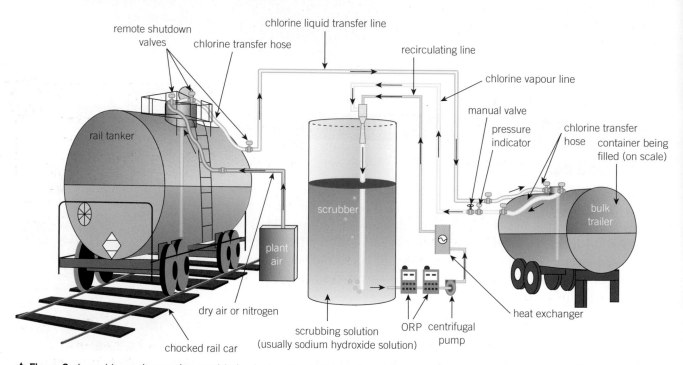

▲ **Figure 3** *A scrubber unit ensuring no chlorine leaks when it is transferred*

Storing chlorine

Chlorine may also be transported and then stored in cylinders. Workers at the chemical plant meet regulations regarding handling the cylinders by carefully moving the cylinders using a hoist to avoid damage to the outside (Figure 4). One method of routine checking of stored cylinders is to take a stick with cloth soaked in concentrated ammonia solution over the end. If a cylinder is leaking then a white cloud of ammonium chloride will be seen.

▲ **Figure 4** *Moving a chlorine cylinder carefully with a hoist*

Uses and benefits of chlorine

In spite of the risks of its transport and storage, chlorine is used in many ways to make our lives safer and more comfortable. About 50 million tonnes of chlorine are produced worldwide annually. The best known use is in water treatment, where it is added to the water to kill bacteria and other pathogens. After its introduction for chlorination of water in the early twentieth century there was a rapid decline in the number of deaths from typhoid. Chlorine is also used in household bleach products, to kill bacteria on surfaces, or to remove stains from clothing. The bleach, which is an oxidising agent, removes stains by breaking bonds in coloured chemicals to form colourless products.

Determining the concentration of sodium chlorate in bleach

Chlorine reacts with sodium hydroxide to make sodium chlorate(I), NaClO. A solution of about 12% NaClO by mass is used in some water purifying plants to kill bacteria. A solution of about 5% is used in household bleach products. One way to determine the accurate concentration of sodium chlorate(I) is to carry out an iodine–thiosulfate titration.

Iodine–thiosulfate titrations

Not all titrations are acid–base reactions. Iodine–thiosulfate titrations involve redox reactions. They are used to find the concentration of a chemical that is a strong enough oxidising agent to oxidise iodide ions.

In the case of bleach, excess iodide ions are added to the chlorate(I) ions. The following redox reaction occurs:

$$ClO^- + 2I^- + 2H^+ \rightarrow I_2 + Cl^- + H_2O$$
$$\text{brown}$$

The iodine produced can be titrated using thiosulfate ions, $S_2O_3^{2-}$, in the following reaction:

$$2S_2O_3^{2-} + I_2 \rightarrow 2I^- + S_4O_6$$
$$\text{brown} \quad \text{pale yellow}$$

The end point of the titration can be clearly identified by adding starch solution. The end point is determined when the final trace of blue/black colour is no longer visible.

Activity ES 5.1

In this activity you can work with other members of your class to investigate the cost-effectiveness of different comercially available bleaches.

Synoptic link

You met acid–base titrations in Chapter 1, Elements of life.

Synoptic link

Detail of how to carry out an iodine–thiosulfate titration can be found in Techniques and procedures.

 Worked example: The concentration of chlorate(I)

In a titration, 25.00 cm³ of sodium chlorate(I) solution was pipetted into a conical flask before excess potassium iodide and sulfuric acid were added. A 0.10 mol dm⁻³ solution of sodium thiosulfate was then run into the conical flask. The end point was reached when 22.0 cm³ of sodium thiosulfate had been added. Calculate the concentration of chlorate(I) ions in the original solution to three significant figures.

Step 1: Convert all cm³ readings to dm³

$$\frac{25.00 \, \text{cm}^3}{1000} = 25.00 \times 10^{-3} \, \text{dm}^3$$

$$\frac{22.00 \, \text{cm}^3}{1000} = 22.0 \times 10^{-3} \, \text{dm}^3$$

Step 2: Write the ionic equations for the titation.

$$ClO^- + 2I + 2H^+ \rightarrow I_2 + Cl^- + H_2O \quad \textbf{Equation 1}$$

$$I_2 + 2S_2O_3^{2-} \rightarrow 2I^- + S_4O_6^{2-} \qquad \textbf{Equation 2}$$

Step 3: Calculate the amount in moles of sodium thiosulfate.

$$\text{moles } S_2O_3^{2-} = \text{concentration } c \times \text{volume } v = 0.10 \times 22.00 \times 10^{-3} = 2.20 \times 10^{-3} \text{ moles}$$

Step 4: Calculate the amount in moles of iodine.

Looking at Equation 2, two moles of thiosulfate react with one mole of iodine. This is the same as one mole of thiosulfate reacting with 0.5 moles of iodine.

$$\text{moles } I_2 = 2.20 \times 10^{-3} \times 0.5 = 1.10 \times 10^{-3}$$

Step 5: Calculate the amount in moles of chlorate(I) ions.

Looking at Equation 1, one mole of chlorate ions produces one mole of iodine.

$$\text{moles } ClO^- = 1.10 \times 10^{-3}$$

Step 6: Calculate the concentration of chlorate(I) ions.

$$\text{concentration of } ClO^- = \frac{\text{moles } ClO^-}{\text{volume of } ClO^- \text{ used}} = \frac{1.10 \times 10^{-3}}{25.00 \times 10^{-3}} = 0.044 \, \text{mol dm}^{-3}$$

Summary questions

1 Calculate the amount of sodium thiosulfate in the following solutions writing the answer in standard form and to 3 s.f.
   a 20.0 cm^3 of solution with a concentration of 1.00 mol dm^{-3}. *(2 marks)*
   b 24.6 cm^3 of solution with a concentration of 0.0100 mol dm^3. *(2 marks)*

2 In a titration, 10.00 cm^3 of sodium chlorate(I) solution was pipetted into a conical flask before excess potassium iodide and sulfuric acid were added. A 0.500 mol dm^{-3} solution of sodium thiosulfate was then run into the conical flask. The end point was reached when 11.2 cm^3 of sodium thiosulfate had been added. Calculate the concentration of the sodium chlorate(I) solution to 3 s.f. *(3 marks)*

3 When chlorine gas reacts with water, it makes dilute hydrochloric acid and dilute chloric(I) acid, HClO. This is a reversible reaction. Write a balanced equation with state symbols. *(2 marks)*

4 Household bleach is diluted by making 10.0 cm^3 of bleach up to 100 cm^3 in a volumetric flask. To a 10.0 cm^3 aliquot of this, excess acid and potassium iodide were added. A mean titre of 9.80 cm^3 of 0.0100 mol dm^{-3} sodium thiosulfate solution was required to change the starch indicator to colourless. Calculate the concentration of the undiluted bleach to 3 s.f. *(4 marks)*

ES 6 Hydrogen chloride in industry and the laboratory

Specification reference: ES(a), ES(l), ES(m)

Making hydrochloric acid

Hydrochloric acid can be made by a variety of methods. A simple method for its production is to start by making hydrogen chloride gas directly from the elements (Figure 1). This would be possible at a plant producing chlorine from brine by electrolysis, as described in Topic ES 3, as both chlorine and hydrogen are produced in the process.

$$H_2(g) + Cl_2(g) \rightarrow 2HCl(g)$$

This is a good example of an atom economy of 100%.

Chemical ideas: Chemistry in industry 15.2

Atom economy

These days, much more household waste is recycled, and less sent to landfill. The movement towards waste reduction is mirrored in the chemical industry, and is widely referred to as green chemistry. When deciding which reactions to use in a chemical plant, the percentage of reactant atoms ending up in the desired product is one factor that is taken into consideration. This percentage is called the **atom economy** and the greater the atom economy, the less the waste.

The following equation shows how to calculate the atom economy in a reaction.

$$\% \text{ atom economy} = \frac{\text{relative formula mass of the desired product}}{\text{relative formula mass of all reactants used}} \times 100$$

 Worked example: Calculating atom economy

What is the atom economy for the production of hydrogen chloride?

Step 1: Write the equation for the reaction.

$$H_2(g) + Cl_2(g) \rightarrow 2HCl(g)$$

Step 2: From the equation, work out the moles of reactants and products.

$$H_2(g) + Cl_2(g) \rightarrow 2HCl(g)$$
$$1\,\text{mol} + 1\,\text{mol} \rightarrow 2\,\text{mol}$$

Step 3: Calculate the mass of one mole for each substance using the relative molecular mass.

$M_r(H_2)$: 2; Mass of one mole = 2.0 g

$M_r(Cl_2)$: 71; Mass of one mole = 71.0 g

$M_r(HCl)$: 36.5; Mass of one mole = 2 × 36.5 = 73.0 g

Step 4: Calculate the percentage economy.

$$\% \text{ atom economy} = \frac{73.0}{2.0 + 71.0} \times 100 = 100\%$$

Learning outcomes

Demonstrate and apply knowledge and understanding of:

→ the concept of amount of substance in performing calculations involving atom economy; the relationship between atom economy and the efficient use of atoms in a reaction

→ the preparation of HCl; the preparation of HBr and HI by using the halide and phosphoric acid; the action of sulfuric acid on chlorides, bromides, and iodides

→ the properties of the hydrogen halides: different thermal stabilities, similar reaction with ammonia and acidity, different reactions with sulfuric acid.

▲ **Figure 1** *A production unit for manufacturing hydrochloric acid*

Synoptic link

You met percentage yield in Chapter 1, Elements of life. Percentage yield is another way of considering the efficiency of a chemical reaction. However, a reaction might have a large percentage yield but a low atom economy.

Activity ES 6.1

This activity gives you the opportunity to calculate some atom economies.

Synoptic link

Atom economy is one aspect of the green chemistry principles. You will find out more about them in Chapter 5, What's in a medicine?

Making hydrochloric acid as a co-product

A large proportion of the hydrochloric acid that is made is a co-product from the chlorination of organic compounds. For example, the first stage in the manufacture of poly(chloroethene), also called poly (vinyl chloride) (PVC), is the reaction of ethene with chlorine. The 1,2-dichloroethane that is formed undergoes thermal cracking to give chloroethene and hydrogen chloride.

$$CH_2ClCH_2Cl \rightarrow CH_2CHCl + HCl$$

The hydrogen chloride can then be converted to hydrochloric acid by passing it through water. A solution of high concentration can be produced easily because hydrogen chloride has a very high solubility in water. Hydrogen chloride gas is made up of covalent molecules – when dissolved in water it forms the hydrated ions $H^+(aq)$ and $Cl^-(aq)$.

Chemical ideas: The periodic table 11.3b

Hydrogen halides

Preparing hydrogen halides

In Topic ES 1 you learnt that fluorine was the strongest oxidising agent in Group 7. Fluorine atoms have the greatest tendency to be reduced or gain electrons. In doing so, fluorine atoms become fluoride ions.

$$F_2 + 2e^- \rightarrow 2F^-$$

Fluoride ions have a low tendency to lose electrons and turn back into atoms. Fluoride ions are difficult to oxidise and so are poor reducing agents.

$$2F^- \rightarrow F_2 + 2e^-$$

stronger oxidising agents $\longleftarrow$

| F_2 | Cl_2 | Br_2 | I_2 |
| F^- | Cl^- | Br^- | I^- |

$\longrightarrow$ stronger reducing agents

Sodium fluoride and sodium chloride

Sodium fluoride and sodium chloride both react with concentrated acid to make hydrogen fluoride or hydrogen chloride gas. In these experiments you see white fumes of hydrogen chloride as it meets the moist air. Tiny droplets of hydrochloric acid are being made.

$$NaCl(s) + H_2SO_4(aq) \rightarrow NaHSO_4(aq) + HCl(g)$$

This is not a redox reaction.

Sodium bromide

Sodium bromide first reacts with concentrated sulfuric acid to make hydrogen bromide.

$$NaBr(s) + H_2SO_4(aq) \rightarrow NaHSO_4(aq) + HBr(g)$$

However, the bromide ions produced are strong enough reducing agents to reduce the sulfuric acid to sulfur dioxide.

$$2H^+(aq) + 2Br^-(aq) + H_2SO_4(aq) \rightarrow SO_2(g) + Br_2(l) + 2H_2O(l)$$

Br	−1		0		increase in oxidation state – reducing agent
S		+6	+4		decrease in oxidation state – reduced

This means that adding concentrated sulfuric acid to sodium bromide would not be a good way to make hydrogen bromide gas because it won't be pure. The gas made will be a mixture of hydrogen bromide, sulfur dioxide, and bromine vapour (since the reaction is exothermic).

Sodium iodide

Sodium iodide first of all reacts with concentrated sulfuric acid to make hydrogen iodide.

$$NaI(s) + H_2SO_4(aq) \rightarrow NaHSO_4(aq) + HI(g)$$

However the iodide ions produced are even stronger reducing agents than the bromide ions above. The sulfuric acid is this time reduced further to make hydrogen sulfide gas.

$$8H^+(aq) + 8I^-(aq) + H_2SO_4(aq) \rightarrow H_2S(g) + 4I_2(s) + 4H_2O(l)$$

I	−1		0		increase in oxidation state – reducing agent
S		+6	−2		decrease in oxidation state – oxidising agent

With bromide ions, the oxidation state of sulfur decreased by two. With iodide ions, the oxidation state of sulfur is reduced by eight. (Iodide is a stronger reducing agent than bromide.) So adding concentrated sulfuric acid to sodium iodide would not be a good way to make hydrogen iodide gas because it won't be pure. The gas made will be a mixture of hydrogen iodide (white fumes) and hydrogen sulfide (smells of rotten eggs).

When preparing the hydrogen halides in the lab, the appropriate sodium halide is used. Concentrated sulfuric acid can be added to the sodium chloride when making hydrogen chloride. To make pure hydrogen bromide or hydrogen iodide concentrated phosphoric acid is used instead. Unlike sulfuric acid, the concentrated phosphoric acid will not be reduced and so a pure hydrogen halide can be collected.

Similarities and differences in the properties of hydrogen halides

Thermal stability

The **thermal stability** of the hydrogen halides decreases as you go down Group 7 (Table 1) – hydrogen iodide, HI, is broken down into its elements at a lower temperature than hydrogen chloride, HCl. This is because the bond strength between hydrogen and the halogen decreases as you go down Group 7. Less energy is needed to break the bond for hydrogen iodide.

Synoptic link

You first met bond enthalpies in Chapter 2, Developing fuels.

▼ **Table 1** Bond enthalpies for hydrogen–halogen bonds

Bond	Average bond enthalpy / kJ mol^{-1}	Bond length / nm
H–F	568.0	0.092
H–Cl	432.0	0.127
H–Br	366.3	0.141
H–I	298.3	0.161

When the hydrogen halides are heated in a laboratory:

- hydrogen fluoride isn't broken down into hydrogen and fluorine
- hydrogen chloride isn't broken down into hydrogen and chlorine
- some brown bromine gas is made when hydrogen bromide is strongly heated

$$2HBr(g) \rightarrow H_2(g) + Br_2(g)$$

- large amounts of purple gaseous iodine are made if a red hot needle is plunged into hydrogen iodide.

$$2HI(g) \rightarrow H_2(g) + I_2(g)$$

Acidity

In solution, the very soluble hydrogen halides are all acidic. Apart from HF they are strongly acidic. For HCl, HBr, and HI there is almost 100% **dissociation**. Remember that all acidic solutions have $H^+(aq)$ ions in. Another way of representing this is the **oxonium ion** $H_3O^+(aq)$:

$$HCl(aq) \rightarrow H^+(aq) + Cl^-(aq) \text{ or } H_2O(l) + HCl(aq) \rightarrow H_3O^+(aq) + Cl^-(aq)$$

Reaction with ammonia

All of the hydrogen halides react with ammonia to make salts. If a glass rod dipped in concentrated ammonia solution is placed in the hydrogen halide, a white cloud of ammonium halide is made. The following reaction is typical of the hydrogen halides:

$$NH_3(g) + HCl(g) \rightarrow NH_4Cl(s)$$

Reaction with sulfuric acid

The reactions of hydrogen halides with concentrated sulfuric acid are different. This is due to the increasing strength of the halide ions as reducing agents. Compare this with the reactions of solid halides with sulfuric acid.

- hydrogen fluoride, HF, and hydrogen chloride, HCl, do not react
- hydrogen bromide, HBr, makes sulfur dioxide, SO_2
- hydrogen iodide, HI, makes hydrogen sulfide, H_2S.

> **Synoptic link**
>
> You found out about acidic solutions in Chapter 1, Elements of life.

> **Activity ES 6.2**
>
> In this activity you will prepare and carry out reactions with hydrogen halides.

Summary questions

1 Propanol can be dehydrated to produce propene and water:

$$CH_3CH_2CH_2OH \rightarrow CH_3CH=CH_2 + H_2O$$

   a Calculate the relative formula mass of the starting material, propanol. (*1 mark*)

   b Calculate the relative formula mass of the useful product, propene. (*1 mark*)

   c Calculate the atom economy of this reaction. (*1 mark*)

2 1-bromobutane, C_4H_9Br, will react (rather slowly) with water to produce butan-1-ol, C_4H_9OH, and hydrogen bromide.
   a Write the equation for this reaction.   (*1 mark*)
   b Calculate the atom economy of this reaction.   (*2 marks*)

   This reaction can be sped up by using sodium hydroxide, NaOH, instead of water. In this case, the waste product of the reaction is not hydrogen bromide, HBr, but sodium bromide, NaBr.

   c Write an equation for this reaction.   (*1 mark*)
   d What effect would changing the reactant in this way have on the atom economy?   (*1 mark*)

3 Write balanced equations with state symbols for the reaction of hydrogen iodide with the following:
   a ammonia   (*2 marks*)
   b concentrated sulfuric acid.   (*2 marks*)

4 1,2-dichloroethane undergoes thermal cracking to give chloroethene.

$$CH_2ClCH_2Cl \rightarrow CH_2{=\!\!=}CHCl + HCl$$

   a Calculate the percentage yield of this process if 10.0 tonnes of the 1,2-dichloroethane yield 2.0 tonnes of chloroethene.   (*2 marks*)
   b Use the percentage yield and atom economy of this reaction to calculate how much in tonnes of the 1,2-dichloroethane is actually converted into chloroethene.   (*2 marks*)

5 Explain why pure hydrogen chloride can be prepared by the addition of concentrated sulfuric acid to sodium chloride but the same method cannot be used to prepare hydrogen bromide from sodium bromide. Include any equations which help your explanation.   (*4 marks*)

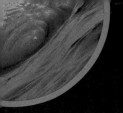

ES 7 The Deacon process solves the problem again

Specification reference: ES(q)

Hydrogen chloride gas has been produced by industrial chemical processes throughout history. Because it is highly toxic, it was important that early chemists could find a way of breaking it down or turning it into something useful. The solution has been built on and improved and is still being used today.

The Deacon process

With increased industrialisation in Britain in the 1800s, industrial pollution became a real problem particularly in areas such as Widnes in Cheshire. Demand for alkalis used to make soap and glass was high but unfortunately hydrogen chloride gas leaving the chimneys of the chemical plants producing the alkalis was devastating the land and killing farmers' crops. Parliament passed the first Alkali Act in 1863 and Victorian inspectors travelled the country to check the fumes from industrial chimneys. At first the hydrogen chloride was just dissolved in water and put into rivers where it killed all the fish. By 1874, Henry Deacon had developed what came to be known as the Deacon process. Hydrogen chloride was mixed with oxygen and passed over a catalyst. The products were chlorine and steam. Chlorine was in demand for bleaching paper and also fabrics.

▲ **Figure 1** *An industrial landscape from the nineteenth and early twentieth century*

$$4HCl(g) + O_2(g) \rightleftharpoons 2Cl_2(g) + 2H_2O(g) \ \Delta H = -114 \, kJ \, mol^{-1}$$

A high yield of chlorine requires that the equilibrium position is to the right. In theory this means that the best yield would be obtained using a high pressure, low temperature, and excess oxygen. In reality conditions in chemical plants are a compromise since factors such as rate of reaction, cost, and safety are taken into account.

The Deacon process today

The Deacon process is still important today. The Japanese Sumitoto Chemical Company have developed and improved the Deacon process so that it can produce almost pure chlorine whilst operating at lower temperature and at low cost.

In the production of the monomer for PVC (chloroethene) the following steps are involved.

Step 1 $\qquad CH_2{=}CH_2 \, (g) + Cl_2 \, (g) \rightarrow CH_2ClCH_2Cl \, (l)$

Step 2 $\qquad CH_2ClCH_2Cl \, (l) \rightarrow CH_2{=}CHCl(g) + HCl(g)$

Chlorine is used as a feedstock and hydrogen chloride is produced as a co-product. The hydrogen chloride can be used to make hydrochloric acid but this requires a great enough demand for it. The improved Deacon process developed by Sumitoto has been licensed to other chemical firms to solve the problem of hydrogen chloride as a product in polymer manufacture. For instance the hydrogen chloride gas made in Step 2 could be oxidised to chlorine using the Deacon process and recycled to use in Step 1. The costs for producing chlorine by this method are less than electrolysis of salt solution and since the chlorine recycling unit is on site the risks associated with the transport of liquid chlorine are eliminated.

Chemical ideas: Equilibrium in chemistry 7.4

Le Chatelier's principle

By studying data from many reactions, in 1888 Henri Le Chatelier (Figure 2) was able to propose a rule that enabled chemists to make qualitative predictions about the effect of a change on a system at equilibrium. He said that if a system is at equilibrium and a change is made in any of the conditions then the system will oppose the change. This is now known as **Le Chatelier's principle**.

Consider the reaction between ethanoic acid, CH_3COOH, and ethanol, C_2H_5OH. If the system was allowed to reach equilibrium, then the concentration of ethanol was increased, the equilibrium position will shift.

$$CH_3COOH(l) + C_2H_5OH(l) \rightleftharpoons CH_3COOC_2H_5(l) + H_2O(l)$$

The change in the equilibrium can be explained using le Chatelier's principle.

1   ethanol added – concentration of ethanol increased

2   The system opposes the change.

3   The system changes to decrease the concentration of ethanol

4   The forward reaction rate increases – ethanol reacts with ethanoic acid

5   More ethyl ethanoate, $CH_3COOC_2H_5$, and water are made

6   The equilibrium position moves to the right

This is exactly the same as the conclusion deduced using the equilibrium constant K_c in Topic ES 4.

Changing the concentration of any of the reactants or products will affect the equilibrium position.

- *Increasing* the concentration of *reactants* causes the equilibrium position to move to the *product* side.

- *Increasing* the concentration of *products* causes the equilibrium position to move to the *reactant* side.

- *Decreasing* the concentration of *reactants* causes the equilibrium position to move to the *reactant* side.

- *Decreasing* the concentration of *products* causes the equilibrium position to move to the *product* side.

▲ **Figure 2** *Henri Le Chatelier*

Using Le Chatelier's principle for the effect of changes in pressure

Many important industrial processes involve reversible reactions that take place in the gas phase. For these processes it is essential that conditions are identified that ensure that the equilibrium is shifted as far to the right (the products) as possible. From the study of equilibria in gas-phase reactions, the following conclusions have been reached:

- increasing the pressure moves the equilibrium to the side of the equation with fewer gas molecules as this tends to reduce the pressure

- decreasing the pressure moves the equilibrium to the side of the equation with more gas molecules as this tends to increase the pressure.

In each case, the position of equilibrium shifts so as to oppose the change in pressure.

 Worked example: The effect of changes in pressure

In the first stage in the steam reforming of methane to make methanol, methane reacts with steam to form carbon monoxide, CO, and hydrogen, H_2. What would happen in this reaction if you *reduced* the pressure?

Step 1: Write the equation for the reaction.

$$CH_4(g) + H_2O(g) \rightleftharpoons CO(g) + 3H_2(g)$$

Step 2: Work out the number of molecules on each side of the equation.

$$CH_4(g) + H_2O(g) \rightleftharpoons CO(g) + 3H_2(g)$$
$$1 + 1 = 2 \qquad\qquad 1 + 3 = 4$$
$$2 \text{ molecules} \qquad 4 \text{ molecules}$$

Step 3: Identify how the system will change.

Pressure is reduced. The system opposes the change, so the system increases the pressure.

Step 4: Identify which way the equilibrium position needs to shift to cause the system change identified in Step 3.

There are more molecules on the right-hand side of the equation than on the left-hand side of the equation. Therefore, making more carbon monoxide and hydrogen will increase the pressure. The equilibrium position moves to the right.

Using Le Chatelier's principle for the effect of changes in temperature

Heating a reaction makes it go faster. However, how fast is not the same as how far. For a reversible reaction, if the forward reaction is exothermic then the reverse reaction will be endothermic to the same extent, and vice versa.

Synoptic link

You used the fact that if a forward reaction is exothermic then the reverse reaction is endothermic when you studied Hess' law in Chapter 2, Developing fuels.

Worked example: The effect of changes in temperature

Nitrogen dioxide, NO_2, is a dark brown gas that exists in equilibrium with its colourless dimer, dinitrogen tetraoxide, N_2O_4.

$$2NO_2(g) \rightleftharpoons N_2O_4(g) \qquad \Delta H \text{ is negative}$$
$$\text{brown} \qquad \text{colourless} \quad \text{(i.e., exothermic)}$$

The forward reaction forms $N_2O_4(g)$ and is exothermic, releasing thermal energy to the surroundings. The reverse reaction forms $NO_2(g)$ and is endothermic. Thermal energy is taken in from the surroundings.

If a sealed container of the brown equilibrium mixture is placed in iced water, it becomes paler. How can you explain this using the rule?

Step 1: Identify the temperature change – temperature is decreased.

Step 2: Identify how the system will change.

The system opposes the change. The system will act to increase the temperature.

Step 3: Identify which reaction will increase the temperature.

The forward reaction releases heat energy so will increase the temperature of the surroundings. This will increase the temperature of the sealed container. More N_2O_4 formed. The equilibrium position moves to the right. The colour becomes paler brown as N_2O_4 is colourless.

If the system in the worked example is put in a beaker of boiling water, the system would turn a darker brown. This is because the system would oppose the change of the increase in temperature, so the reverse reaction (the formation of the brown NO_2) would be favoured as this reaction takes in thermal energy. This reduces the temperature of the system.

- Heating a reversible reaction at equilibrium shifts the reaction in the direction of the endothermic reaction.

- Cooling a reversible reaction at equilibrium shifts the reaction in the direction of the exothermic reaction.

Summary questions

1 Which element is oxidised and which element is reduced in the Deacon process? *(2 marks)*

2 Ethanol is produced industrially at about 70 atmospheres pressure and 300 °C by the following reaction. The reaction needs a catalyst.

$$C_2H_4(g) + H_2O(g) \rightleftharpoons C_2H_5OH(g) \qquad \Delta H = -46\,kJ\,mol^{-1}$$

Which of the following would move the position of equilibrium to the right?

A Increasing the temperature.
B Increasing the concentration of steam.
C Decreasing the pressure.   (*1 mark*)

3 State the direction in which the position of equilibrium of each system would move (if at all) if the pressure was increased by compressing the reaction mixture. Give your answer as 'to the left' or 'to the right', or 'no change'.

a $2NO(g) + O_2(g) \rightleftharpoons 2NO_2(g)$ (*1 mark*)
b $C_2H_6(g) \rightleftharpoons C_2H_4(g) + H_2(g)$ (*1 mark*)
c $2HI(g) \rightleftharpoons H_2(g) + I_2(g)$ (*1 mark*)
d $2NO_2(g) \rightleftharpoons N_2O_4(g)$ (*1 mark*)
e $2CO(g) + O_2(g) \rightleftharpoons 2CO_2(g)$ (*1 mark*)

4 Consider the reaction between hydrogen and oxygen to produce steam.

a Write an equation for the reaction with state symbols.   (*2 marks*)
b Write an expression for K_c.   (*1 mark*)
c Describe and explain how the equilibrium position is affected by:
   i   an increase in temperature   (*2 marks*)
   ii an increase in the total pressure.   (*2 marks*)

5 For the Deacon process explain why the following changes in conditions would increase the yield of chlorine.

$$HCl(g) + O_2(g) \rightleftharpoons 2Cl_2(g) + 2H_2O(g) \qquad \Delta H = -114\,kJ\,mol^{-1}$$

a adding excess oxygen   (*2 marks*)
b decreasing the pressure   (*2 marks*)
c decreasing the temperature   (*2 marks*)

6 This equilibrium exists in bleach:

$$Cl_2(aq) + 2NaOH(aq) \rightleftharpoons NaCl(aq) + NaOCl(aq) + H_2O(l)$$

Explain why you should never use another cleaning product that is acidic alongside the bleach.   (*3 marks*)

Practice questions

1   Look at the two reactions of chlorine with ethene:

$$C_2H_4 + Cl_2 \rightarrow C_2H_4Cl_2 \qquad \textbf{Reaction 1}$$
$$C_2H_4 + Cl_2 \rightarrow C_2H_3Cl + HCl \qquad \textbf{Reaction 2}$$

Which of the following rows is correct about the atom economies of these reactions in terms of the organic product?

	Reaction 1	Reaction 2
A	80%	50%
B	100%	37%
C	100%	63%
D	40%	100%

(1 mark)

2   Which row of the table is correct for the electrolysis of aqueous potassium iodide?

	Cathode	Anode
A	potassium	iodine
B	hydrogen	oxygen
C	iodine	potassium
D	hydrogen	iodine

(1 mark)

3   Which row in the table contains the correct half-equations for the electrolysis of molten sodium chloride?

	Cathode	Anode
A	$Na^+ + e^- \rightarrow Na$	$Cl^- \rightarrow Cl + e^-$
B	$2H^+ + 2e^- \rightarrow H_2$	$Cl^- \rightarrow \frac{1}{2}Cl_2 + e^-$
C	$Na^+ \rightarrow Na + e^-$	$Cl_2 \rightarrow 2Cl^- + 2e^-$
D	$Na^+ + e^- \rightarrow Na$	$2Cl^- \rightarrow Cl_2 + 2e^-$

(1 mark)

4   Which row in the table correctly describes the halogen elements at room temperature?

	Chlorine	Bromine	Iodine
A	green gas	brown gas	purple gas
B	colourless solution	brown solution	brown solution
C	green gas	red liquid	grey solid
D	yellow-green gas	brown liquid	purple solid

(1 mark)

5   Sodium bromide is reacted with silver nitrate solution.

The result is:

A   a white precipitate that is soluble in dilute ammonia solution

B   a cream precipitate that is soluble in concentrated ammonia solution

C   a yellow precipitate that is insoluble in ammonia solution

D   a white precipitate that is soluble in concentrated ammonia solution.   *(1 mark)*

6   Which of the following will react with sodium iodide to produce the purest sample of hydrogen iodide?

A   concentrated sulfuric acid

B   dilute sulfuric acid

C   dilute hydrochloric acid

D   phosphoric acid   *(1 mark)*

7   An aqueous solution of chlorine is added to an aqueous solution of sodium iodide. Some cyclohexane is added, forming the upper layer. Which of the following is the correct observation.

A   There is no reaction.

B   The upper layer goes purple and the aqueous layer goes brown.

C   Both layers go brown.

D   The lower layer is brown and the upper layer is yellow.   *(1 mark)*

8   When sulfuric acid reacts with a bromide, which of the following are correct?

1   hydrogen bromide is produced

2   sulfur dioxide is produced

3   bromine is produced

A   1, 2, and 3 correct

B   1 and 2 are correct

C   2 and 3 are correct

D   Only 1 is correct   *(1 mark)*

9 Which of the following are true about hydrogen chloride, hydrogen bromide and hydrogen iodide?

1 they all react with ammonia

2 they are all acidic

3 they all reduce sulfuric acid

A 1, 2, and 3 correct

B 1 and 2 are correct

C 2 and 3 are correct

D Only 1 is correct   (*1 mark*)

10 A solution of copper sulfate is electrolysed with copper electrodes. Which of the following is true?

1 Copper is transferred from the anode to the cathode.

2 Copper is plated on the anode.

3 Sulfur dioxide is produced at the anode.

A 1, 2, and 3 correct

B 1 and 2 are correct

C 2 and 3 are correct

D Only 1 is correct   (*1 mark*)

11 In the manufacture of bromine from sea water:

Step 1: Chlorine is bubbled through sea water containing a very dilute bromide solution to release bromine.

Step 2: Air is blown through to produce bromine vapour.

Step 3: The vapour is mixed with sulfur dioxide and passed into water:

$$Br_2 + SO_2 + ... H_2O \rightarrow 2HBr + ...$$

Equation 11.1

Step 4: Steam and chlorine are blown through to release bromine from the hydrogen bromide.

Step 5: The bromine is dried using concentrated sulfuric acid.

a (i) Write an ionic equation for the reaction that occurs in both Steps 1 and 4.   (*1 mark*)

(ii) Which property of the halogens does this reaction illustrate?   (*1 mark*)

b Suggest why it is necessary to produce bromine in Step 1 and then again in Step 4.   (*1 mark*)

c (i) On what property of bromine does Step 2 depend?   (*1 mark*)

(ii) Suggest the appearance of the gas stream after Step 2.   (*1 mark*)

d Use oxidation states to complete and balance Equation 11.1 and explain your reasoning.   (*3 marks*)

e Chlorine is made by electrolysing an aqueous solution of sodium chloride. Give the half-equations for the reactions at the positive and negative electrodes during this electrolysis.   (*2 marks*)

f Chlorine reacts with cold aqueous sodium hydroxide as follows:

$$Cl_2 + H_2O \rightleftharpoons HCl + HClO \quad \textbf{Equation 11.2}$$

(i) Write the oxidation states under the chlorine atoms in Cl_2, HCl, and HClO in Equation 11.2.

What is being reduced and what is being oxidised in this equation?   (*3 marks*)

(ii) Give the systematic name of HClO.   (*1 mark*)

12 Hydrogen chloride is made industrially by the reaction:

$$H_2(g) + Cl_2(g) \rightleftharpoons 2HCl(g) \quad \textbf{Equation 12.1}$$

a This reaction can reach dynamic equilibrium. Explain the meaning of the term *dynamic equilibrium*.   (*2 marks*)

b (i) Write the equation for the equilibrium constant K_c of the reaction in Equation 12.1.   (*1 mark*)

(ii) $K_c = 2 \times 10^{33}$ for this reaction at 298 K. What conclusion can be drawn about the composition of an equilibrium mixture at 298 K?   (*1 mark*)

(iii) The reaction in Equation 12.1 is exothermic. Discuss the effect on an equilibrium mixture of, separately, changing the hydrogen concentration, varying the temperature and varying the pressure.   (*6 marks*)

13 The chemist Max Bodenstein investigated the equilibrium shown below:

$$H_2(g) + I_2(g) \rightleftharpoons 2HI(g) \qquad \textbf{Equation 13.1}$$

He allowed mixtures of known masses of hydrogen and iodine to react in sealed tubes at high temperatures until equilibrium had been established. Then he rapidly cooled the tubes and analysed the iodine present with sodium thiosulfate.

a Suggest why the flasks are rapidly cooled. *(1 mark)*

b Describe how a sealed tube could be investigated to measure the mass of iodine it contains. *(4 marks)*

c From such an experiment at 500 K, the masses of substances in a 100 cm³ tube were found to be:

Substance	Mass / g
H_2	0.20
I_2	25.38
HI	161.15

Calculate a value for K_c for the reaction in Equation 13.1 at 500 K. *(4 marks)*

d 20.0 cm³ of a solution of 0.002 mol dm⁻³ KIO_3 is reacted with excess iodide ions in the presence of acid.

$$KIO_3 + 5I^- + 6H^+ \rightarrow 3I_2 + 3H_2O$$

$$\textbf{Equation 13.2}$$

(i) Give the systematic name of KIO_3. *(1 mark)*

(ii) Calculate the volume of 0.50 mol dm⁻³ $Na_2S_2O_3$ that would react with the iodine formed. Give your answer to a *suitable* number of significant figures. *(4 marks)*

(iii) A teacher tells a student that this is not a very satisfactory titration result. Suggest why the teacher says this and suggest what the student could do to improve the titration result without changing the apparatus used. *(3 marks)*

Chapter 3 Summary

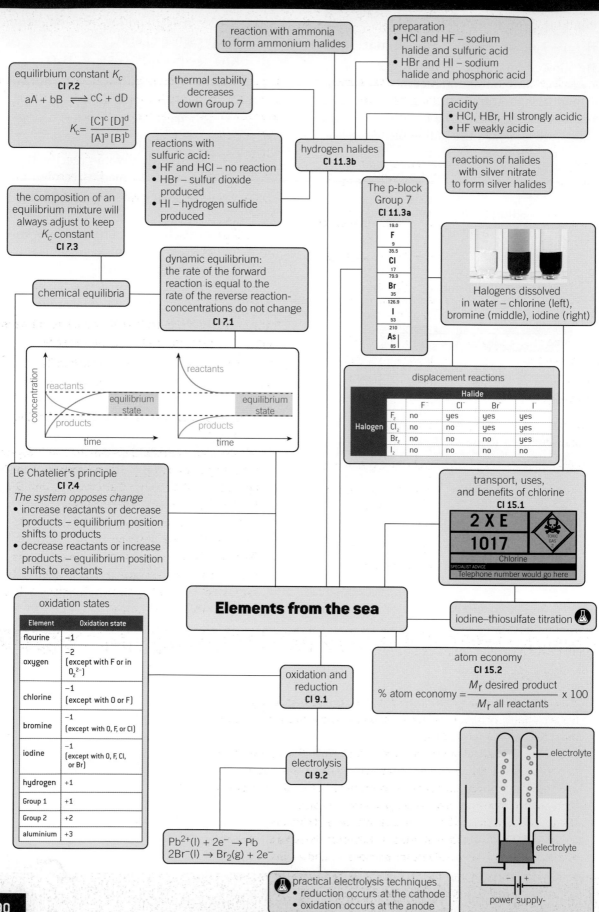

reaction with ammonia to form ammonium halides

preparation
- HCl and HF – sodium halide and sulfuric acid
- HBr and HI – sodium halide and phosphoric acid

equilibrium constant K_c
CI 7.2
$$aA + bB \rightleftharpoons cC + dD$$
$$K_c = \frac{[C]^c\,[D]^d}{[A]^a\,[B]^b}$$

thermal stability decreases down Group 7

acidity
- HCl, HBr, HI strongly acidic
- HF weakly acidic

reactions with sulfuric acid:
- HF and HCl – no reaction
- HBr – sulfur dioxide produced
- HI – hydrogen sulfide produced

hydrogen halides
CI 11.3b

reactions of halides with silver nitrate to form silver halides

the composition of an equilibrium mixture will always adjust to keep K_c constant
CI 7.3

The p-block Group 7
CI 11.3a

19.0	
F	
9	
35.5	
Cl	
17	
79.9	
Br	
35	
126.9	
I	
53	
210	
As	
85	

dynamic equilibrium: the rate of the forward reaction is equal to the rate of the reverse reaction- concentrations do not change
CI 7.1

chemical equilibria

Halogens dissolved in water – chlorine (left), bromine (middle), iodine (right)

concentration / time graphs

reactants — products — equilibrium state — time

displacement reactions

Halogen		Halide			
		F⁻	Cl⁻	Br⁻	I⁻
	F₂	no	yes	yes	yes
	Cl₂	no	no	yes	yes
	Br₂	no	no	no	yes
	I₂	no	no	no	no

Le Chatelier's principle
CI 7.4
The system opposes change
- increase reactants or decrease products – equilibrium position shifts to products
- decrease reactants or increase products – equilibrium position shifts to reactants

transport, uses, and benefits of chlorine
CI 15.1

2 X E
1017
Chlorine
SPECIALIST ADVICE
Telephone number would go here

iodine–thiosulfate titration

Elements from the sea

oxidation states

Element	Oxidation state
flourine	−1
oxygen	−2 (except with F or in $O_2{}^{2-}$)
chlorine	−1 (except with O or F)
bromine	−1 (except with O, F, or Cl)
iodine	−1 (except with O, F, Cl, or Br)
hydrogen	+1
Group 1	+1
Group 2	+2
aluminium	+3

oxidation and reduction
CI 9.1

atom economy
CI 15.2
$$\% \text{ atom economy} = \frac{M_r \text{ desired product}}{M_r \text{ all reactants}} \times 100$$

electrolysis
CI 9.2

electrolyte

$$Pb^{2+}(l) + 2e^- \rightarrow Pb$$
$$2Br^-(l) \rightarrow Br_2(g) + 2e^-$$

electrolyte

practical electrolysis techniques
- reduction occurs at the cathode
- oxidation occurs at the anode

power supply-

Interhalogens

Interhalogens are compounds containing two or more different halogen atoms. There are many known interhalogens.

Interhalogen	Formula	Notes
chlorine monofluoride	ClF	colourless gas and the lightest interhalogen compound
iodine monofluoride	IF	gas which decomposes at 0 °C into iodine, I_2, and iodine pentafluoride
iodine monochloride	ICl	red crystalline substance that melts at 27°C
astatine monoiodide	AtI	heaviest diatomic interhalogen compound
bromine trifluoride	BrF_3	yellow liquid that is an electrical conductor
chlorine trifluoride	ClF_3	colourless gas
iodine trifluoride	IF_3	crystalline yellow solid
chlorine pentafluoride	ClF_5	colourless gas which reacts violently with water
iodine pentafluoride	IF_5	colourless highly reactive liquid

The three interhalogens containing chlorine and fluorine atoms are highly reactive oxidising and fluorinating agents. Examples of their reactions with metals are given below.

$$W + 6\ ClF \rightarrow WF_6 + 3\ Cl_2 \qquad \text{Reaction A}$$
$$U + 3\ ClF_3 \rightarrow UF_6 + 3\ ClF \qquad \text{Reaction B}$$

A dot-and-cross diagram of chlorine pentafluoride is shown below.

Some interhalogen molecules dissociate into halogens or other interhalogen molecules, as shown below.

$$ClF_3 \rightleftharpoons ClF + F_2 \qquad \text{Reaction C}$$
$$5\ IF \rightleftharpoons IF_5 + 2\ I_2 \qquad \text{Reaction D}$$

1 What is the oxidation state of chlorine in ClF, ClF_3, and ClF_5?

2 Use oxidation states to show that the interhalogens are acting as oxidising agents in Reactions A and B.

3 Comment on the number of electrons surrounding the chlorine atom in chlorine pentafluoride.

4 Write an expression for K_c for the dissociation of chlorine trifluoride (Reaction C).

5 Disproportionation occurs when different atoms of the same element are oxidised and reduced in the same reaction. Identify which of Reactions A–D is a disproportionation reaction.

Extension

1 Research further examples of equilibria in the chemical industry. Explain how careful control of the reaction conditions in each case can maximise the yield of product.

2 Prepare a summary of redox reactions. Include definitions of oxidation and reduction, and describe identification of oxidising and reducing agents, balancing equations, naming compounds, electrolysis, redox titrations and industrial applications of redox reactions.

3 There have been calls to ban the use of chlorine and chlorine compounds for environmental reasons. Research the risks and benefits of these chemicals and evaluate the opinions for and against a ban.

CHAPTER 4
The ozone story

Topics in this chapter

Why a chapter on The ozone story?

The chemical and physical processes going on in the atmosphere have a profound influence on life on Earth. They involve a highly complex system of interrelated reactions, and yet much of the underlying chemistry is essentially simple. The focus of this chapter is change – change in the atmosphere brought about by human activities and the potential effects on life. The influences of human activities in causing the depletion of the ozone layer in the upper atmosphere are explored, together with the role of chemists in recognising and explaining the phenomenon. The success of the Montreal Protocol to limit ozone-depleting chemicals is discussed and the current state of the ozone layer is investigated.

Some important chemical principles are introduced and developed in considering the ozone story. In particular the effect of radiation on matter, the factors that affect the rate of a chemical reaction, the formation and reactions of radicals, and the idea of intermolecular bonding will be discussed, as well as the specific chemistry of species that are met in the context of atmospheric chemistry – such as oxygen, carbon dioxide, methane, and organic halogen compounds.

Knowledge and understanding checklist

From your Key Stage 4 study you will have studied the following. Work through each point, using your Key Stage 4 notes and the support available on Kerboodle.

☐ Factors affecting the rate of a chemical reaction.

☐ How catalysts work.

☐ Covalent bonding.

You will learn more about some ideas introduced in earlier chapters:

☐ the use of moles and quantitative chemistry (**Developing fuels**)

☐ the electromagnetic spectrum (**Elements of life**)

☐ electronic structure (**Elements of life**)

☐ the chemistry of simple organic molecules (**Developing fuels**)

☐ enthalpy changes and bond enthalpies (**Developing fuels**)

☐ oxidation states (**Elements from the sea**)

☐ catalysis (**Developing fuels**).

Maths skills checklist

In this chapter, you will need to use the following maths skills. You can find support for these skills on Kerboodle and through MyMaths.

☐ Recognise and make use of appropriate units in calculations.

☐ Recognise and use expressions in decimal and ordinary form.

☐ Use percentages.

☐ Understand and use the symbols $=, <, \ll, \gg, >, \propto, \sim, \rightleftharpoons$.

☐ Translate information between graphical and numerical forms.

☐ Plot variables for experimental data.

MyMaths.co.uk
Bringing Maths Alive

The **atmosphere** is a relatively thin layer of gas that surrounds the Earth's surface. It extends about 100 km above the Earth's surface. If the world were a blown-up balloon, the rubber would be thick enough to contain nearly all the atmosphere. In Figure 1, the atmosphere is the thin blue haze you can see surrounding the Earth. Thin though it is, this layer of gas has an enormous influence on the Earth.

Structure and composition of the atmosphere

A simplified picture of the lower and middle parts of the atmosphere is shown in Figure 2. The two most chemically important regions are the **troposphere** and the **stratosphere**. In fact, 90% of all the molecules in the atmosphere are in the troposphere, and the atmosphere becomes less dense the higher you go. Figure 2 also shows the way that temperature changes with altitude. Mixing is easy in the troposphere because hot gases can rise and cold gases can fall. The reverse temperature gradient in the stratosphere means that mixing is much more difficult in the vertical direction. However, horizontal circulation is rapid in the stratosphere, particularly around circles of latitude.

▲ **Figure 1** *The Earth as seen from space – the blue haze is the atmosphere of the Earth*

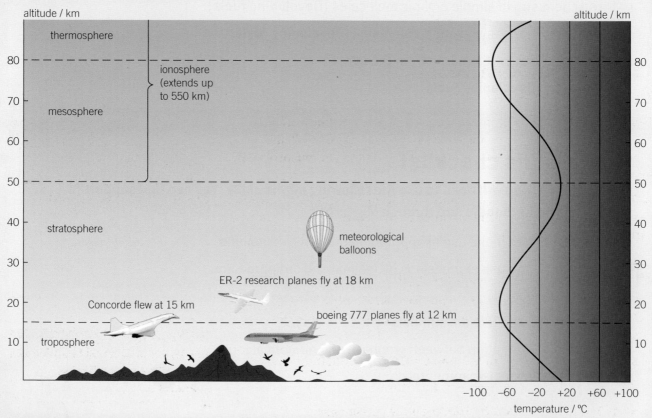

▲ **Figure 2** *The structure of the atmosphere and the change in temperature with altitude*

Table 1 shows the average composition by volume of dry air from an unpolluted environment. This is typical of the troposphere.

▼ **Table 1** *Composition by volume of dry tropospheric air from an unpolluted environment. The concentrations of some of these gases are measured in parts per million (ppm) by volume. Gases marked with an asterisk are found naturally in the atmosphere, but their concentration is increased by human activity*

Gas	Concentration by volume
nitrogen, N_2	78%
oxygen, O_2	21%
argon, Ar	1%
carbon dioxide, CO_2	399 ppm *
neon, Ne	18.2 ppm
helium, He	5.2 ppm
methane, CH_4	1.8 ppm *
krypton, Kr	1.1 ppm
hydrogen, H_2	0.5 ppm
dinitrogen oxide, N_2O	0.3 ppm *
carbon monoxide, CO	0.1 ppm *
xenon, Xe	0.09 ppm
nitrogen monoxide, NO, and nitrogen dioxide, NO_2 (NO_x)	0.003 ppm *

Chemical ideas: Measuring amounts of substance 1.5b

Calculations involving gases

Gas concentrations

When dealing with a gas it is sometimes more useful to know its volume than its mass. When the concentration of a gas is small, it is more convenient to express the concentration as parts per million (ppm) by volume.

In Table 1 carbon dioxide has a concentration of 399 ppm. This means that of one million particles in a sample of air, 399 of them will be carbon dioxide molecules.

 Worked example: Calculating percentage composition

The concentration of carbon dioxide in a sample of air is 399 ppm. Calculate the percentage composition of carbon dioxide in this sample.

Step 1: Divide number of CO_2 molecules by 1 000 000.

$$\frac{399}{1\,000\,000} = 3.99 \times 10^{-4}$$

Step 2: Multiply answer by 100 to get the percentage.

$$3.99 \times 10^{-4} \times 100 = 0.04\%$$

> **Synoptic link**
>
> You have already looked at different types of gas calculations in Chapter 2, Developing fuels.

> **Study tip**
>
> This is the same as dividing by 10 000.

From the worked example, you can see that converting from per cent to parts per million involves a factor of 10 000. A concentration of 1% is equivalent to 10 000 ppm.

 Worked example: Calculating concentration in parts per million

A sample of air is 78% nitrogen. Calculate the concentration of nitrogen in ppm in this sample.

Step 1: Multiply the percentage by 10 000.

$$78 \times 10\,000 = 780\,000 \text{ ppm}$$

The early atmosphere

The atmosphere hasn't always had the same composition. The first atmosphere was lost during the upheavals in the early years of the Solar System. The next atmosphere consisted of compounds such as carbon dioxide, methane, and ammonia, which bubbled out of the Earth itself.

3000 million years ago there was very little oxygen in the atmosphere. When the first simple plants appeared they began to produce oxygen through photosynthesis. For more than 1000 million years very little of this oxygen reached the atmosphere. It was used up quickly as it oxidised sulfur and iron compounds, and other chemicals in the Earth's crust. It wasn't until this process was largely complete that oxygen began to collect in the atmosphere.

 The atmosphere past, present, and future

Researchers looking at the phenomenon of oxidative weathering believe that the Earth's atmosphere may have contained significant amounts of oxygen up to three billion years ago, earlier than previous estimates.

Their research involved studying the oxidation of two isotopes of chromium, chromium-53 and chromium-52. When oxidised, ^{53}Cr becomes slightly more soluble than ^{52}Cr. Rain water washes the heavier ^{53}Cr isotope from the soil into the sea more readily than ^{52}Cr. Consequently, soils should become depleted in ^{53}Cr and sea-bed sediments should become richer in ^{53}Cr.

This prediction was borne out when the researchers analysed ancient soils and sediments in Kwazulu-Natal Province, South Africa. The age of the sediments and the ratio of ^{52}Cr to ^{53}Cr indicates that 2.95 billion years ago the atmosphere could have had an oxygen concentration of 63 ppm.

Mauna Loa

From 1958 carbon dioxide levels in the atmosphere have been measured by continuous atmospheric monitoring devices at the Mauna Loa Observatory in Hawaii, USA. In 1958 the concentration of carbon dioxide was recorded as 315 ppm.

In early 1988 the level was recorded as 350 ppm, and on 10 May 2013 daily averages were recorded as 400 ppm. It is thought that this is the highest level for the last three million years of Earth's history. There is a natural annual variation of 3 to 9 ppm due to plants' growing seasons, but there has been a steady upward rise in average carbon dioxide levels.

March 2012	394.36 ppm
March 2013	397.27 ppm
March 2014	399.47 ppm

There is widespread concern at this increase in carbon dioxide levels, as they are linked to global warming.

1. State the number of protons, neutrons, and electrons in an atom of ^{52}Cr and an atom of ^{53}Cr.
2. Calculate the percentage composition of oxygen in the early atmosphere (63 ppm).
3. Suggest how plants' growing seasons affect the atmospheric concentration of carbon dioxide.

Synoptic link

You will learn more about increasing carbon dioxide levels and global warming in Chapter 8, Oceans.

When the oxygen concentration reached about 10%, there was enough for the first animals to evolve using oxygen for respiration. Eventually there was enough respiration and other processes going on to remove the oxygen as fast as it was formed. Since then, the oxygen concentration has remained at about 21%.

Look again at Table 1. All the gases listed are produced as a result of natural processes. Human activities add more gases to the atmosphere. Some of them, like carbon dioxide, are already present, but we increase their concentration. These gases are marked by an asterisk in Table 1. Their main sources as a result of human activities are shown in Table 2. Other gases in the atmosphere, like chlorofluorocarbons (CFCs) and hydrofluorocarbons (HFCs), are produced only as a result of human activity.

Synoptic link

You studied the origins of nitrogen oxides, carbon dioxide, and carbon monoxide in Chapter 2, Developing fuels.

▼ Table 2 *Sources of some of the gases in the atmosphere produced as a result of human activities*

Gas	Main source as a result of human activities
carbon dioxide	combustion of hydrocarbon fuels (e.g., in power stations, motor vehicles); deforestation
methane	cattle farming; landfill sites; rice paddy fields; natural gas leakage
nitrous oxide	fertilised soils; changes in land use (e.g., from the soil when land is ploughed up)
carbon monoxide	incomplete combustion of hydrocarbons (e.g., from car exhausts)
nitrogen oxides	internal combustion engines (from the reaction of N_2 and O_2 at high temperatures)

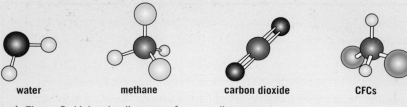

water methane carbon dioxide CFCs

▲ **Figure 3** *Molecular diagrams of some pollutant gases*

Given time, gases mix together and this natural diffusion process is greatly speeded up in the atmosphere by air currents and prevailing winds. So, in time, pollutant gases spread throughout the atmosphere. **Atmospheric pollution** is a global problem – it affects us all. In this chapter, we shall be concentrating on the depletion of the ozone layer in the stratosphere, looking at how chemists identified the problem, and at the international agreements to limit the damage.

Summary Questions

1 State the main gases in unpolluted air. *(1 mark)*

2 State three human activities that add gases to the air. *(1 mark)*

3 a Calculate how many parts per million by volume of argon are in a typical sample of tropospheric air. *(1 mark)*
  b Calculate the percentage of neon in a typical sample of tropospheric air. *(1 mark)*

4 a Calculate the volume of methane present in 1 dm^3 of tropospheric air. *(1 mark)*
  b Calculate the percentage of methane molecules in the sample. *(1 mark)*

OZ 2 Screening the sun
Specification reference: OZ(r), OZ(s), OZ(t), OZ(u)

The sunburn problem

Until the 1920s a suntan was considered undesirable, as it showed that you had to work outdoors in the Sun. The clothes designer Coco Chanel made sunbathing fashionable after a cruise on the yacht belonging to the Duke of Westminster. As you discover more about the effects of the Sun's **electromagnetic radiation** on the chemical bonds in living material, sunbathing will seem like less of a good idea.

The Sun radiates a spectrum of electromagnetic radiation. Part of this spectrum corresponds to the energy required to break chemical bonds, including those in molecules such as DNA. This can cause damage to genes and lead to skin cancer. On a less serious level, sunlight can damage the proteins within skin, so that years of exposure can make people look wrinkly and leathery. Brief exposure to the Sun may irritate the blood vessels in the skin, making it look red and sunburnt.

▲ **Figure 1** The trend-setting clothes designer Coco Chanel

Chemical sunscreens

Figure 3 shows the effects of different parts of the Sun's spectrum on the skin.

The most damaging part of this spectrum to the skin is the **ultraviolet** part. Fortunately, there are chemicals which absorb much of this radiation. You can sit by a window, or in a greenhouse, for hours on a sunny day without burning because the glass in the window lets through visible light but absorbs the high-energy ultraviolet radiation, so it never reaches your skin. On the other hand, water does let through some ultraviolet, so it is possible to burn under water.

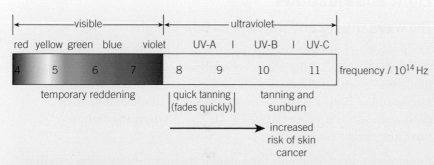

▲ **Figure 3** The effects of sunlight on the skin

Learning outcomes

Demonstrate and apply knowledge and understanding of:

→ the principal radiations of the Earth and the Sun in terms of the following regions of the electromagnetic spectrum: infrared, visible, ultraviolet

→ calculation of values for frequency, wavelength, and energy of electromagnetic radiation from given data

→ the formation and destruction of ozone in the stratosphere and troposphere; the effects of ozone in the atmosphere, including:

- ozone's action as a sunscreen in the stratosphere by absorbing high-energy UV (and the effects of such UV, including on human skin)

- the polluting effects of ozone in the troposphere, causing problems including photochemical smog

→ the effect of UV and visible radiation promoting electrons to higher energy levels, sometimes causing bond-breaking.

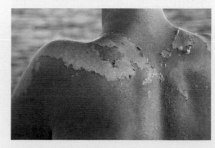

▲ **Figure 2** Over-exposure to sunlight causes severe sunburn and peeling skin, and increases the risk of developing skin cancers

Activity OZ 2.1

This activity helps you develop your skills planning and investigating sunscreens.

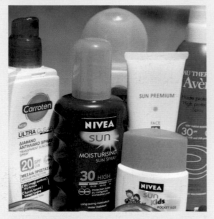

▲ **Figure 4** *The Sun protection factor (SPF) number on sunscreens indicates the time it will take for the Sun to produce a certain effect on your skin*

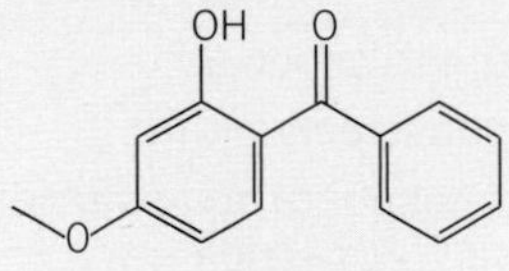

oxybenzone

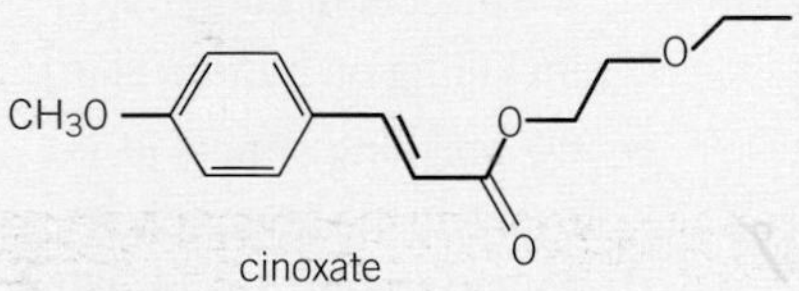

cinoxate

▲ **Figure 5** *Some examples of the molecules in sunscreen that are effective at absorbing ultraviolet*

Chemists have developed sunscreens which absorb high-energy ultraviolet radiation, and millions of pounds are spent on them in the UK every summer. The molecules in sunscreen often contain benzene rings or alternating double and single bonds (Figure 5). When ultraviolet light is absorbed, the electrons in the π orbitals in these bonds are promoted to higher energy levels.

However the best sunscreen of all is not made by chemists. It has always been with us – it is the atmosphere.

Why is the atmosphere such a good sunscreen?

Certain atmospheric gases absorb ultraviolet radiation strongly. They act as a global sunscreen, preventing much of the Sun's harmful radiation from reaching the Earth.

Most of this absorption goes on in the region of the upper atmosphere called the stratosphere (see Figure 2 in Topic OZ 1). Particularly important is the gas **ozone**, which is a form of oxygen with the formula O_3. It absorbs ultraviolet radiation in the region $10.1 \times 10^{14} - 14.0 \times 10^{14}$ Hz. This includes the UV-B and UV-C regions (Figure 3) which can damage DNA potentially leading to skin cancer, damage eyes leading to cataracts, and damage crops.

Although ozone in the stratosphere protects us from high energy ultraviolet radiation, ozone at ground level in the troposphere is a significant pollutant. Ozone is involved in reactions producing photochemical smog that causes haziness and reduced visibility, and irritation and respiratory problems for many people.

There is no life in the stratosphere because the high energy ultraviolet radiation would break down the delicate molecules of living things. Even simple molecular substances are broken down. Some of the covalent bonds break to give fragments of molecules called radicals.

Higher up in the atmosphere the radiation is powerful enough to knock electrons out of atoms, molecules, and radicals. Ions are produced, giving that part of the atmosphere its name – the ionosphere.

Chemical ideas: Radiation and matter 6.1b

Light and electrons

The wave theory and particle nature of light

The behaviour of light can be described using the wave model or the particle model. Like all electromagnetic radiation, light behaves like a wave with a characteristic wavelength λ and frequency ν. The speed of light c is the same for all kinds of electromagnetic radiation. It has a value of $3.00 \times 10^8\,m\,s^{-1}$ when the light is travelling in a vacuum.

Frequency and wavelength are related. If you multiply wavelength and frequency together you get a constant – the speed of light.

$$\text{speed of light } c\ (m\,s^{-1}) = \text{wavelength } \lambda\ (m) \times \text{frequency } \nu\ (s^{-1})$$

Synoptic link

In Chapter 10, Colour by design, you will learn more about how and why molecules absorb ultraviolet or visible radiation.

Study tip

You should use the term *high energy* ultraviolet radiation, rather than harmful UV radiation.

The behaviour of light can also be explained in some situations by thinking of it not as waves but as particles called photons.

Bringing the wave and particle models of light together

The two theories of light – the wave and photon models – are linked by a relationship:

the energy of a photon E (J) = Planck constant h (Js) × frequency v (s^{-1})

The energy of a photon is equal to the frequency of the light multiplied by the Planck constant. This is has a value of 6.63×10^{-34} Js.

Worked example: Calculations using the particle theory

Calculate the frequency associated with a photon of red light with an energy of 3.00×10^{-19} J.

Step 1: Rearrange the equation the energy of a photon
E (J) = Planck constant h (Js) × frequency v (s^{-1})
to make frequency the subject.

$$v = \frac{E}{h}$$

Step 2: Calculate the frequency of the photon.

$$\frac{3.00 \times 10^{-19}}{6.63 \times 10^{-34}} = 4.52 \times 10^{14}\,\text{s}^{-1}$$

Chemical ideas: Radiation and matter 6.2

What happens when radiation interacts with matter?

Energy interacts with matter

Electromagnetic radiation can interact with matter, transferring energy to the chemicals involved. This can cause changes in the chemicals, depending on the chemical and the amount of energy involved.

A molecule has energy associated with several different aspects of its behaviour, including:

- translation (the molecule moving around as a whole)
- rotation (of the molecule as a whole)
- vibration of the bonds
- electron energy.

These different kinds of energetic activities involve different amounts of energy as shown in Figure 6.

Electrons can occupy definite energy levels. The electronic energy of an atom or molecule changes when an electron moves from one level to another. Electronic energy is **quantised**; it has fixed levels. But

Synoptic link

You first encountered the wave theory of light and defined wavelength, frequency, and speed in Topic EL 2, How do we know so much about outer space?

Activity OZ2.2

You can check your understanding of the electromagnetic spectrum in this activity.

Study tip

If you are asked to calculate an energy value per mole of photon, remember to multiply the value per photon by the Avagadro constant N_A.

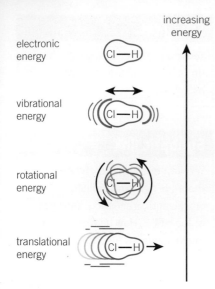

electronic
energy

vibrational
energy

rotational
energy

translational
energy

increasing
energy

▲ **Figure 6** *An HCl molecule has energy associated with different aspects of its behaviour*

here is a crucial point – *all* the other types of energy (translational, rotational, and vibrational) are quantised too.

Different energy changes for different parts of the spectrum

The spacing between vibrational energy levels corresponds to the infrared part of the spectrum. You sense **infrared radiation** as heat. The radiation makes bonds in the chemicals in your skin vibrate more energetically. The molecules have more kinetic energy and this is why you feel warmer.

Making molecules rotate requires less energy than making their bonds vibrate. Therefore, changes in rotational energy correspond to a lower energy, a lower frequency part of the electromagnetic spectrum, namely the microwave region. The spacing between translational energy levels is even smaller.

However, making electronic changes occur in a molecule requires *higher* energy than for vibrational changes. Exciting electrons to higher electronic energy levels requires energy corresponding to the **visible** and ultraviolet parts of the spectrum. See Table 1.

▼ **Table 1** *Summary of molecular energy changes*

Change occurring	Size of energy change / J	Type of radiation absorbed
change of rotational energy level	1×10^{-22} to 1×10^{-20}	microwave
change of vibrational energy level	1×10^{-20} to 1×10^{-19}	infrared
change of electronic energy level	1×10^{-19} to 1×10^{-16}	visible and ultraviolet

Activity OZ2.3 and OZ2.4

You can check your understanding of how ozone and other substances act as a sunscreen in these activities.

The table gives *ranges* of energy. The particular value of the energy change depends on the substance involved. For example, the C—F bond is stronger than the C—Br bond, so it takes infrared of a higher energy to make a C—F bond vibrate than to make a C—Br bond vibrate. The electromagnetic radiation reaching the Earth's atmosphere is mainly in the visible and ultraviolet part of the electromagnetic spectrum. Radiation emitted by the Earth's surface is mainly in the infrared region.

What kind of electronic changes occur when molecules absorb ultraviolet radiation?

The electrons in a molecule, such as chlorine, Cl_2, occupy definite energy levels. The outer shell electrons are in the highest energy levels so can move most easily to higher levels.

When a chlorine molecule absorbs radiation, one of three things can happen depending on the amount of energy involved:

● Electrons may be excited to a higher energy level. Chlorine owes its colour to this process – Cl_2 absorbs visible light of such a frequency that the remaining, unabsorbed light looks green.

● If higher energy radiation is used the molecule may absorb so much energy that the bonding electrons can no longer bond the atoms together. This is **photodissociation** and **radicals** are formed. Radicals are molecules or atoms with at least one unpaired electron. They are usually very reactive. Their formation may lead to further chemical reactions. You will find out more about photodissociation and radicals later in this module.

• With very high energy photons, the molecules may acquire so much energy that an electron is able to leave it – the molecule is *ionised*.

Figure 8 illustrates these three possibilities:

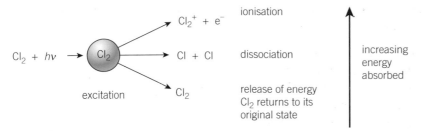

▲ **Figure 8** *When chlorine molecules absorb radiation ($h\nu$), they become excited – the excited molecules may then ionise, dissociate, or just release the energy*

▲ **Figure 7** *Chlorine gas*

Summary questions

1 A beam of infrared radiation has an energy of 3.65×10^{-20} J per photon.
  Calculate the frequency of the radiation. *(1 mark)*

2 A beam of X-rays has a frequency of 2.60×10^{17} s^{-1}.
  Calculate the wavelength of the X-rays. *(1 mark)*

3 Carbon dioxide is a greenhouse gas. It absorbs some of the infrared radiation given off from the Earth's surface.
  a Explain why carbon dioxide molecules absorb only certain frequencies of infrared radiation. *(2 marks)*
  b Explain why absorbing infrared radiation makes the atmosphere warmer. *(2 marks)*

4 To change 1 mole of molecular HCl from the lowest vibrational energy level (ground state) to the next vibrational level requires 32.7 kJ.
  a Calculate the energy, in joules, gained by *one molecule* of HCl when energy is absorbed in this way. [Avogadro constant, $N_A = 6.02 \times 10^{23}$ mol^{-1}] *(2 marks)*
  b Calculate the corresponding frequency of radiation. State the type of radiation this corresponds to. *(2 marks)*
  c Calculate the wavelength of this radiation in metres. *(1 mark)*

5 You want to heat up a cup of coffee in a microwave cooker. The cooker uses radiation of frequency 2.45×10^9 Hz. The cup contains 150 cm^3 of coffee, which is mainly water.
  a Calculate the energy needed to raise the temperature of the water by 30 °C. [Specific heat capacity of water = 4.18 J g^{-1} K^{-1}] *(2 marks)*
  b Calculate the energy transferred to the water by each photon of microwave radiation. *(2 marks)*
  c Calculate the energy transferred by one mole of photons. *(1 mark)*
  d Calculate how many moles of photons are needed to supply the energy calculated in part **a**. *(1 mark)*

If all the ozone in the atmosphere were collected and brought to the Earth's surface at atmospheric pressure, it would form a layer only 3 mm thick.

It isn't really surprising that there is so little ozone in the atmosphere – it reacts so quickly with other substances and gets destroyed. In fact, you might ask why the ozone in the atmosphere hasn't run out. Some reactions must be producing it too.

Many of the ideas in this part of the storyline are concerned with the formation and reactions of radicals.

Chemical ideas: Radiation and matter 6.3

Radiation and radicals

Ways of breaking bonds

In a covalent bond, a pair of electrons is shared between two atoms. For example, in the HCl molecule.

$$H \text{---} \overset{..}{\text{.}} \text{---} Cl$$

During a reaction bonds break and get remade. Bond-breaking is sometimes called **bond fission** and it involves the redistribution of the electrons in the covalent bond. There are two ways this can happen – by **heterolytic fission** or by **homolytic fission**.

Heterolytic fission

In this type of fission, both of the shared electrons go to just one of the atoms when the bond breaks. This atom becomes negatively charged because it has one more electron than it has protons. The other atom becomes positively charged.

In the case of HCl:

$$H \text{---} \overset{..}{\text{.}} \text{---} Cl \rightarrow H^+ + \text{:}Cl^-$$

Heterolytic fission is common when a bond is already **polar**. You will find out more about polar bonds in Topic OZ 5.

Homolytic fission

In this type of bond fission, one of the two shared electrons goes to each atom, as in the case of Br_2.

$$Br \text{---} \overset{..}{\text{.}} \text{---} Br \rightarrow Br^\bullet + Br^\bullet$$

The dot beside each atom shows the *unpaired electron* that the atom has inherited from the shared pair in the bond. The atoms have no overall charge because they have the electronic structure they had before they shared their electrons to form the bond. Sometimes the dot is omitted and the radical simply represented as, for example, Br.

The unpaired electron has a strong tendency to pair up again with another electron from another substance. Radicals are most commonly formed when the bond being broken is non-polar, but many polar bonds can break this way too – particularly when the reaction is taking place in the gas phase and in the presence of light.

The amount of energy it takes depends on the bond enthalpy of the bond being broken. For example, a pattern can be seen in the bond enthalpies of carbon–halogen bonds, with C—F strongest through to C—I being weakest. Photodissociation of a C—F bond requires higher energy (shorter wavelength) light than photodissociation of a C—I bond.

Radical reactions involving ozone

Ozone, O_3, is formed when an oxygen atom (an example of a radical) reacts with a dioxygen molecule.

$$O + O_2 \rightarrow O_3$$

One way to make oxygen atoms is by dissociating dioxygen molecules. This requires quite a lot of energy ($+498\,kJ\,mol^{-1}$), which can be provided by ultraviolet radiation or by an electric discharge.

You can often smell the sharp odour of ozone near electric motors, photocopiers, or ultraviolet lamps used to kill bacteria in food shops. The electric discharges make some of the dioxygen molecules in the air dissociate into atoms, which then react to make ozone.

Some of the ozone in the troposphere is formed in reactions taking place in photochemical smogs. In this case, oxygen atoms are produced by the action of sunlight on the pollutant gas nitrogen dioxide.

In the stratosphere, oxygen atoms are formed by the photodissociation of dioxygen molecules when ultraviolet radiation of the right frequency (indicated by hv) is absorbed.

The reaction can be summarised as:

$$O_2 + hv \rightarrow O + O \qquad \textbf{Reaction 1}$$

The resulting oxygen atoms may collide and react with an O_2 molecule, another O atom or an O_3 molecule in reactions 2–4:

$$O + O_2 \rightarrow O_3 \qquad \textbf{Reaction 2}$$
$$O + O \rightarrow O_2 \qquad \textbf{Reaction 3}$$
$$O + O_3 \rightarrow O_2 + O_2 \qquad \textbf{Reaction 4}$$

Reaction 2 is the one that produces ozone.

When the ozone absorbs radiation in the 10.1×10^{14} to $14.0 \times 10^{14}\,Hz$ region, some molecules undergo photodissociation and split up again:

$$O_3 + hv \rightarrow O_2 + O \qquad \textbf{Reaction 5}$$

This reaction is responsible for the vital screening effect of ozone, since it absorbs the radiation responsible for sunburn.

▲ **Figure 1** *Helium-filled balloons are sent up into the stratosphere to measure ozone concentrations*

Synoptic link

You first came across photochemical smogs in Chapter 2, Developing fuels.

Other radical reactions

Radical polymerisation uses initiators to produce radicals. The initiator radical attaches to the monomer, forming a new larger radical, which then attacks another monomer molecule. Consequently the polymer chain grows. Polymerisation stops when a termination step occurs, or when all of the monomer is used up.

Combustion is a series of radical chain reactions. Flammable materials require a lower concentration of radicals before initiation and propagation reactions cause combustion of the material. Termination reactions cause the flame to extinguish. Lead additives to petrol reduced propagation and promoted termination, thereby reducing auto-ignition.

▲ **Figure 2** *Combustion*

Radicals have been implicated in *age-related biological effects*. Radicals interact with biological molecules, removing an electron. This often causes the molecule to become a radical, resulting in changes to the biological molecules:

- Chains of DNA can become cross-linked, which may lead to cancer.
- Fats and proteins can become cross-linked, leading to wrinkles in the skin.
- Low-density lipoproteins can become oxidised, leading to arterial plaques which can cause heart disease or strokes.

Antioxidants such as vitamin C can reduce this damage because they give electrons to radicals but remain stable.

1 Suggest why radical polymerisation tends to produce highly-branched irregular polymer chains.
2 Suggest how initiation, propagation, and termination reactions may explain the process of combustion.

Radicals are reactive

Filled outer electron shells are more stable than unfilled ones. Radicals are reactive because they tend to try to fill their outer shells by grabbing an electron from another atom or molecule.

When $Cl^\bullet$ collides with H_2, the chlorine grabs an electron from the pair of electrons in the bond between the hydrogen atoms. This makes a new bond between the chlorine and hydrogen atoms.

▲ **Figure 3** *The reaction mechanism of a chlorine radical with hydrogen*

The **curly arrow** indicates the movement of an electron. The 'tail' of the arrow shows where the electron starts and the 'head' shows where it finishes. Look carefully at the head of the arrow – it is drawn this way to show the movement of a *single* electron. Full-headed curly arrows indicate movement of a *pair* of electrons. Half-headed curly arrows indicate movement of a single electron (Figure 4).

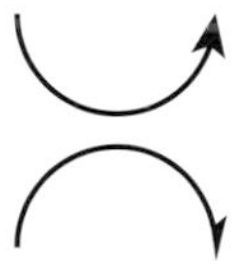

A hydrogen radical is formed in this reaction. This is also highly reactive and it will combine with another molecule, once again creating a new radical – and so it goes on. It is a **radical chain reaction** and it has three key stages:

Initiation   Chlorine radicals are initially formed by the photodissociation of chlorine molecules:

$$Cl_2 + hv \rightarrow Cl^\bullet + Cl^\bullet$$

$Cl^\bullet$ are so reactive that they soon react with something else – they *initiate* the reaction.

Propagation   Chlorine radicals react with hydrogen molecules. The hydrogen radicals go on to react with chlorine molecules:

$$Cl^\bullet + H_2 \rightarrow HCl + H^\bullet$$
$$H^\bullet + Cl_2 \rightarrow HCl + Cl^\bullet$$

These reactions produce new radicals which keep the reaction going – they *propagate* the reaction.

Termination   Occasionally two radicals collide with each other. When this happens the reaction is *terminated* because the radicals have been removed:

$$H^\bullet + H^\bullet \rightarrow H_2$$
$$Cl^\bullet + Cl^\bullet \rightarrow Cl_2$$
$$H^\bullet + Cl^\bullet \rightarrow HCl$$

The overall effect is to convert hydrogen and chlorine into hydrogen chloride:

$$H_2 + Cl_2 \rightarrow 2HCl$$

▲ **Figure 4** *Full-headed (top) and half-headed (bottom) curly arrows*

Synoptic link

You were first introduced to the use of curly arrows to represent the movement of electrons in Chapter 2, Developing fuels.

Activity OZ 3.1

This activity helps you check your understanding of radical mechanisms.

Methane and chlorine

Alkanes are generally considered to be unreactive, although they will react with halogens in the presence of light. This is another radical chain reaction, and has the same three key stages.

Initiation Chlorine radicals are formed from chlorine molecules:

$$Cl_2 + hv \rightarrow Cl^\bullet + Cl^\bullet$$

Propagation Chlorine radicals react with methane molecules. The resulting methyl radicals can then react with chlorine molecules:

$$Cl^\bullet + CH_4 \rightarrow HCl + CH_3^\bullet$$
$$CH_3^\bullet + Cl_2 \rightarrow CH_3Cl + Cl^\bullet$$

These two steps *propagate* the reaction.

Termination The reaction ends when two free radicals combine:

$$Cl^\bullet + Cl^\bullet \rightarrow Cl_2$$
$$CH_3^\bullet + Cl^\bullet \rightarrow CH_3Cl$$
$$CH_3^\bullet + CH_3^\bullet \rightarrow C_2H_6$$

The effect is to produce hydrogen chloride, chloromethane, and small amounts of ethane. Further substitution may occur to form dichloromethane and trichloromethane.

Study tip

You should be able to identify whether radical reactions are initiation, propagation, or termination.

Activity OZ 3.2 and OZ 3.3

These activities involve worked with radical reactions in the laboratory.

Summary questions

1   State whether or not each of the following species is a radical. You may need to draw electron dot–cross diagrams to help you to decide.
    a F   (*1 mark*)   c H_2O (*1 mark*)   e NO_2 (*1 mark*)
    b Ar   (*1 mark*)   d OH   (*1 mark*)   f CH_3 (*1 mark*)

2   The hydroxyl radical, $HO^\bullet$, is an important species in atmospheric chemistry. Reaction **A** shows one process in which $HO^\bullet$ is produced. The reaction is brought about by radiation with a wavelength below 190 nm.

    Reaction A $H_2O + hv \rightarrow H^\bullet + HO^\bullet$

    Hydroxyl radicals are very reactive and act as scavengers in the atmosphere. One set of reactions which involve stratospheric ozone is:

    Reaction B $HO^\bullet + O_3 \rightarrow HO_2^\bullet + O_2$
    Reaction C $HO_2^\bullet + O_3 \rightarrow HO^\bullet + 2O_2$

    a State whether reaction A is a homolytic or a heterolytic process.   (*1 mark*)
    b Explain whether reactions **A**, **B**, and **C** are initiation, propagation, or termination.   (*1 mark*)
    c i Write an equation which shows the overall result of reactions **B** and **C**.
       ii State the role of $HO^\bullet$ in this process.   (*2 marks*)

3 The creation of nitrogen monoxide from human activities is of concern because it is thought to lead to a loss of ozone from the stratosphere. Reactions D and E show how this loss can occur.

Reaction D $NO^\bullet + O_3 \longrightarrow NO_2^\bullet + O_2$ $\Delta H = -100\,kJ\,mol^{-1}$

Reaction E $NO_2^\bullet + O \longrightarrow NO^\bullet + O_2$ $\Delta H = -192\,kJ\,mol^{-1}$

  a State one human activity which leads to the production of a significant amount of NO. *(1 mark)*

  b i Deduce the the overall effect of reactions D and E.

    ii State the role of NO in this process.

    iii Calculate the value of ΔH for the overall process. *(3 marks)*

4 Reactions F to K show various processes involving ethane.

Reaction F $C_2H_6 \longrightarrow 2CH_3^\bullet$

Reaction G $CH_3^\bullet + C_2H_6 \longrightarrow CH_4 + C_2H_5^\bullet$

Reaction H $C_2H_5^\bullet \longrightarrow C_2H_4 + H^\bullet$

Reaction I $H^\bullet + C_2H_6 \longrightarrow H_2 + C_2H_5^\bullet$

Reaction J $2C_2H_5^\bullet \longrightarrow C_2H_4 + C_2H_6$

Reaction K $2C_2H_5^\bullet \longrightarrow C_4H_{10}$

        (C_2H_4 is ethene, $CH_2{=}CH_2$)

  a State whether these reactions are initiation, propagation, or termination. *(3 marks)*

  b For most of these reactions, you would need information about bond enthalpies in order to decide whether the process is exothermic or endothermic. But two of the reactions can be classified by inspection.

    i Explain which reaction is endothermic and which reaction is exothermic.

    ii Explain your answer to i. *(4 marks)*

  c State the names and formulae of the chemical species in the reaction sequence which are radicals. *(3 marks)*

5 The radical chain reaction of methane (CH_4) with chlorine in the presence of sunlight is given below.

$$CH_4 + Cl_2 \longrightarrow CH_3Cl + HCl$$

  a Suggest a mechanism for the reaction showing clearly which reactions correspond to the initiation, propagation, and termination stages. *(5 marks)*

  b Explain why the reaction product also contains some CH_2Cl_2, $CHCl_3$, and CCl_4. *(3 marks)*

OZ 4 Ozone – here today and gone tomorrow

Specification reference: OZ(e), OZ(f)

Learning outcomes

Demonstrate and apply knowledge and understanding of:

→ activation enthalpy and enthalpy profiles

→ the effect of concentration and pressure on the rate of a reaction, explained in terms of the collision theory

→ use of the concept of activation enthalpy and the Boltzmann distribution to explain the qualitative effect of temperature changes and catalysts on rate of reaction

→ techniques and procedures for experiments in reaction kinetics.

Reactions 1–5 in Topic OZ 3 show that ozone is being made and destroyed all the time. Left to themselves, these reactions would reach a point where ozone was being made as fast as it was being used up. This is called a steady state.

It's like the situation in Figure 1 when you are running water into a basin with the plug out of the waste pipe. When water is running out as fast as it's running in, the level of water stays constant. If you turned the tap on more, the water would rise, but it would run out faster because of the higher pressure. Before long you would reach a steady state again, but this time with more water in the basin.

The water coming out of the tap is like the reactions producing ozone, and the water going down the waste pipe is like the reactions that destroy it.

Chemists have studied the rates of the reactions that produce and destroy ozone and can write mathematical expressions for all the reactions involved. From these expressions, they can work out what the concentration of ozone should be in different circumstances. In the Ideas section below you will find out some of the ways that chemists calculate the rates of reactions.

▲ **Figure 1** *A steady state situation*

▲ **Figure 2** *Decomposition of hydrogen peroxide*

Chemical ideas: Rates of reactions 10.1

Factors affecting reaction rates

Some chemical reactions, such as burning fuel in a car engine or precipitating silver chloride from solution, go very fast. Others, such as the souring of milk or the rusting of iron, are much slower. The study of the **rate of reaction**, or **reaction kinetics**, helps chemists to find ways of controlling and predicting chemical reactions, and to understand the mechanisms of chemical reactions. When you talk about the rate of something, you mean the rate at which some quantity changes. When you talk about the rate of reaction, you mean the rate at which reactants are converted into products.

In the decomposition of hydrogen peroxide, H_2O_2, the rate of the reaction means the rate at which $H_2O(l)$ and $O_2(g)$ are formed, which is the same as the rate at which $H_2O_2(aq)$ is used up.

$$2H_2O_2 (aq) \rightarrow H_2O(l) + O_2 (g)$$

You could measure the rate of this reaction in moles of water or oxygen formed per second, or moles of hydrogen peroxide used up per second.

Measuring rate of reaction

To determine the rate of a reaction you need to measure a property that changes during the reaction and that is proportional to the concentration of a particular reactant or product. The rate can then be calculated using:

$$\text{rate of reaction} = \frac{\text{change in property}}{\text{time taken}}$$

Measuring volumes of gases evolved

The reaction between calcium carbonate and hydrochloric acid produces carbon dioxide which can be collected. The volume produced can be used to follow the reaction rate. This method can be used for a range of experiments where a gas is evolved.

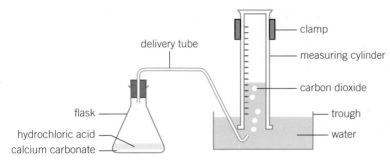

▲ **Figure 3** *Measuring the rate of a gas produced*

Measuring mass changes

A different way of monitoring the same reaction is by recording the mass lost (in the form of carbon dioxide) from the reaction.

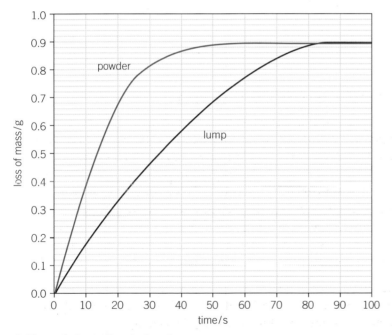

▲ **Figure 4** *Recording the loss in mass*

Synoptic link

You first encountered pH in Chapter 1, Elements of life.

Synoptic link

You will learn about rate equations in Chapter 9, Developing metals.

pH measurement

As the above reaction proceeds, the hydrochloric acid concentration will fall, as it reacts with the calcium carbonate. Measuring the pH of the reaction mixture is another way of following the reaction. Again, this method is suitable for a range of reactions.

▲ **Figure 5** *When zinc reacts with copper sulfate solution the blue colour of CuSO₄ fades*

Synoptic link 🧪

How to use a colorimeter can be found in Techniques and procedures.

Colorimetry

A **colorimeter** measures the change in colour of a reaction. When zinc reacts with aqueous copper(II) sulfate (Figure 5) the blue coloration of the copper sulfate solution decreases and the reaction can be followed using a colorimeter.

$$Zn(s) + CuSO_4(aq) \rightarrow ZnSO_4(aq) + Cu(s)$$

Chemical analysis

All the techniques described so far have not interfered with the progress of the reaction. Chemical analysis, however, involves taking samples of the reaction mixture at regular intervals, and stopping the reaction in the sample (quenching it), before analysis.

For example, iodine and propanone react in the presence of an acid catalyst. A sample can be extracted from the reaction mixture and quenched by the addition of sodium hydrogen carbonate. This neutralises the acid catalyst, effectively stopping the reaction. The amount of unreacted iodine remaining can then be determined by titration.

What affects the rate of a reaction?

The rate of a chemical reaction may be affected by:

- The *concentration* of the reactants. In solutions, concentration is measured in $mol\,dm^{-3}$ – in gases, the concentration is proportional to the *pressure*.

- The *temperature*. Nearly all reactions go faster at higher temperatures.

- The *intensity of radiation*, if the reaction involves radiation. Dissociation of oxygen molecules happens faster when the intensity of the radiation increases.

- The *particle size* of a solid. A powder reacts faster than a lump of solid because there is a much larger *surface area* of solid exposed for reaction to take place on.

- The presence of a *catalyst*.

The collision theory of reactions

You can explain the effect of these factors using the **collision theory**. The idea is that reactions occur when particles of reactants collide with a certain minimum kinetic energy. An ozone molecule and a chlorine atom in the stratosphere must collide in order to react. This will happen more often if there are more particles in a given volume – so increasing concentration speeds up the reaction.

But not every collision causes a reaction. As the particles approach and collide. When a plot is drawn of reaction progress against enthalpy, this is known as an **enthalpy profile** (Figure 6).

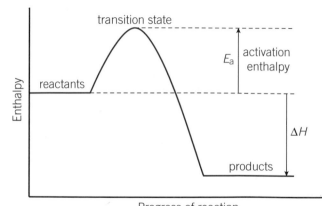

▲ **Figure 6** *Enthalpy profile for an exothermic reaction*

Existing bonds start to stretch and break and new bonds start to form. Only those pairs with enough combined energy on collision to overcome the energy barrier, or **activation enthalpy**, for the reaction will go on to produce products. At higher temperatures, a much larger proportion of colliding pairs have enough energy to react.

Enthalpy profiles

Plots like the one in Figure 6 are a useful way of picturing the energy changes that take place as a reaction proceeds. In going from reactants to products, the highest point on the pathway corresponds to the **transition state**, where old bonds stretch and new bonds start to form. This state exists for only a very short time. The curve in Figure 6 applies to a simple one-step reaction. Many reactions actually take place in a series of steps and there will be many curves – one for each step.

Catalysts provide an alternative reaction pathway of lower activation energy, thereby increasing the rate.

Synoptic link

You first encountered enthalpy in Chapter 2, Developing fuels.

Activation enthalpy

Activation enthalpy is the minimum kinetic energy required by a pair of colliding particles before reaction will occur.

Synoptic link

You were introduced to catalysis in Chapter 2, Developing fuels.

Chemical ideas: Rates of reactions 10.2

The effect of temperature on rate

Temperature has an important effect on the rate of chemical reactions. If you've measured the rates of reactions at different temperatures, you'll know that for many reactions the rate is roughly doubled by a temperature rise of just 10 °C.

The distribution of energies

At any temperature, the speeds of the molecules in a substance are distributed over a wide range. Just as the walking speeds of people in a street, some are moving slowly and some quickly, but the majority are moving at moderate speeds.

This distribution of kinetic energies in a gas at a given temperature is shown in Figure 7. The pattern is called the **Maxwell–Boltzmann distribution**.

Activity OZ 4.1

This activity you will have the opportunity to measure the rate of reactions at different temperatures.

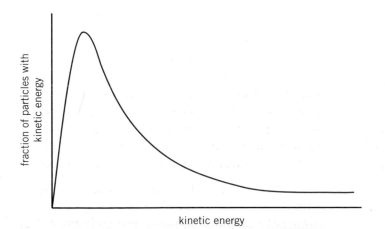

▲ **Figure 7** *Distribution curve for molecular kinetic energies in a gas*

As the temperature increases, more molecules move at higher speeds and have higher kinetic energies. Figure 8 shows how the distribution of energies changes when you increase the temperature from 300 to 310 K. There is still a spread of energies, but now a greater proportion of molecules have higher energies. The area under the curve is the same, but the distribution of energies is different.

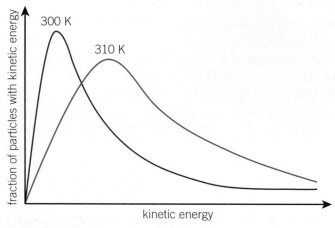

▲ **Figure 8** *Distribution curves for molecular kinetic energies in a gas at 300 K and 310 K*

Now let's look at the significance of this for reaction rates. Imagine a reaction whose activation enthalpy E_a is + 50 kJ mol^{-1}, a value typical of many reactions.

Figure 9 shows the number of collisions with energy greater than 50 kJ mol^{-1} for the reaction at 300 K – it's given by the shaded area underneath the curve. Only those collisions with energies in the shaded area can lead to a reaction.

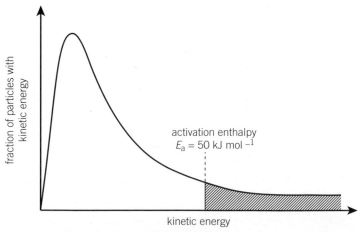

▲ **Figure 9** *Distribution curve showing collisions with energy 50 kJ mol^{-1} and above*

Now look at the graph in Figure 10, which shows the curves for both 300 K and 310 K. At the higher temperature a significantly higher proportion of molecules have energies above 50 kJ mol^{-1}.

Reactions go faster at higher temperatures because a larger proportion of the colliding molecules have the minimum activation enthalpy needed to react.

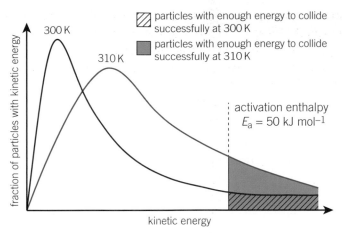

▲ **Figure 10** *Distribution curves showing the effect of changing the temperature, from 300 K to 310 K, on the proportion of collisions with energy 50 kJ mol^{-1} and above*

For many reactions, the rate is roughly doubled by a 10 °C temperature rise. It is only a rough rule, and in fact it only works exactly for reactions with an activation enthalpy of +50 kJ mol^{-1} and for a temperature rise from 300 to 310 K, but it is a reasonable rough guide. The greater the activation enthalpy, the greater is the effect of increasing the temperature on the rate of the reaction.

Catalysts provide an alternative pathway for a reaction with a lower activation energy. This means that at any given temperature, there will be a larger proportion of particles that on collision produce a successful collision (Figure 11).

Synoptic link

You will find out more about the role of catalysts in Chapter 6, The chemical industry.

Synoptic link

You first came across catalysts in Chapter 2, Developing fuels

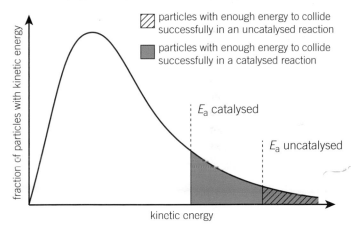
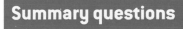

▲ **Figure 11** *Distribution curve showing the effect of a catalyst on the proportion of particles with energy for successful collisions*

Summary questions

1 Use the collision theory to explain:
   a why coal burns faster when it is finely powdered than when it is in a lump *(1 mark)*
   b why nitrogen and oxygen in the atmosphere do not normally react to form nitrogen oxides *(1 mark)*
   c why reactions between two solids take place very slowly *(1 mark)*
   d why flour dust in the air can ignite with explosive violence. *(1 mark)*

2 The collision theory assumes that the rate of a reaction depends on:
  A the rate at which reactant molecules collide with one another
  B the proportion of reactant molecules that have enough energy to react once they have collided.

Which, out of **A** and **B**, explains each of the following observations?

  a Reactions in solution go faster at higher concentration.   (*1 mark*)
  b Solids react faster with liquids or gases when their surface area is greater.   (*1 mark*)
  c Catalysts increase the rate of reactions.   (*1 mark*)
  d Increasing the temperature increases the rate of a reaction.   (*1 mark*)

3 For each of the following reactions, state which of the following factors might affect the rate.
  A temperature
  B total pressure of gas
  C concentration of solution
  D surface area of solid.

  a The reaction of magnesium with hydrochloric acid

$$Mg(s) + 2HCl(aq) \rightarrow MgCl_2(aq) + H_2(g)$$ (*1 mark*)

  b The reaction of nitrogen with hydrogen in the presence of an iron catalyst

$$N_2(g) + 3H_2(g) \rightarrow 2NH_3(g)$$ (*1 mark*)

  c The decomposition of aqueous hydrogen peroxide

$$2H_2O_2(aq) \rightarrow 2H_2O(l) + O_2(g)$$ (*1 mark*)

4 A mixture of hydrogen and oxygen doesn't react until it is ignited by a spark – then it explodes. The mixture also explodes if you add some powdered platinum.
  a The energy of a spark is tiny, yet it is enough to ignite any quantity of a hydrogen/oxygen mixture, large or small. Suggest an explanation for this.   (*2 marks*)
  b Explain why platinum makes the hydrogen/oxygen reaction occur at room temperature.   (*2 marks*)

5 The activation enthalpy for the decomposition of hydrogen peroxide to oxygen and water is $+36.4\,kJ\,mol^{-1}$ in the presence of an enzyme catalyst and $+49.0\,kJ\,mol^{-1}$ in the presence of a very fine colloidal suspension of platinum. The overall reaction is exothermic.
  a Sketch, on the same enthalpy diagram, the enthalpy profiles for both catalysts.   (*2 marks*)
  b How will the rate of the decomposition of hydrogen peroxide differ for the two catalysts at room temperature? Explain your answer.   (*2 marks*)

When chemists compare their calculated concentrations of ozone from the expected reaction rates with the actual measured values, they find that the measured values are a lot lower than expected – ozone is being removed faster than expected. Going back to the analogy of the basin and the running tap from Topic OZ 4, it's as if the waste pipe had been made larger. But by what?

You have seen that ozone is very reactive and reacts with oxygen atoms. But oxygen atoms aren't the only radicals to be found in the stratosphere.

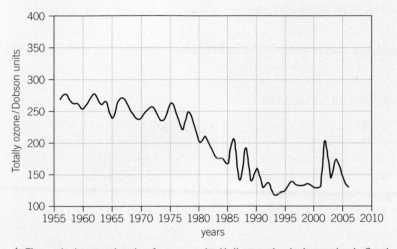

▲ **Figure 1** *Average levels of ozone at the Halley station in Antarctica in October between 1955 and 2006*

Chemical ideas: Rates of reaction 10.4

How do catalysts work?

In a chemical reaction, existing bonds in the reactants must first stretch and break. Then new bonds can form as the reactants are converted to products.

Bond breaking is an endothermic process. A pair of reacting particles must collide with an energy greater than the activation energy before any reaction can occur.

Catalysts speed up reactions by providing an alternative reaction pathway with a lower activation enthalpy for the breaking and remaking of bonds. Now that the energy barrier is lower, more pairs of molecules can react when they collide, so the reaction proceeds more quickly.

The enthalpy change ΔH is the same for the catalysed and uncatalysed reactions. Figure 2 shows the enthalpy profiles for an uncatalysed and a catalysed reaction. Figure 3 shows the Maxwell–Boltzmann distribution for an uncatalysed and a catalysed reaction.

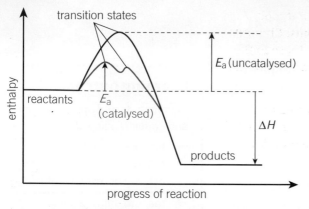

▲ **Figure 2** *The effect of a catalyst on the enthalpy profile for a reaction*

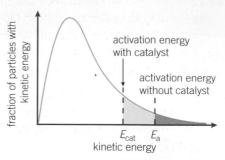

▲ **Figure 3** *A catalyst provides an alternative route with a lower activation energy, allowing a greater proportion of molecules to react*

Catalysts and equilibrium

Catalysts do not affect the position of equilibrium in a reversible reaction. They alter the *rate* at which the equilibrium is attained, but not the *composition* of the equilibrium mixture.

Homogeneous catalysts

Homogeneous catalysts are in the same physical state as the reactants in the reaction. They normally work by forming an intermediate compound with the reactants (sometimes called a transition state). That is why the enthalpy profile for the catalysed reaction in Figure 2 has two humps – one for each step. The intermediate compound then breaks down to give the product and reform the catalyst.

An example of homogeneous catalysis is the destruction of ozone in the stratosphere by chlorine and bromine atoms.

How do chlorine and bromine atoms delete ozone?

Small amounts of chloromethane, CH_3Cl, and bromomethane, CH_3Br, reach the stratosphere as a result of natural processes. Once in the stratosphere, their molecules are split up by solar radiation to give chlorine atoms and bromine atoms. CH_3Cl and CH_3Br are examples of compounds called haloalkanes. In the following reactions the unpaired electron of the radicals is not shown.

Chlorine atoms react with ozone like this:

$$Cl + O_3 \rightarrow ClO + O_2 \quad \textbf{Reaction 1}$$

and bromine atoms react in the same way.

The ClO formed is another reactive radical and can react with oxygen atoms:

$$ClO + O \rightarrow Cl + O_2 \quad \textbf{Reaction 2}$$

So now you have two radicals (Topic OZ 3) competing with each other to remove ozone from the stratosphere:

$$O + O_3 \rightarrow O_2 + O_2 \quad \textbf{Reaction 3}$$
$$Cl + O_3 \rightarrow ClO + O_2 \quad \textbf{Reaction 1}$$

The concentration of Cl atoms in the stratosphere is much less than the concentration of O atoms. So how significant is Reaction 1?

The reaction of O_3 with Cl atoms would not matter much if it took place a lot more slowly than the reaction of O_3 with O atoms. Chemists have shown that, at temperatures and pressures similar to those in the stratosphere, the reaction of O_3 with Cl atoms is more than 1500 times faster than the reaction of O_3 with O atoms.

What's more, the Cl atoms used in Reaction 1 are regenerated in Reaction 2 – and can then go on to react with more O_3. Adding together the equations for Reactions 1 and 2 gives the equation for the overall reaction.

$$Cl + O_3 \rightarrow ClO + O_2 \quad \textbf{Reaction 1}$$
and
$$ClO + O \rightarrow Cl + O_2 \quad \textbf{Reaction 2}$$
overall reaction:
$$O + O_3 \rightarrow O_2 + O_2$$

The chlorine atoms act as a homogeneous catalyst for this reaction. By going through the catalytic cycle many times, a single chlorine atom can remove about one million ozone molecules. So you can see why even low concentrations of chlorine atoms can be devastating.

A similar catalytic cycle involves bromine atoms and, although the concentration of bromine atoms is much lower, bromine is about 100 times more effective in destroying ozone than chlorine.

Other ways ozone is removed

Chlorine and bromine atoms aren't the only radicals present in the stratosphere which can destroy ozone in a catalytic cycle in this way.

If you represent the radical by the generic symbol X, you can rewrite Reactions 1 and 2 as a general catalytic cycle:

$$X + O_3 \rightarrow XO + O_2$$
and
$$XO + O \rightarrow X + O_2$$
overall reaction:
$$O + O_3 \rightarrow O_2 + O_2$$

Two other important radicals (the hydroxyl radical and nitrogen monoxide) which can destroy ozone in this way are described below.

Hydroxyl radicals, HO

These are formed by the reaction of oxygen atoms with water in the stratosphere. They react with ozone like this:

$$HO + O_3 \rightarrow HO_2 + O_2$$

The HO_2 radicals then go on to react with oxygen atoms to reform the HO radicals:

$$HO_2 + O \rightarrow HO + O_2$$

So this is another example of a catalytic cycle, and the HO radicals released can go on to react with more O_3 molecules.

Nitrogen monoxide, NO

Nitrogen monoxide reacts with ozone to form nitrogen dioxide, NO_2, and dioxygen. Nitrogen dioxide can then react with oxygen atoms to release nitrogen monoxide and dioxygen to complete the catalytic cycle.

NO and NO_2 are both radicals. They are unusual radicals because they are relatively stable molecules and they can be prepared and collected like ordinary molecular substances. (It is important to remember that not all radicals are highly reactive.)

The radicals mentioned in this section (Cl, Br, HO, and NO) are important, but they are only part of the whole picture. Hundreds of reactions have been suggested which affect the gases in the stratosphere.

Many of these have been going on since long before there were humans on Earth. However human activities can have a serious effect on certain key reactions, and so lead to dramatic changes in the concentration of ozone in the stratosphere.

Summary questions

1 The enthalpy profiles **A–D** represent four different reactions – all the diagrams are drawn to the same scale.

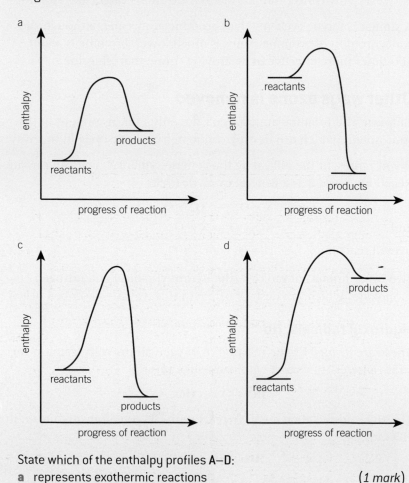

State which of the enthalpy profiles **A–D**:

a represents exothermic reactions   (*1 mark*)
b represents endothermic reactions   (*1 mark*)
c shows the largest activation enthalpy   (*1 mark*)
d shows the smallest activation enthalpy   (*1 mark*)
e represents the most exothermic reaction   (*1 mark*)
f represents the most endothermic reaction.   (*1 mark*)

2 The reaction of chlorine atoms with ozone in the upper atmosphere (stratosphere) is thought to be responsible for the destruction of the ozone layer:

$$Cl + O_3 \longrightarrow ClO + O_2$$

The chlorine atoms are produced by the action of high energy solar radiation on CFCs.

  a Use collision theory to explain how an increase in the concentration of chlorine atoms increases the rate of reaction with ozone.   *(2 marks)*

  b The activation enthalpy for the reaction of chlorine atoms with ozone is relatively small. Explain what effect you would expect a change of temperature to have on the rate of this reaction.   *(2 marks)*

3 The diagram below shows the energy distribution for collisions between Cl atoms and O_3 molecules at temperature T_1. E_a is the activation enthalpy for the reaction of Cl atoms and O_3 molecules.

$$Cl + O_3 \longrightarrow ClO + O_2$$

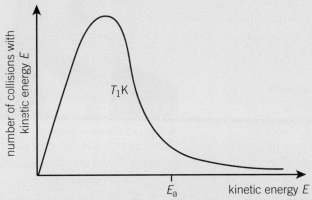

  a Copy the diagram and shade the area of the graph to indicate the number of collisions with sufficient energy to lead to a reaction.
   (1 mark)

  b On your graph, indicate the energy distribution for collisions between Cl atoms and O_3 molecules when the temperature of the reactants is increased by 10 K to T_2. Shade (using a different colour) the area under the graph to indicate the number of collisions with sufficient energy to lead to a reaction at T_2.   *(2 marks)*

4 The breakdown of ozone in the stratosphere is catalysed by chlorine atoms. The overall reaction is:

$$O_3 + O \longrightarrow 2O_2$$

  a The Cl atoms act as a homogeneous catalyst in this process. State the meaning of the term homogeneous.   *(1 mark)*

  b The enthalpy profiles for the catalysed and uncatalysed reactions are shown in Figure 1.   *(2 marks)*

    i Explain why the enthalpy profile for the catalysed reaction has two humps.

   ii Describe what is happening at the peaks and troughs on the two curves.

Learning outcomes

Demonstrate and apply knowledge and understanding of:

→ recognition of and formulae for examples of haloalkanes including systematic nomenclature

→ the term electronegativity

- qualitative electronegativity trends in the Periodic Table

- use of relative electronegativity values to predict bond polarity in a covalent bond

- relation of overall polarity of a molecule to its shape and the polarity of its individual bonds

→ intermolecular bonds: instantaneous dipole–induced dipole bonds (including dependence on branching and chain length of organic molecules and M_r), permanent dipole–permanent dipole bonds

→ the relative boiling points of substances in terms of intermolecular bonds

→ the characteristic properties of haloalkanes, comparing fluoro-, chloro-, bromo-, and iodo-compounds by considering boiling points and their dependence on intermolecular bonds.

Synoptic link

CFCs have similar structures to alkanes which you studied in Chapter 2, Developing fuels, with some hydrogen atoms substituted by halogen (Group 7) atoms.

CFCs: very handy compounds

In 1930 the American engineer Thomas Midgley, Jr demonstrated a new refrigerant to the American Chemical Society. He inhaled a lungful of dichlorodifluoromethane, CCl_2F_2, and used it to blow out a candle.

Midgley was flamboyantly demonstrating the lack of toxicity and lack of flammability of CCl_2F_2. Up to that time, ammonia had been the main refrigerant in use, but unfortunately it is toxic and very smelly.

Midgley had found what appeared to be the ideal replacement for ammonia. CCl_2F_2 belongs to a family of compounds called **chlorofluorocarbons** (CFCs) that contain chlorine, fluorine, and carbon. CFCs were used as:

- refrigerants and in air conditioning units

- aerosol propellants

- blowing agents for expanded plastics such as polystyrene

- dry cleaning solvents.

By the early 1970s, industry was producing about a million tonnes of CFCs a year.

▲ Figure 1 *Thomas Midgley (1889–1944) developed CFCs and pioneered the use of lead compounds as 'anti-knock' agents in petrol*

CFC-11   CFC-12

▲ Figure 2 *The structures of two common CFCs*

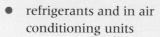
Chemical ideas: Organic chemistry: modifiers 13.2a

Haloalkanes

Properties and naming of haloalkanes

Organic halogen compounds have one or more halogen atoms (F, Cl, Br, or I) attached to a hydrocarbon backbone. Their occurrence in nature is limited, but they are very useful in chemical synthesis.

The halogen atom modifies the properties of the relatively unreactive hydrocarbon chain. The simplest examples are the **haloalkanes** (sometimes called halogenoalkanes), with the halogen atom attached to an alkane chain (Figure 3).

The haloalkanes are named after the parent alkanes, using the same basic rules as for naming alcohols – except that the halogen atom is added as a prefix to the name of the parent alkane.

Therefore, $CH_3CH_2CH_2Cl$ is called 1-chloropropane whilst CH3CHClCH2Cl is called 1,2-dichloropropane. CH3CHBrCH2CH2Cl is called 3-bromo-1-chlorobutane.

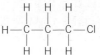
1-chloropropane

▲ **Figure 3** *1-chloropropane is a typical haloalkane*

▼ **Table 1** *Naming haloalkanes*

Full structural formula	Skeletal formula	Name
		1-chloropropane
		1,2-dichloropropane
		3-bromo-1-chlorobutane

The prefixes bromo- and chloro- are listed in alphabetical order. The numbers used to show the positions of the bromine and chlorine atoms are the lowest ones possible, for example, 3 and 1 rather than 2 and 4 in the case of 3-bromo-1-chlorobutane.

Physical properties of haloalkanes

The properties and reactions of haloalkanes are largely determined by the fact that carbon–halogen bonds are polar. Understanding the bond polarity helps explain the boiling points, intermolecular bonds, and reactions of haloalkanes.

Chemical ideas: Structure and properties 5.3

Bonds between molecules: temporary and permanent dipoles

Polar bonds

Figure 4 shows the way a hydrogen molecule is bonded.

The atoms are held together because both nuclei are attracted to the shared electrons between them. Both atoms are identical so the electrons are shared equally.

With larger atoms, it is the atomic cores (everything except the outer shell electrons) that are attracted to the shared electrons. When the two atoms bonded together are different sizes, the core of the smaller atom is closer to the shared electrons and exerts a stronger pull on them (Figure 5a). A similar situation arises when the atoms are from different groups in the periodic table and have different core charges (Figure 5b).

Synoptic link
You learnt about covalent bonding in Chapter 1, Elements of life, and Chapter 2, Developing fuels.

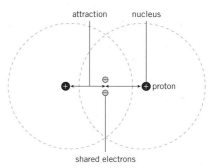
▲ **Figure 4** *A hydrogen molecule – the electrons are held an equal distance from each of the two hydrogen nuclei because they are identical*

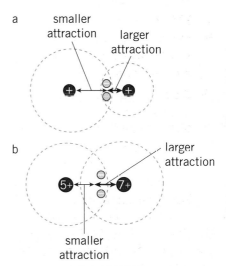
▲ **Figure 5** *a) Shared electrons are attracted more strongly by the core of the smaller atom, which is closer* *b) Shared electrons are attracted more strongly by the atom with the greater core charge*

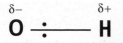

O ⋮ H

$\delta-$ ⬩ $\delta+$

O ⋮ H

More negative	More positive
end of the bond	end of the bond

because O has greater
share of electrons

▲ **Figure 6** *The O—H covalent bond is polar – the two electrons in the bond are drawn closer to the oxygen atom than the hydrogen atom*

Study tip

$\delta+$ and $\delta-$ are referred to as *partial charges*.

Electronegativity

Electronegativity is a measure of the ability of an atom in a molecule to attract electrons in a chemical bond to itself.

▼ **Table 2** *Boiling points of some organic halogen compounds*

Compound	State at 298 K	Boiling point / K
CH_3F	gas	195
CH_3Cl	gas	249
CH_3Br	gas	277
CH_3I	liquid	316
CH_2Cl_2	liquid	313
$CHCl_3$	liquid	335
CCl_4	liquid	350
C_6H_5Cl	liquid	405

Activity OZ 6.1

In this activity you can investigate bond polarity.

When atoms attract bonding electrons unequally, one gets a slight negative charge because it has a greater share of the bonding electrons. The other atom consequently becomes slightly positively charged. Bonds like this are called **polar bonds** (Figure 6).

The small amounts of electrical charge are shown by $\delta-$ and $\delta+$ where δ (lower case Greek letter delta) means a small amount of.

Some bonds are polar and some are not. The O—H bond is a particularly important example of a polar covalent bond. The polarity of the O—H bond has many consequences for the chemistry of water molecules, as you will learn in Topic OZ 7. For example, polar substances tend to be soluble in water whilst non-polar substances tend not to dissolve.

Electronegativity

To decide the polarity of a covalent bond, you need a measure of each atom's attraction for bonding electrons – its **electronegativity**. The electronegativity values in Figure 8 are derived from a method suggested by Linus Pauling.

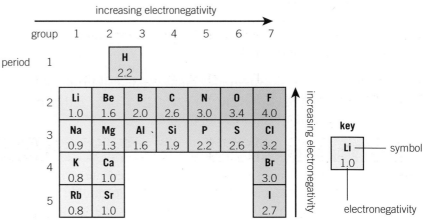

▲ **Figure 8** *Pauling electronegativity values for some main group elements in the Periodic Table, showing trends in groups and periods*

You can now predict how polar a particular covalent bond will be. In the C—F bond, fluorine has a higher electronegativity value (4.0) than carbon (2.6), so it attracts the shared electrons more strongly and the polarity of the bond is:

$$\overset{\delta+}{C} - \overset{\delta-}{F}$$

Electronegativity and haloalkanes

Carbon–halogen bonds are polar, but not enough to make a big difference to the physical properties of the compounds. For example, all haloalkanes are immiscible with water, like alkanes. Their boiling points depend on the halogen atoms present – the bigger the halogen atom and the more halogen atoms there are, Δthe higher the boiling point.

Intermolecular bonds

In a liquid or a solid there are **intermolecular bonds** causing molecules to be *attracted to one another*, otherwise they would move apart and become a gas. When a solid melts and boils (Figure 9), it is the intermolecular bonds that are broken, so boiling point gives a good indication of the strength of bonds *between* the molecules. Any covalent bonds *within* the molecules remain intact.

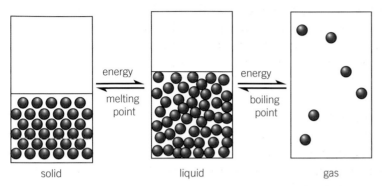

solid   liquid   gas

▲ **Figure 9** *Energy must be supplied to break intermolecular bonds. Covalent bonds within the molecules remain intact*

The low boiling points of noble gases (Table 3) show that the bonds *between the atoms* are very weak.

The boiling point trend of alkanes (Table 4) shows that the longer molecules have stronger bonds between them.

▼ **Table 3** *Boiling points for the noble gases*

Element	Boiling point / K
Helium	4
Neon	27
Argon	87
Krypton	121
Xenon	161

▼ **Table 4** *Boiling points of some alkanes*

Alkane	Structural formula	Boiling point / K	Alkane	Skeletal formula	Boiling point / K
methane	CH_4	111	hexane		342
ethane	CH_3CH_3	184	3-methylpentane		336
propane	$CH_3CH_2CH_3$	231	2-methylpentane		333
butane	$CH_3CH_2CH_2CH_3$	273	2,3-dimethylbutane		331
pentane	$CH_3CH_2CH_2CH_2CH_3$	309	2,2-dimethylbutane		323

For alkanes, the longer the chain, the stronger the intermolecular forces and the higher the boiling point. More energy is needed to break these stronger bonds and separate the molecules.

The table shows some branched isomers of hexane, C_6H_{14}. In straight-chain alkanes there are more contacts between molecules and more opportunities for intermolecular bonds to form, so straight-chain alkanes have higher boiling points than their branched isomers.

Dipoles

A **dipole** is a molecule (or part of a molecule) with a positive end and a negative end, because of its polar bonds. When a molecule has a dipole we say it is **polarised**. There are several ways a molecule can become polarised.

Permanent dipoles

Permanent dipoles occur when the two atoms in a bond have substantially different electronegativities. Hydrogen chloride has a permanent dipole because chlorine is much more electronegative than hydrogen, so attracts the shared electrons more.

When the atoms that form a molecule have only a small difference in electronegativity, any dipole will be very small. It is also possible for a molecule to have no overall dipole even though the bonds are polar. For example in CCl_4 each chlorine atom carries a small negative charge and the central carbon is positive.

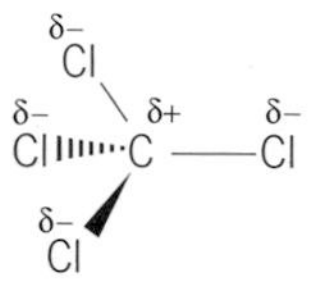

Because the chlorine atoms are distributed tetrahedrally around the carbon, the centre of negative charge is midway between all the chlorines. It is at the centre of the molecule and is superimposed on the positive charge on the carbon. The dipoles cancel due to the symmetry of the molecule and there is no overall dipole – the molecule is **non-polar** (Figure 10).

Instantaneous dipoles

Chlorine, Cl_2, does not have a permanent dipole, but a *temporary* dipole can arise. The electrons within the Cl—Cl bond are in constant motion, and at a particular instant they may not be evenly distributed. Then one end of the molecule has a greater negative charge than the other end – an **instantaneous dipole** (Figure 11).

Electron cloud evenly distributed–no dipole

At some instant, more of the electron cloud happens to be at one end of the molecule than the other–molecule has an instantaneous dipole

▲ **Figure 11** *How a dipole forms in a chlorine molecule*

After an instant the electrons change position, changing the dipole.

Induced dipoles

An **induced dipole** occurs if an unpolarised molecule is next to a dipole. The dipole attracts or repels electrons in the unpolarised molecule, *inducing* a dipole in it (Figure 12).

In Figure 12, the dipole has been induced by a *permanent* dipole. A dipole can also be induced by an *instantaneous* dipole, so a whole series of dipoles can be set up in a substance that contains no permanent dipoles.

$$\delta-$$
$$Cl$$
$$\delta- \quad \backslash \delta+ \quad \delta-$$
$$Cl \text{|||·····} C —— Cl$$
$$\delta- \diagup$$
$$Cl$$

tetrachloromethane

▲ **Figure 10** *The chlorine atoms are distributed symmetrically around the carbon atom, and CCl_4 has a tetrahedral shape, like that of methane. CCl_4 has polar bonds, but no overall dipole*

Study tip

Bond polarity depends on electronegativity differences.

Molecular dipoles depend on electronegativity differences *and* the shape of the molecule. In symmetrical molecules such as CO_2 or CCl_4 the dipoles cancel.

Study tip

In the diagram for Figures 11, 12, and 13 red areas indicated electron deficiency ($\delta+$) and blue areas indicated areas of higher electron density ($\delta-$).

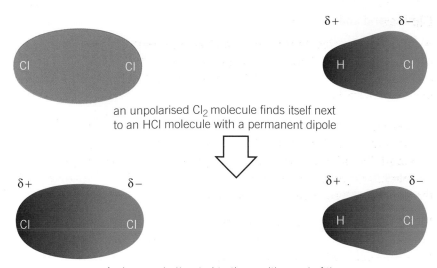

an unpolarised Cl_2 molecule finds itself next to an HCl molecule with a permanent dipole

electrons get attracted to the positive end of the HCl dipole, inducing a dipole in the Cl_2 molecule

▲ **Figure 12** *How a dipole can be induced in a chlorine molecule*

Dipoles and intermolecular bonds

All intermolecular bonds arise from the attractive forces between dipoles. There are three kinds of bond:

- **instantaneous dipole–induced dipole bonds**, for example, in chlorine. These are the weakest type of intermolecular bonding. They act between *all* molecules because instantaneous dipoles can arise in molecules that already have a permanent dipole. However, they are most noticeable in non-polar substances. The more electrons the atoms have the greater the instantaneous dipole–induced dipole effect. This can be seen in the increased boiling points for the noble gases. The larger the atoms, the more electrons, the greater the instantaneous dipole-induced dipole bonds.

- **permanent dipole–permanent dipole bonds**, for example, in hydrogen chloride. These are relatively strong and such molecules are more likely to be liquids or solids.

- **permanent dipole–induced dipole bonds**, for example, between hydrogen chloride and chlorine molecules.

In the liquid state, the molecular dipoles are constantly moving. Sometimes opposite charges are next to each other causing attraction (Figure 13a) and sometimes like charges are next to each other causing repulsion (Figure 13b).

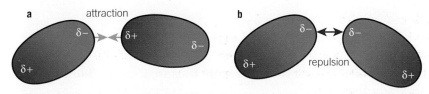

▲ **Figure 13** *Molecular dipoles in the liquid state*

> ## Study tip
>
> Make sure that you can identify which substances have which type of intermolecular bond. You should be able to identify these in a range of different substances.

Melting and boiling points of the halogens

In the case of halogens (which form diatomic, non-polar molecules) the strongest type of intermolecular bond that can form is an instantaneous dipole–induced dipole. The smaller the atoms in the molecule the weaker the instantaneous dipole–induced dipole is.

Fluorine has the weakest instantaneous dipole–induced dipole bonds, having the fewest electrons, and therefore has the lowest melting and boiling points of the halogens.

▼ **Table 5** *The melting and boiling points and physical appearance of the halogens*

Halogen	Melting point / K	Boiling point / K	Appearance at room temperature
fluorine	53	85	pale yellow gas
chlorine	172	239	green gas
bromine	266	332	dark red volatile liquid
iodine	387	457	black shiny solid

As the molecules get bigger and have more electrons, going down Group 7, the instantaneous dipole–induced dipole bonds increase and the melting and boiling points increase correspondingly.

Damage to the ozone layer

Chemists' concerns were raised in the early 1970s that nitrogen oxides from jet aircraft may be destroying the ozone layer. It turned out that this wasn't a significant problem because the number of aircraft was relatively small.

The issue arose again in 1974 when American professor Sherry Rowland published a paper predicting that CFCs would damage the ozone layer if they reached the stratosphere. For this work, Professor Rowland, together with Professors Mario Molina and Paul Crutzen, was awarded the Nobel Prize for Chemistry in 1995.

Chlorine reservoirs

Left to themselves, chlorine atoms would quickly destroy most of the ozone in the stratosphere, but fortunately there are other molecules there that react with chlorine atoms.

Methane, CH_4, is an important example. It removes chlorine atoms by reacting with them.

$$CH_4 + Cl \rightarrow CH_3 + HCl$$

This reaction is an example of a radical propagation reaction (Topic OZ 3).

HCl is a **chlorine reservoir molecule** because it stores chlorine in the stratosphere.

Another important reaction which produces a chlorine reservoir molecule is the reaction of nitrogen dioxide and chlorine monoxide.

$$NO_2 + ClO \rightarrow ClONO_2$$

These reservoir molecules are both soluble in water and can be removed in raindrops. Most, however, remain in the stratosphere with serious consequences.

Summary questions

1   Name the following haloalkanes.
    a $CHCl_3$ (*1 mark*)   **d** CH_3—$CHCl$—CF_3 (*1 mark*)
    b $CH_3CHClCH_3$ (*1 mark*)   **e** CH_3—$CHCl$—CBr_2—CH_3 (*1 mark*)
    c CF_3CCl_3 (*1 mark*)

2   Explain why noble gases have very low boiling points.   (*1 mark*)

3   Draw skeletal formulae showing how two molecules of pentane can approach close to one another. Now do the same for both of its structural isomers, 2-methylbutane, and 2,2-dimethylpropane.
    a Explain the boiling points of the isomers in terms of the strength of the intermolecular bonds present.   (*3 marks*)

Isomer	Boiling point / K
pentane	309
2-methylbutane	301
2,2-dimethylpropane	283

    b Explain the differences in strengths of the intermolecular bonds.   (*2 marks*)

4   Explain which of the following molecules possesses a permanent dipole, considering the shape of the molecule and the electronegativity of its atoms.
    a CO_2 (*1 mark*)   **d** CH_3OH (*1 mark*)
    b $CHCl_3$ (*1 mark*)   **e** $(CH_3)_2CO$ (*1 mark*)
    c C_6H_{12} (cyclohexane) (*1 mark*)   **f** benzene   (*1 mark*)

5   Silane, SiH_4, boils at 161 K whereas hydrogen sulfide, H_2S, boils at 213 K.
    a i State the number of electrons in each molecule.
       ii Explain the strengths of instantaneous dipole–induced dipole bonds in the two compounds.
       iii Explain whether either molecule possesses a permanent dipole.
                                                                      (*6 marks*)
    b Use your answers in part **a** to explain the different boiling points of these two compounds.   (*3 marks*)

▲ **Figure 1** *Polar stratospheric clouds over the Antarctic. Years in which the temperature is low, with extensive formation of polar stratospheric clouds, correspond to particularly wide and deep ozone holes*

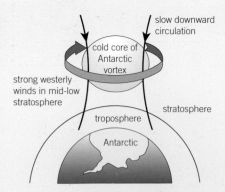

▲ **Figure 2** *The winter vortex over the Antarctic. The cold core is almost isolated from the rest of the atmosphere*

Why does the ozone hole develop over the poles?

Satellite measurements have shown that, whilst ozone concentrations have decreased in other parts of the globe too, the effect is particularly dramatic in the Antarctic spring.

The reasons for this are associated with the special weather conditions occurring in the Antarctic. Firstly, the very low temperatures (below −80 °C) that occur in the polar winter lead to the formation of polar stratospheric clouds. These clouds are made up of tiny solid particles – some are mainly particles of ice, others are rich in nitric acid, HNO_3. These particles provide surfaces on which chemical reactions can occur.

Secondly, during the Antarctic winter a vortex of circulating air forms, which effectively isolates the air at the centre of the vortex. This cold core turns into a giant sealed reaction vessel (Figure 2).

Chlorine reservoir molecules – HCl and chlorine nitrate, $ClONO_2$ – are adsorbed onto the surface of the solid particles in the polar stratospheric clouds, where they react together:

$$ClONO_2 + HCl \rightarrow Cl_2 + HNO_3$$

The HNO_3 remains dissolved in the ice particle, but the chlorine molecules are released as a gas trapped in the isolated core of the vortex. In the Antarctic spring the vortex starts to break up and the chlorine molecules undergo homolytic fission to form chlorine atoms (Figure 3).

Ozone depletion also occurs over the Arctic. The effect is more variable here because temperatures are not usually as low as those above the Antarctic so polar stratospheric clouds are less abundant and a stable polar vortex does not form. However, ozone depletion here extended over more densely populated areas including Canada, the USA, and Europe, and is a cause for concern.

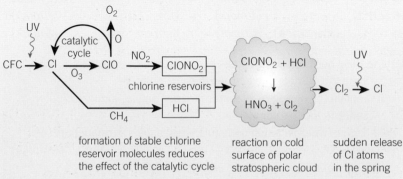

▲ **Figure 3** *How reactions in polar stratospheric clouds contribute to the dramatic loss of ozone in the Antarctic spring*

Ice particles are critical in the process above. Water has some interesting properties which can be explained by the bonds between its molecules.

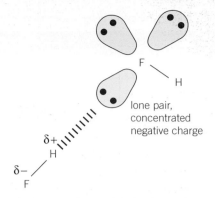

Bonds between molecules

Hydrogen bonding

Hydrogen bonding is the strongest type of intermolecular bond and can be thought of as a special case of permanent dipole–permanent dipole bonding.

What is special about hydrogen bonding?

For hydrogen bonding to occur, the molecules involved must have the following three features:

- a *large dipole* between a hydrogen atom and a highly electronegative atom such as oxygen, nitrogen, or fluorine

- a *small hydrogen atom* which can get very close to oxygen, nitrogen, or fluorine atoms in nearby molecules

- a *lone pair of electrons* on the oxygen, nitrogen, or fluorine atom that the positively charged hydrogen atom can line up with.

In hydrogen fluoride, HF, the hydrogen atoms have a strong positive charge because they are bonded to the highly electronegative fluorine atom. This positive charge lines up with a lone pair on another fluorine atom. The hydrogen and fluorine atoms can get very close, and therefore attract very strongly, because the hydrogen atom is so small.

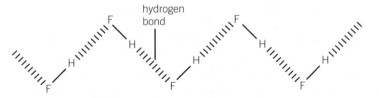

▲ Figure 4 *Hydrogen bonding in liquid hydrogen fluoride*

Dipole–dipole bonds are present in all the hydrogen halides (HF, HCl, HBr, and HI) but the greater strength of the hydrogen bonds in HF can be seen from a comparison of the boiling points. In Table 1, hydrogen fluoride has the highest boiling point despite having the lowest relative molecular mass. This is because the electronegativity of the halogens decreases down the group F>Cl>Br>I.

Water molecules can form twice as many hydrogen bonds as hydrogen fluoride. The oxygen atom possesses two lone pairs of electrons and there are twice as many hydrogen atoms as oxygen atoms (Figure 6).

Water is unique in this respect. In hydrogen fluoride, the fluorine has three lone pairs but there are only as many hydrogen atoms as fluorine atoms – only one-third of the available lone pairs can be used. In ammonia, NH_3, another hydrogen-bonded substance, there is only one lone pair on the nitrogen so, on average, only one of the three hydrogen atoms can form hydrogen bonds.

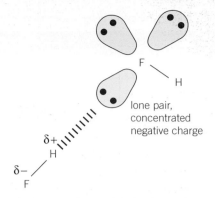

▲ Figure 5 *The positively charged hydrogen atom lines up with a lone pair on a fluorine atom*

Activity OZ 7.1

This activity visibly demonstrates the link between viscosity and hydrogen bonding.

▼ Table 1 *Boiling points of the hydrogen halides*

Compound	Boiling point / K
HF	292
HCl	188
HBr	206
HI	238

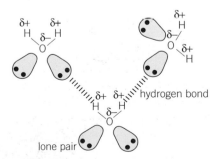

▲ Figure 6 *The positively charged hydrogen atoms line up with the lone pairs on the oxygen atoms*

▼ Table 2 *The effect of hydrogen bonding on boiling point*

Compound	Formula	Relative molecular mass	Hydrogen bonding?	Boiling point / K
Propane	$CH_3CH_2CH_3$	44	no	231
Ethanol	CH_3CH_2OH	46	yes	351
Heptane	$CH_3(CH_2)_5CH_3$	100	no	371
Glycerol	$CH_2(OH)CH(OH)CH_2(OH)$	92	yes	563

Synoptic link

Many covalent substances have N—H and O—H bonds in their molecules. Amino acids and proteins have both types of bonds and hydrogen bonding plays a major role in determining their properties. You will find out more about amino acids and proteins in Chapter 7, Polymers and life. You will also find out more about the properties of water in Chapter 8, The ocean.

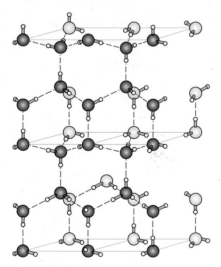

▲ Figure 7 *The arrangement of water molecules in ice*

In Table 2 you can see than the compounds with hydrogen bonding have higher boiling points that compounds with similar molecule masses that do not have hydrogen bonding. This is because more energy is needed to break hydrogen bonds than instantaneous dipole–induced dipole bonds.

Liquids that have hydrogen bonding have a high **viscosity** – viscosity is a measure of how easily a liquid flows. For a liquid to flow the molecules must be able to move past each other and this requires the constant breaking and forming of intermolecular bonds. The stronger the bonds between the molecules, the more difficult this becomes.

Substances with hydrogen bonding are also often soluble in water. Hydrogen bonds can form between water molecules and molecules of the substance, helping the dissolving process. When water freezes it forms ice crystals. The open structure, with four groups around each oxygen atom, maximises the hydrogen bonding between them (Figure 7)

Summary questions

1　a　Explain why the O—H bond is polar.

　　b　State which of the covalent bonds in the following list will be significantly polar. In cases where there is a polar bond, show which atom is positive and which is negative using the $\delta+$ and $\delta-$ convention.

a	C—F	*(1 mark)*	e	H—N	*(1 mark)*
b	C—H	*(1 mark)*	f	S—Br	*(1 mark)*
c	C—S	*(1 mark)*	g	C—O	*(1 mark)*
d	H—Cl	*(1 mark)*			

2　Hydrogen bonding can occur between different molecules in mixtures. Draw diagrams to show where the hydrogen bonds form in the following mixtures.

　　a　NH_3 and H_2O　　　　　　　　　　　　　　　　*(2 marks)*

　　b　CH_3CH_2OH and H_2O　　　　　　　　　　　*(2 marks)*

3　Explain why water exhibits a greater degree of hydrogen bonding than other substances.　　　　　　　　　　　　　　*(3 marks)*

Sherry Rowland's predictions come true

Science is all about making predictions, then testing them experimentally. The problem with Sherry Rowland's predictions (see OZ 6) was that they involved a long time scale. However in 1985, scientists examining the atmosphere above the Antarctic made a momentous discovery. Since the mid-1980s ongoing measurements have confirmed the presence of the 'hole' over the Antarctic.

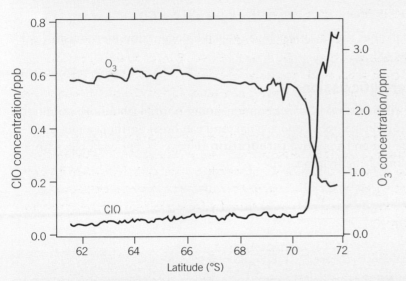

▲ **Figure 1** *These graphs show measurements of ClO radicals (in parts per billion) and ozone (in parts per million) at 18 km altitude. The measurements provided convincing evidence that chlorine radicals are involved in ozone depletion*

Could the trouble with CFCs have been foreseen?

When Thomas Midgley and other chemists developed CFCs, they found a family of compounds that had many useful applications. The trouble is that they are too unreactive.

In the 1930s, questions about environmental consequences of CFCs were simply not on the agenda. Atmospheric chemistry was in its infancy and sensitive instruments that could detect minute concentrations of compounds in the air did not exist.

Laboratory studies and computer modelling allowed scientists to run simulations and make predictions. In the 1980s and 1990s this contributed to an explosion of knowledge of the chemistry of the atmosphere.

It is now known that most organic compounds are broken down in the troposphere by species such as HO radicals and never reach the stratosphere. However the stability of CFCs, initially seen as a huge advantage, proved to be their downfall.

Learning outcomes

Demonstrate and apply knowledge and understanding of:

→ the recognition of and formulae for examples of members of the following homologous series: amines

→ the characteristic properties of haloalkanes, comparing fluoro-, chloro-, bromo-, and iodo-compounds, considering the following aspects:

  • nucleophilic substitution with water and hydroxide ions to form alcohols, and with ammonia to form amines

→ the terms: substitution and nucleophile

→ the use of the S_N2 mechanism as a model to explain nucleophilic substitution reactions of haloalkanes using 'curly arrows' and partial charges

→ the possible dependence of the relative reactivities of the haloalkanes on either bond enthalpy or bond polarity and how experimental evidence determines that the bond enthalpy is more important

→ the ease of photodissociation of the haloalkanes (fluoroalkane to iodoalkane) in terms of bond enthalpy.

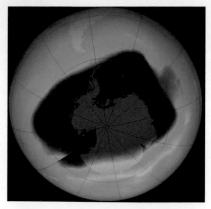

▲ **Figure 2** *The ozone hole over the Antarctic as recorded on 25th September 2006. This was the biggest recorded hole in the ozone layer. Blue represents areas of low ozone levels and green represents areas of high ozone levels*

Chemical ideas: Organic chemistry: modifiers 13.2b

Haloalkanes

Chemical reactions of haloalkanes

Reactions of haloalkanes involve breaking the carbon–halogen bond. The bond can break homolytically or heterolytically (Topic OZ 3).

Homolytic fission

Homolytic fission of haloalkanes occurs when haloalkanes reach the stratosphere, where they are exposed to intense ultraviolet radiation. This is how ozone-depleting chlorine radicals are formed.

$$CH_3-Cl + h\nu \rightarrow CH_3{}^\bullet + Cl^\bullet$$

Haloalkanes can be *formed* by a radical halogenation mechanism (Topic OZ 3).

Heterolytic fission

Heterolytic fission is more common under normal laboratory conditions. The carbon–halogen bond is polar, and can break forming a negative halide ion and a positive **carbocation** (Figure 3). For example, with 2-chloro-2-methylpropane:

$$
\begin{array}{ccc}
& CH_3 & & CH_3 \\
& | & & | \\
H_3C-C-Cl & \longrightarrow & H_3C-C^+ & + & Cl^- \\
& | & & | \\
& CH_3 & & CH_3 \\
\text{2-chloro-2-methylpropane} & & \text{carbocation} & & \text{chloride ion}
\end{array}
$$

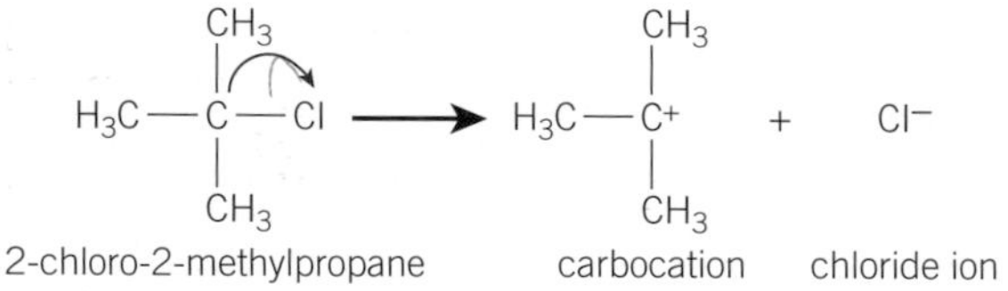

▲ **Figure 3** *2-chloro-2-methylpropane breaks down to a carbocation and a chloride ion*

Sometimes, heterolytic fission is caused by a negatively charged substance reacting with the positively polarised carbon atom, causing a **substitution** reaction.

Nucleophilic substitution reactions of haloalkanes

You have already seen radical reactions for haloalkanes. Another type of reaction haloalkanes take part in are substitution reactions. The halogen atom is substituted by another group. This group is called a **nucleophile**.

Consider the reaction between hydroxide ions, OH⁻, and 1-bromobutane. The haloalkane is heated under reflux with ethanolic sodium hydroxide

- The nucleophile, OH⁻, attacks the electron deficient carbon atom in the C—Br bond.

- The OH⁻ donates two electrons to form a new dative covalent bond.

- The C—Br bond breaks heterolytically and the bromine atom receives two electrons, producing a bromide ion. In this case the bromide ion is called the **leaving group**.

Activity OZ 8.1

This activity provides an opportunity to develop your chemical literacy skills.

Substitution

A reaction in which one atom or group in a molecule is replaced by another atom or group.

Nucleophile

A nucleophile is a molecule or negatively charged ion with a lone pair of electrons that it can donate to a positively charged atom to form a covalent bond.

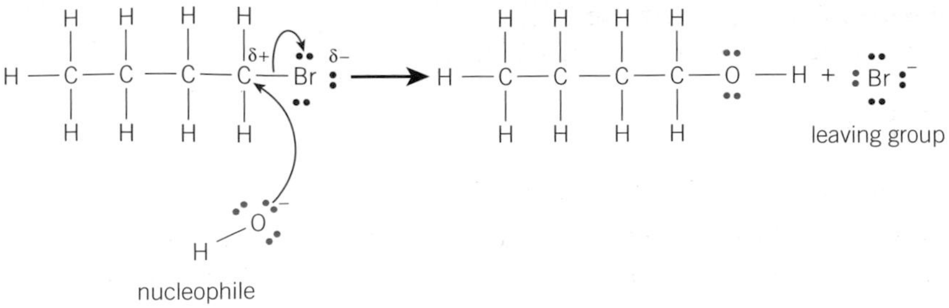

▲ **Figure 4** *The reaction mechanism of bromobutane and hydroxide ions*

Synoptic link

You first came across the idea of reaction mechanisms when studying electrophilic addition in Chapter 2, Developing fuels.

Because the overall reaction involves a nucleophile replacing another group, this type of reaction is called a **nucleophilic substitution**.

This reaction involves heterolytic fission. A free carbocation is not formed because the OH^- ion attacks at the same time as the C–Br bond breaks.

Curly arrows show the movement of electrons. In this case a full-headed arrow shows the movement of *a pair* of electrons. Table 2 shows a number of nucleophiles. each nucleophile in Table 2 has a lone pair of electrons.

▼ **Table 2** *Some common nucleophiles*

Name and formula	Structure showing lone pairs
hydroxide ion, OH^-	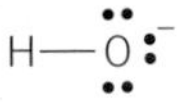
cyanide ion, CN^-	
ethanoate ion, CH_3COO^-	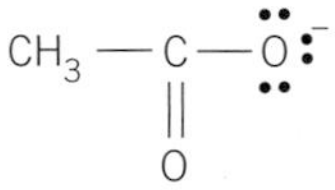
ethoxide ion, $C_2H_5O^-$	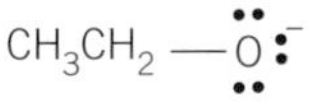
water, H_2O	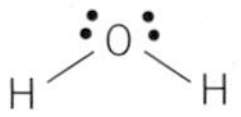
ammonia, NH_3	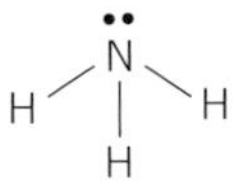

Study tip

Nucleophile means nucleus loving and can help you remember that nucleophiles are attracted to areas of positive charge.

Using Nu^- as a general symbol for any nucleophile and X as a symbol for a halogen atom, the nucleophilic substitution process can be described by:

$$R\text{—}X + Nu^- \rightarrow R\text{—}Nu + X^-$$

The curly arrow moves from the lone pair of electrons to the electron deficient carbon.

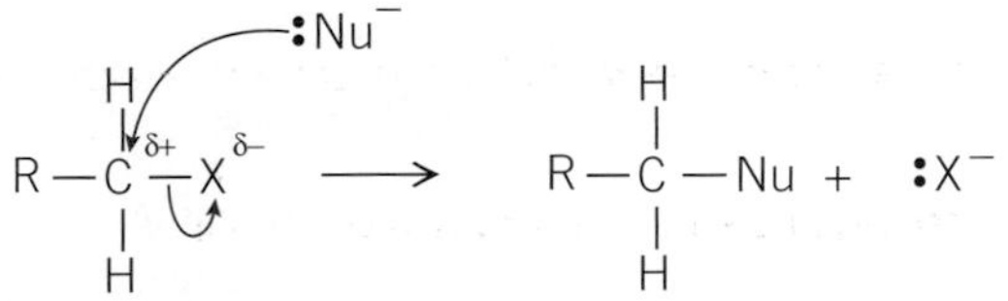

Synoptic link

You will study other types of organic reaction mechanisms in Chapter 10, Colour by design.

▲ **Figure 5** *General reaction mechanism and equation for nucleophilic substitution of haloalkanes.*

Water as a nucleophile

A water molecule has a lone pair on the oxygen atom, so water can act as a nucleophile – though this reaction is slower than the reaction with OH^- ions. When the two are heated together under reflux, water attacks the haloalkane.

Then the resulting ion loses H^+ to form an alcohol.

The general equation is known as a *hydrolysis* reaction:

$$R—X + H_2O \rightarrow R—OH + H^+ + X^-$$

Ammonia as a nucleophile

Ammonia, NH_3, can act in a similar way to water with the lone pair of electrons on the nitrogen atom attacking the haloalkane. The haloalkane is heated in a sealed tube with concentrated ammonia solution. The product is an **amine** with an NH_2 group. The overall equation is:

$$R—X + NH_3 \rightarrow R—NH_3^+ X^- \rightleftharpoons R—NH_2 + H^+ + X^-$$

Amines have the general formula $R—NH_2$. They are nitrogen analogues of alcohols ($R—OH$).

Some examples of amines are shown in Figure 6.

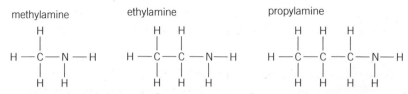

▲ **Figure 6** *Some examples of amines*

Using nucleophilic substitution to make haloalkanes

You can use the *reverse* of a hydrolysis reaction to produce a haloalkane from an alcohol. This time the nucleophile is a halide ion (X^-). The reaction is done in the presence of a strong acid, and the first step involves bonding between H^+ ions and the oxygen atom on the alcohol. For example:

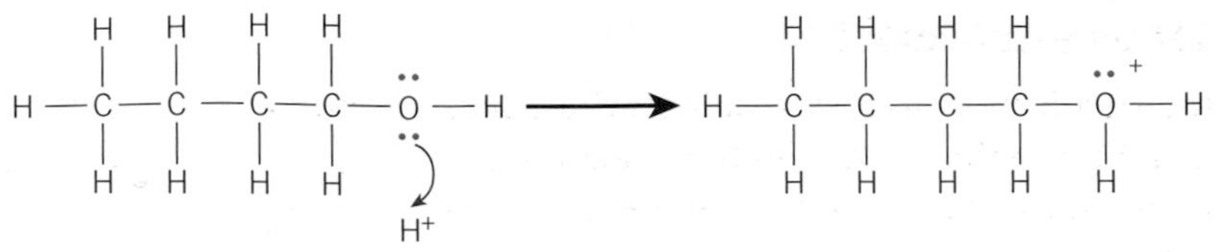

This gives the carbon atom to which the oxygen is attached a higher partial positive charge. It is now more readily attacked by halide ions:

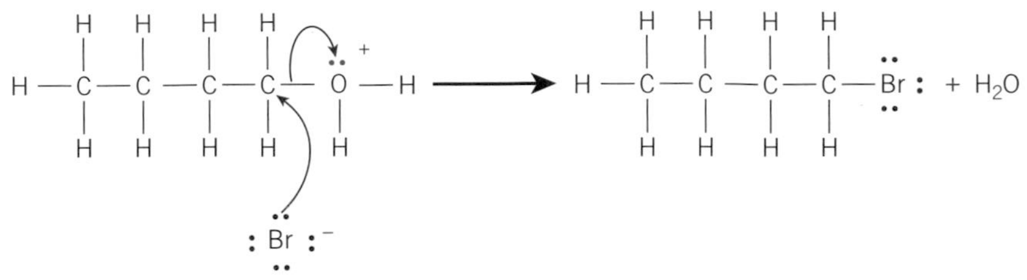

The overall equation for the reaction is:

$$CH_3CH_2CH_2CH_2OH + H^+ + Br^- \rightarrow CH_3CH_2CH_2CH_2Br + H_2O$$

Different halogens, different reactivity

Table 1 gives the bond enthalpies of the four different types of C—Hal bond.

▼ **Table 1** *Bond enthalpies of carbon–halogen bonds*

Bond	Bond enthalpy /kJ mol^{-1}
C—F	467
C—Cl	346
C—Br	290
C—I	228

You might imagine that a large bond polarity in a carbon–halogen bond would result in it breaking easily, for example, C—F bonds would break more easily than C—Cl bonds. However, it has been shown experimentally that bond enthalpy is the overriding factor in determining reactivity. Bond strength decreases in the order C—F > C—Cl > C—Br > C—I. On this basis, you would expect the C—I bond to be hydrolysed most easily because it is the weakest.

Bond polarity decreases in the order C—F > C—Cl > C—Br > C—I. On this basis, you would expect the C—F bond to be hydrolysed most easily because it is the most polar. Experimentally you find that C—I bonds are most easily hydrolysed, so bond strength is the most important factor.

The strength of the C—F bond makes it very difficult to break, so fluoro compounds are very unreactive. As you go down the halogen group the carbon–halogen bond gets weaker, so the compounds get more reactive. Bromo- and iodo-compounds are fairly reactive, so they are useful intermediates in organic synthesis.

Study tip

You need to be able to write balanced equations and reaction mechanisms for nucleophilic substitution reactions.

Activity OZ 8.3

You can investigate the relative reactivities of some haloalkanes in this activity.

What is the state of the ozone layer now?

It was clear in the 1980s that removing ozone from the stratosphere would have serious consequences, but it was not known what the full extent of these would be. One thing was certain – the numbers of cases of skin cancers and eye cataracts were increasing as the ozone was destroyed.

Another problem was whether increased ultraviolet radiation could affect species such as plankton in the oceans? That could affect other organisms involved in the food chain. Furthermore, could changes in the amount of radiation reaching the Earth affect its temperature and weather?

▲ **Figure 7** *Bioluminescent plankton. There were concerns raised that plankton such as these could be directly affected by increases in UV radiation*

The Montreal Protocol

In 1987, Governments realised that it was not worth risking the global experiment needed to find the answers to these questions. In Montreal a procedure was agreed for restricting the release of CFCs into the atmosphere. This was known as the Montreal Protocol. Since then a series of amendments have strengthened the protocol, with the aim of eliminating emissions of ozone-depleting substances as a result of human activity.

By 1998 the developed nations had almost phased out their use of CFCs except for some specialised appliances such as asthma inhalers. A fund was set up to help developing countries to move away from CFCs by 2010.

CFC replacements

The chemical industry rapidly found replacements for CFCs that would have no significant damaging effect on the ozone layer. In the short term, replacement compounds were hydrochlorofluorocarbons (HCFCs), such as $CHClF_2$. The hydrogen–carbon bonds mean that HCFCs are

Synoptic link

You will learn more about greenhouse gases and global warming in Chapter 8, The oceans.

broken down in the troposphere. Nonetheless, some molecules will make it to the stratosphere where they will photodissociate to release chlorine atoms. It is hoped that HCFCs will be phased out in developed countries by 2020 and in the rest of the world by 2040.

For the longer term, hydrofluorocarbons, HFCs are a better option because they have no ozone-depleting effect, even if they make it to the stratosphere (Table 2).

Sadly, there is no perfect solution. Many of these compounds are greenhouse gases and there is concern about HFCs in the troposphere, since they can produce HF and trifluoroethanoic acid, CF_3COOH, but it is felt that the concentrations are too small to be a problem.

▼ **Table 2** *Properties and uses of some CFCs, HCFCs, and HFCs. The ozone-depleting potential (ODP) is a measure of the effectiveness of the compound in destroying stratospheric ozone. CFC-11 is defined as having an ODP of 1.0*

Compound	Code	Ozone-depleting potential (ODP)	Lifetime / years	Uses
Main culprits				
CCl_3F	CFC-11	1.0	45	refrigeration, air conditioning, foams, cleaning solvents
CCl_2F_2	CFC-12	1.0	100	
$CBrF_3$	Halon-1301	10.0	65	firefighting
$CBrClF_2$	Halon-1211	3.0	16	
CCl_4	–	1.1	26	solvents
CH_3CCl_3	–	0.1	5	
Short-term solutions				
$CHClF_2$	HCFC-22	0.06	12	replaces CFC-11 and CFC-12
$CHCl_2CF_3$	HCFC-123	0.02	1	replaces CFC-11
Long-term solution				
CH_3CHF_2	HFC-152a	0	1.4	replaces CFC-11 and CFC-12

When will the ozone layer recover?

The concentrations of CFCs in the troposphere peaked around the beginning of this century and are now starting to decrease. The concentrations of HCFCs and HFCs are rising. Figure 7 shows the results from monitoring air samples around the world.

The long lifetimes of CFCs mean that recovery of the ozone layer will be slow – and will be subject to variations caused by weather conditions and solar activity. The lowest recorded ozone concentration was in 1992 when the level was 92 'Dobson units' but by 2002 this had risen to 131 units. It seemed that the ozone layer was recovering, but the ozone depletion in September 2006 was the most severe to date. The hole at that point covered 27 million km^2.

In 2013 the ozone concentration had risen to 133 Dobson units, and the size of the hole had receded to 21 million km². It seems that there is an established positive trend, but complete recovery is not expected until 2060–2070.

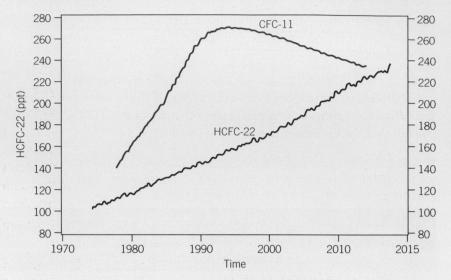

▶ **Figure 7** *Results from air samples – the concentration of CFC-11 is decreasing but HCFC-22 is increasing*

Summary questions

1 When 1-chloropropane is heated under reflux with aqueous sodium hydroxide solution, a nucleophilic substitution reaction occurs, forming propan-1-ol.
  a Write a balanced equation for this reaction. (*1 mark*)
  b Explain why this is a *substitution* reaction. (*1 mark*)
  c Draw the mechanism of the reaction, showing the relevant lone pairs and partial charges. (*3 marks*)

2 Table 6 shows some common nucleophiles. Write balanced equations for each of the following nucleophilic substitution reactions, showing the structures of the reactants and products clearly.
  a iodoethane and OH⁻ ions (*1 mark*)
  b bromoethane and CN⁻ ions (*1 mark*)
  c chlorocyclopentane and OH⁻ ions (*1 mark*)
  d 2-chloro-2-methylpropane and H₂O (*1 mark*)
  e 1,2-dibromoethane and OH⁻ ions (*1 mark*)
  f bromomethane and C₂H₅O⁻ (ethoxide) ions (*1 mark*)
  g 2-chloropropane and CH₃COO⁻ (ethanoate) ions (*1 mark*)

3 Concentrated ammonia solution reacts with 1-bromoethane when heated in a sealed tube.
  a Write a balanced equation for this reaction. (*1 mark*)
  b Using the reaction of haloalkanes with water as a guide, draw out the mechanism for this reaction. (*2 marks*)
  c Write a few sentences to explain this mechanism to a fellow student. (*2 marks*)
  d Give definitions of all the terms in your mechanism. (*2 marks*)

Practice questions

1 The following substances all have similar M_r values.

Which has the highest boiling point?

A $CH_3CH_2CH_2CH_3$

B CH_3CH_2Cl

C $CH_3CH_2CH_2OH$

D CH_3COCH_3 (1 mark)

2 Which of the following molecules would have the greatest overall dipole?

A CCl_4

B BF_3

C NH_3

D C_2H_6 (1 mark)

3 Nitrogen and hydrogen are mixed. Which of the following conditions would result in the equilibrium being set up fastest?

$$N_2 + 3H_2 \rightarrow 2NH_3 \quad \text{exothermic}$$

	Pressure	Temperature
A	high	high
B	high	low
C	low	high
D	low	low

(1 mark)

4 Which of the following are correct statements about the role of CFCs in ozone depletion?

	Troposphere	Stratosphere
A	no ozone to deplete	form Cl radicals that deplete ozone
B	do not form Cl radicals	form Cl radicals that deplete ozone
C	no ozone to deplete	react with ozone to deplete it
D	do not form Cl radicals	react with ozone to deplete it

(1 mark)

5 Which of the following statements is correct about UV radiation in the stratosphere?

A It is reflected from the Earth.

B It comes from the Sun and is all absorbed.

C It breaks down molecules.

D It causes photochemical smog. (1 mark)

6 The names of some bromoalkanes and their boiling points are shown below.

Name	1-bromobutane	2-bromobutane	2-bromo-2-methyl propane
Boiling point / K	375	364	346

Which of the following is a correct statement?

A The dipole of the C—Br bond is largest in 1-bromobutane.

B The instantaneous dipole–induced dipole bonds get weaker as the molecules become more compact.

C The molecules contain different numbers of electrons.

D 1-bromobutane is the most volatile. (1 mark)

7 Hydrogen bonds can form between which of the following pairs of atoms in molecules?

A	Any hydrogen atom.	An oxygen atom in a molecule.
B	A hydrogen atom attached to a nitrogen atom.	A nitrogen atom in a molecule.
C	Any hydrogen atom.	A fluorine atom in a molecule.
D	A hydrogen atom attached to a chlorine atom.	A carbon atom attached to a chlorine atom.

(1 mark)

8 Which of the following is a characteristic of a catalyst?

1 It provides a route of lower activation enthalpy.

2 It is unchanged at the end of the reaction.

3 It does not take part in the reaction.

A 1, 2, and 3 correct

B 1 and 2 are correct

C 2 and 3 are correct

D Only 1 is correct (1 mark)

9 Which of the following are characteristics of a nucleophile?

1 It must have negative (or partial negative) charge.

2 It attacks positively charged atoms.

3 It has a lone pair of electrons.

A 1,2, and 3 correct

B 1 and 2 are correct

C 2 and 3 are correct

D Only 1 is correct (*1 mark*)

10 Which of the following are termination steps in the radical chlorination of methane?

$$CH_4 + Cl_2 \rightarrow CH_3Cl + HCl$$

1 $2CH_3 \rightarrow C_2H_6$

2 $2Cl \rightarrow Cl_2$

3 $CH_3 + Cl \rightarrow CH_3Cl$

A 1,2, and 3 correct

B 1 and 2 are correct

C 2 and 3 are correct

D Only 1 is correct (*1 mark*)

11 The hydroxyl radical OH has been described as 'the detergent of the atmosphere' as it is able to oxidise (and thus remove) most substances in the troposphere.

It is made by the reaction of oxygen atoms with water molecules:

$$O + H_2O \rightarrow 2OH \qquad \textbf{Equation 11.1}$$

a Draw a *'dot-and-cross'* diagram for a hydroxyl radical and explain why OH is called a *radical*. (*2 marks*)

b Oxygen atoms are formed in the troposphere by the photolysis of ozone:

(i) Write an equation for the photolysis of ozone. (*1 mark*)

(ii) 3% of oxygen atoms are removed from the atmosphere by the reaction in **Equation 11.1**. Suggest another equation (related to that from (i)) by which oxygen atoms are removed from the atmosphere. (*1 mark*)

c In the stratosphere, oxygen molecules are broken down by photolysis.

(i) Given that the bond enthalpy of $O{=}O$ is 498 kJ mol^{-1}, calculate the minimum frequency of electromagnetic radiation needed to break this bond. (*3 marks*)

(ii) Calculate the corresponding wavelength to this frequency. (*2 marks*)

(iii) Suggest why this reaction only takes place in the stratosphere, whereas the photolysis of ozone takes place in the troposphere. (*2 marks*)

d Carbon monoxide is one of the major substances removed from the troposphere by hydroxyl radicals.

$$CO + OH \rightarrow CO_2 + H \qquad \textbf{Equation 11.2}$$

Classify the reaction in **Equation 11.2** as initiation, propagation, or termination, giving a reason. (*2 marks*)

12 Some students carry out experiments with 1-chlorobutane, 1-bromobutane, and 1-iodobutane.

a The students wish to discover the relative rates of hydrolysis of the three haloalkanes, using silver nitrate solution and ethanol.

Describe how they would carry out the experiment. Give equations for the reactions that occur and say what is seen. (*6 marks*)

b Before the experiment, one student predicts that the haloalkane with the greatest permanent dipole would react fastest.

(i) Explain what is meant by *permanent dipole* and explain which of the haloalkanes used by the students has the greatest permanent dipole. (*3 marks*)

(ii) The student uses the mechanism of the reaction to justify the prediction.

Draw out the mechanism of the reaction for the hydrolysis of 1-chlorobutane with water, showing curly arrows and partial charges.

(3 marks)

(iii) Indicate **one** *heterolytic* bond-breaking that occurs in the mechanism. *(1 mark)*

(iv) Explain why the haloalkane with the greatest dipole might have reacted fastest. *(1 mark)*

(v) State and explain the more important factor in determining the rate of hydrolysis. *(2 marks)*

c The students carry out a reaction in which they heat 1-bromobutane with concentrated ammonia in a sealed tube. Write an equation for the reaction that will occur and name the functional group in the product. *(2 marks)*

13 When aircraft started flying in or near to the stratosphere, there were concerns that nitrogen oxides in their exhausts would catalyse the breakdown of ozone by the reaction in **Equation 13.1**:

$$O + O_3 \rightarrow 2O_2 \qquad \textbf{Equation 13.1}$$

a The first reaction by which NO breaks down ozone is:

$$NO + O_3 \rightarrow NO_2 + O_2 \qquad \textbf{Equation 13.2}$$

(i) NO is a catalyst in the breakdown of ozone by the reaction in **Equation 13. 1.** Write the equation for the reaction that follows **Equation 13.2**. *(1 mark)*

(ii) What **type** of catalysis is involved here? *(1 mark)*

b In order to be a catalyst, the activation enthalpy of the catalysed reaction must be smaller than that for the uncatalysed reaction.

(i) Explain the meaning of the term *activation enthalpy*. *(1 mark)*

(ii) Label the Boltzmann distribution below to show how adding a catalyst makes a reaction go faster. *(2 marks)*

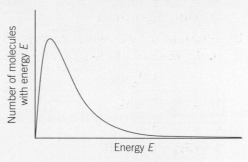

(iii) Complete the enthalpy profile below to show that a catalysed reaction has a lower activation enthalpy than the uncatalysed reaction. *(2 marks)*

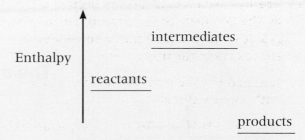

c (i) Why is there concern about the destruction of the ozone layer? *(2 marks)*

(ii) It was later discovered that the effect of NO from aircraft on the ozone layer was very small compared with the effect of another radical from human-derived sources.

Give the name of this radical and its source in the stratosphere. *(2 marks)*

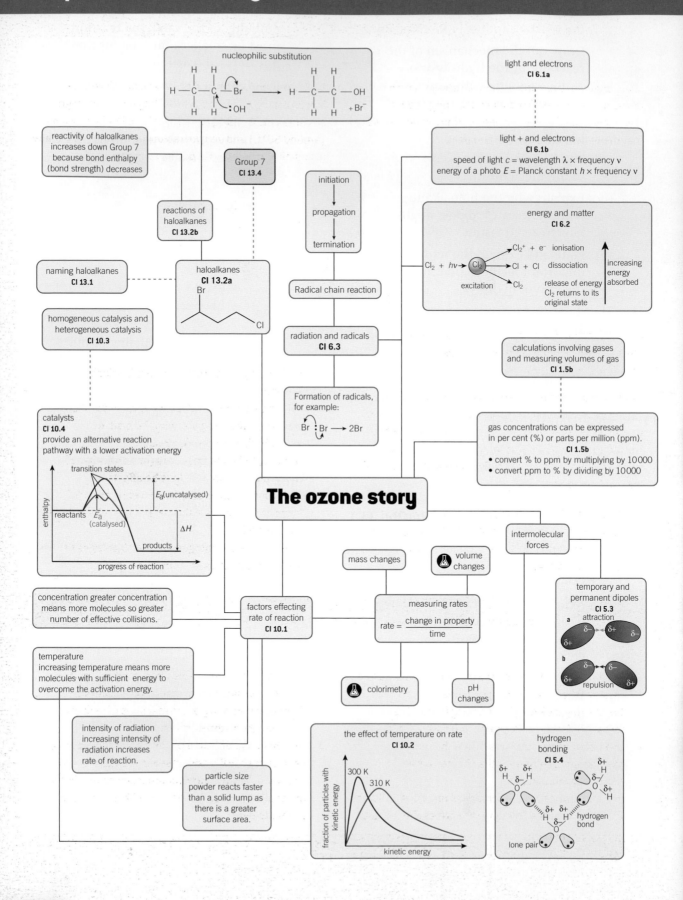

nucleophilic substitution

light and electrons
CI 6.1a

reactivity of haloalkanes increases down Group 7 because bond enthalpy (bond strength) decreases

Group 7
CI 13.4

light + and electrons
CI 6.1b
speed of light c = wavelength λ × frequency ν
energy of a photo E = Planck constant h × frequency ν

reactions of haloalkanes
CI 13.2b

initiation
↓
propagation
↓
termination

energy and matter
CI 6.2
$Cl_2 + h\nu$ → Cl_2
$Cl_2^+ + e^-$ ionisation
$Cl + Cl$ dissociation
Cl_2 release of energy Cl_2 returns to its original state
excitation
increasing energy absorbed

naming haloalkanes
CI 13.1

haloalkanes
CI 13.2a

Radical chain reaction

homogeneous catalysis and heterogeneous catalysis
CI 10.3

radiation and radicals
CI 6.3

calculations involving gases and measuring volumes of gas
CI 1.5b

catalysts
CI 10.4
provide an alternative reaction pathway with a lower activation energy

transition states

E_a(uncatalysed)

reactants E_a (catalysed)

ΔH

products

enthalpy

progress of reaction

Formation of radicals, for example:

Br : Br → 2Br

gas concentrations can be expressed in per cent (%) or parts per million (ppm).
CI 1.5b
• convert % to ppm by multiplying by 10000
• convert ppm to % by dividing by 10000

The ozone story

intermolecular forces

concentration greater concentration means more molecules so greater number of effective collisions.

factors effecting rate of reaction
CI 10.1

mass changes

volume changes

measuring rates

rate = $\dfrac{\text{change in property}}{\text{time}}$

temporary and permanent dipoles
CI 5.3
a attraction
$\delta-$ $\delta+$ $\delta-$
$\delta+$
b
$\delta-$ $\delta-$
$\delta+$ $\delta+$
repulsion

temperature
increasing temperature means more molecules with sufficient energy to overcome the activation energy.

colorimetry

pH changes

intensity of radiation increasing intensity of radiation increases rate of reaction.

particle size powder reacts faster than a solid lump as there is a greater surface area.

the effect of temperature on rate
CI 10.2

fraction of particles with kinetic energy

300 K
310 K

kinetic energy

hydrogen bonding
CI 5.4
$\delta+$ H $\delta+$ H $\delta+$ H
H $\delta-$ H   O $\delta+$ H
O $\delta+$ $\delta+$ hydrogen bond
$\delta+$ $\delta+$ H $\delta-$ H
O
lone pair

Anesthetics

Many anaesthetics contain halogen atoms in their molecules. Anaesthetics are used to bring about a reversible loss of consciousness during operations, so that the patient does not feel any pain.

Isoflurane, enflurane, and halothane are all volatile liquids which readily evaporate. They have quite high M_r values (184–197 g mol^{-1}), but have low boiling points (around 50 °C) and very low solubility in water. Halothane is unstable in light and is packaged in dark bottles.

In contrast to some early anaesthetics such as diethyl ether, $C_2H_5OC_2H_5$, these three molecules are all completely non-flammable.

Isoflurane and enflurane are isomers of each other, and all three molecules contain a chiral carbon. This is a carbon with four different atoms or groups attached. The different positioning of the groups around the chiral carbon can have important biological effects, as you will learn in later topics.

1 What does the low boiling point of these molecules suggest about their intermolecular bonding? Identify the type(s) of intermolecular bonding they would contain.

2 Explain why these molecules are insoluble in water.

3 Write an equation for the complete combustion of diethyl ether. Suggest why the modern anaesthetics are non-flammable.

4 What might be the effect of light on halothane? Which bond is most likely to be affected?

5 Suggest the possible ozone depleting effect of these anaesthetics if they were to be released into the atmosphere.

6 Evaluate the advantages and disadvantages of these anaesthetics for doctors and patients.

Extension

1 Nucleophilic substitution reactions can occur by two main mechanisms, largely dependent on whether the starting molecule is a primary haloalkane or tertiary haloalkane. Research and contrast the two different mechanisms. (Note that only the mechanism described in the chapter will be tested in examinations.)

2 In the extension box, Other radical reactions, in Topic OZ 3, How is ozone formed in the atmosphere?, you

learnt that radicals are involved in polymerisation, combustion, and age-related biological effects. Produce a summary of the reactions of radicals, including these three examples.

3 The challenge of ozone depletion appears to have been successfully addressed by politicians and chemists. Can the same approaches be applied to the challenge of global warming? What role can chemists play?

CHAPTER 5
What's in a medicine?

Topics in this chapter

→ **WM 1 The development of modern ideas about medicine**
reactions of alcohols

→ **WM 2 Identifying the active chemical in willow bark**
the −OH group in different environments and derivatives of carboxylic acids

→ **WM 3 Infrared spectroscopy**
Infrared spectroscopy

→ **WM 4 Mass spectrometry**
Mass spectrometry for compounds

→ **WM 5 The synthesis of salicylic acid and aspirin**
principles of green chemistry

Why a chapter on What's in a medicine?

This chapter introduces the pharmaceutical industry, which not only produces new and more effective medicines but is a net exporter of medicinal products and so contributes to the financial health of the UK. The molecule focussed on in this chapter is aspirin. Aspirin is a relatively simple drug which did not go through the drug testing regime carried out today but which is widely used as an over-the-counter painkiller, as well as being prescribed by doctors for heart disease and other conditions.

During the chapter you will study the application of instrumental methods for determining the structure of molecules and practise organic synthesis and use test-tube reactions to identify functional groups. You will also have the opportunity to see how green principles can be applied in the chemical industry.

The chemistry of alcohols, phenols, aldehydes, ketones, and carboxylic acids is studied in some detail. You will see how alcohols can react with carboxylic acids and acid anhydrides to produce esters, and that esters can also be made from phenols. You will also apply your understanding of acids and bases.

You will learn more about some ideas introduced in earlier chapters:

- [ ] hydrogen bonding (**The ozone story**)
- [ ] alcohols (**Developing fuels**)
- [ ] oxidation (**Elements from the sea**)
- [ ] alkenes (**Developing fuels**)
- [ ] equilibria (**Elements from the sea**)
- [ ] acids (**Elements of life**)
- [ ] the interaction of radiation with matter (**Elements of life and The ozone story**)
- [ ] bond polarity (**The ozone story**)
- [ ] mass spectrometry (**Elements of life**)
- [ ] atom economy (**Elements from the sea**).

Maths skills checklist

In this unit, you will need to use the following maths skills. You can find support for these skills on Kerboodle and through MyMaths.

- [ ] Translate information between graphical, numerical, and algebraic forms.
- [ ] Analyse and use data in a range of contexts.

MyMaths.co.uk
Bringing Maths Alive

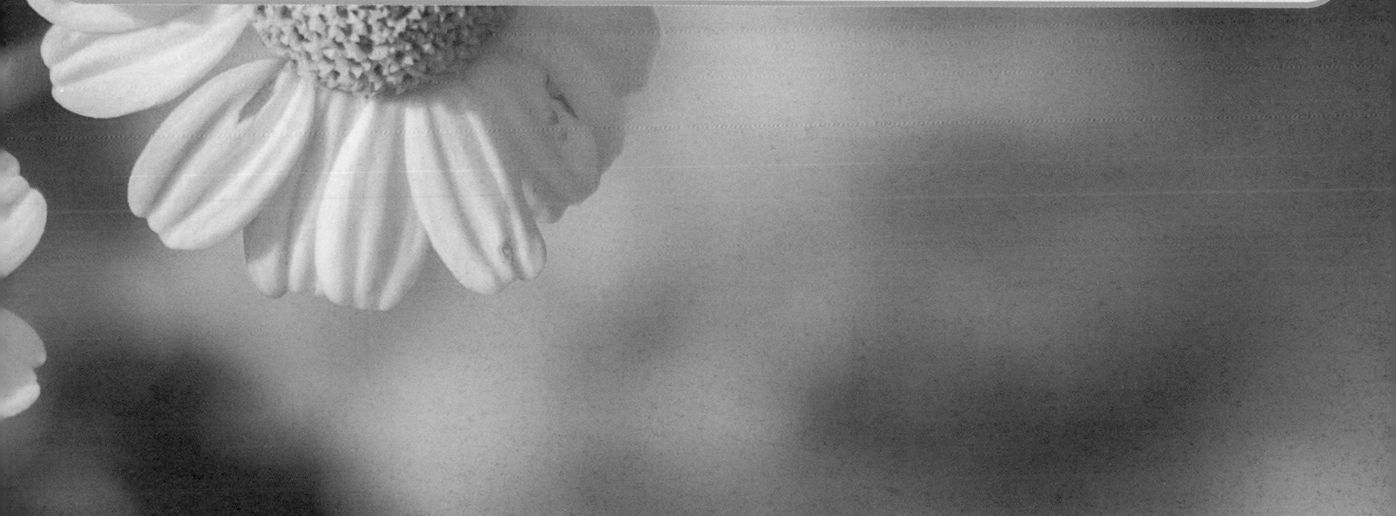

WM 1 The development of modern ideas about medicines

Specification reference: WM(a), WM(b), WM(d), WM(f), WM(h)

Learning outcomes

Demonstrate and apply knowledge and understanding of:

→ the formulae of the following homologous series: carboxylic acids, acid anhydrides, esters, aldehydes, ketones, ethers

→ primary, secondary, and tertiary alcohols in terms of the differences in structures

→ the following reactions of alcohols and two-step syntheses involving these reactions and other organic reactions in the specification:

- with carboxylic acids, in the presence of concentrated sulfuric acid or concentrated hydrochloric acid (or with acid anhydrides) to form esters

- oxidation to carbonyl compounds (aldehydes and ketones) and carboxylic acids with acidified dichromate(VI) solution, including the importance of the condition (reflux or distillation) under which it is done

- dehydration to form alkenes using heated Al_2O_3 or refluxing with concentrated H_2SO_4

- substitution reactions to make haloalkanes

→ techniques and procedures for preparing and purifying a liquid organic product including the use of a separating funnel and of Quickfit or reduced scale apparatus for distillation and heating under reflux

→ the term elimination reaction.

When there is something wrong with your body, a **medicine**, such as aspirin or penicillin, prevents things getting worse and can help to bring about a cure. The active ingredients of medicines are **drugs** – substances that alter the way your body works. If your body is already working normally the drug will not be beneficial, and if the drug throws the body a long way off balance it may even be a poison. Not all drugs are medicines – alcohol and nicotine are not medicines, but they *are* drugs. Some drugs, such as opium, may or may not be medicines depending on your state of health.

The study of drugs and their action is called **pharmacology**. Making and dispensing medicines is called **pharmacy**.

People have been using medicines for thousands of years – most of that time with no idea how they worked. Their effectiveness was discovered by trial and error and sometimes there were disastrous mistakes. Today's medicines are increasingly designed to have specific effects – this is becoming easier as scientists learn more about the body's chemistry and begin to understand the intricate detail of the complex molecules from which people are made. Work at this level comes into the field of molecular pharmacology.

Medicines from nature

Modern pharmacy has its origins in folklore, and the history of medicine abounds with herbal and folk remedies. Many of these can be explained in present day terms and the modern pharmaceutical industry investigates 'old wives' tales' to see if they lead to important new medicines.

▲ Figure 2 Foxglove contains the compound digitalin, which is active against heart disease

▲ Figure 1 Feverfew has been used since ancient times for the treatment of migraine – research in the 1970s confirmed that it was an effective medicine for this disorder

In 400 BC, Hippocrates recommended a brew of willow leaves to ease the pain of childbirth, and in 1763 the Reverend Edward Stone, an English clergyman living in Chipping Norton, Oxfordshire, used a willow bark brew to reduce fevers.

Unknown at the time, it was the substance salicin that was extracted from the willow bark. Salicin has no pharmacological effect by itself. The body converts it by hydrolysis and oxidation into the active chemical, salicylic acid. Salicin can be oxidised to salicylic acid because it contains a primary alcohol group. You need to know about the structure of alcohols, and their reactions, to fully understand this oxidation.

▲ Figure 3 Extracts from willow trees have been used in medicine for thousands of years

Chemical ideas: Organic chemistry: modifiers 13.3

Alcohols

Physical properties of alcohols

Like water molecules, alcohol molecules are polar because of the polarised O—H bond. In both water and alcohols, there is a special sort of strong attractive force between the molecules due to hydrogen bonds.

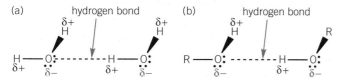

▲ Figure 4 Hydrogen bonding in water (a) and alcohols (b)

Hydrogen bonds are not as strong as covalent bonds, but are stronger than other attractive bonds between covalent molecules. When a liquid boils, these forces must be broken so the molecules escape from the liquid to form a gas. This explains why the boiling points of alcohols are higher than those of corresponding alkanes with similar relative molecular mass M_r. For example, ethanol ($M_r = 46.0$) is a liquid, whilst propane ($M_r = 44.0$) is a gas at room temperature.

Hydrogen bonding between alcohol and water molecules (Figure 5) also explains why the two liquids mix together.

Table 1 shows the solubility of some alcohols in water. As the hydrocarbon chain becomes longer and the molecule becomes larger, the influence of the −OH group on the properties of the molecule becomes less important. So the properties of the higher alcohols get more and more like those of the corresponding alkane.

> **Synoptic link**
> You studied the naming of alcohols in Chapter 2, Developing fuels.

> **Synoptic link**
> You studied hydrogen bonding, as well as other intermolecular bonds, in Chapter 4, The ozone story.

▼ Table 1 Solubility of primary alcohols in water

Name	Formula	Solubility / g per 100 g water
methanol	CH_3OH	fully soluble
ethanol	CH_3CH_2OH	fully soluble
propan-1-ol	$CH_3CH_2CH_2OH$	fully soluble
butan-1-ol	$CH_3CH_2CH_2CH_2OH$	8.0
pentan-1-ol	$CH_3CH_2CH_2CH_2CH_2OH$	2.7
hexan-1-ol	$CH_3CH_2CH_2CH_2CH_2CH_2OH$	0.6

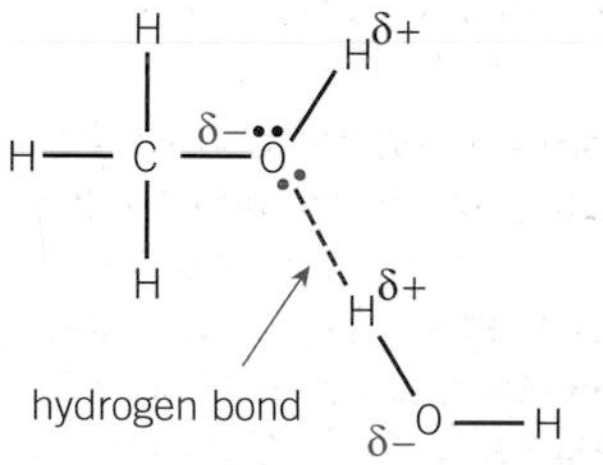

▲ Figure 5 Hydrogen bonding between water molecules and alcohol molecules

Reactions of alcohols

There are of three types of alcohols – **primary alcohols**, **secondary alcohols**, and **tertiary alcohols** – named according to the position of the –OH group (Table 2).

▼ **Table 2** *Primary, secondary, and tertiary alcohols*

Type of alcohol	Position of –OH group	General formula	Example
primary	–OH bonded to a carbon bonded to *one* other carbon atom	RCH_2OH	butan-1-ol
secondary	–OH bonded to a carbon bonded to *two* other carbon atoms	$RCH(OH)R$	butan-2-ol
tertiary	–OH bonded to a carbon bonded to *three* other carbon atoms	$R_2C(OH)R$	2-methylpropan-2-ol

The reactions of alcohols depend on the type of alcohol involved.

Oxidation

The –OH group can be oxidised by strong oxidising agents, such as acidified potassium dichromate(VI), $K_2Cr_2O_7$, solution. The orange dichromate(VI) ion, $Cr_2O_7^{2-}$ (aq), is reduced to green chromate(III) ions, Cr^{3+}(aq), in the reaction. On being oxidised, the –OH group is converted into a carbonyl, C=O, group, and the reaction mixture turns from orange to green.

In this reaction two atoms of hydrogen are being removed – one from the oxygen atom, and one from the carbon atom. Oxidation of the –OH group will not take place unless there is a hydrogen atom on the carbon atom to which the –OH is attached.

The product is a **carbonyl** compound – an aldehyde or a ketone. The type of product you get depends on the type of alcohol you start with and the reaction conditions used.

Primary alcohols, such as ethanol, are initially oxidised to **aldehydes**. The aldehyde is then oxidised to a carboxylic acid, in the presence of excess oxidising agent.

Synoptic link

You encountered oxidation and reduction in Chapter 3, Elements from the sea.

Study tip

$Cr_2O_4^{2-}/H^+$ does not exist so when asked to give an example of such an oxidising agent it should be $K_2Cr_2O_7$(aq)/H^+(aq) *or* acidified potassium chromate(VII) solution.

Study tip

You do not need to know the reaction mechanism for this type of reaction.

If the aldehyde is required it can be distilled in situ out from the reaction mixture before it is oxidised further. In this case, having the alcohol in excess reduces the likelihood of the aldehyde being further oxidised.

Secondary alcohols, such as propan-2-ol, are oxidised to **ketones** by heating under reflux, in the presence of an oxidising agent.

$$H-\overset{\overset{\displaystyle H}{|}}{\underset{\underset{\displaystyle H}{|}}{C}}-\overset{\overset{\displaystyle H}{|}}{\underset{\underset{\displaystyle OH}{|}}{C}}-\overset{\overset{\displaystyle H}{|}}{\underset{\underset{\displaystyle H}{|}}{C}}-H \quad \xrightarrow[\text{reflux}]{[O]} \quad H-\overset{\overset{\displaystyle H}{|}}{\underset{\underset{\displaystyle H}{|}}{C}}-\overset{\overset{\displaystyle}{\|}}{\underset{\underset{\displaystyle O}{}}{C}}-\overset{\overset{\displaystyle H}{|}}{\underset{\underset{\displaystyle H}{|}}{C}}-H \;+\; H_2O$$

propan-2-ol $\qquad\qquad$ propanone (a ketone)

The ketone is not oxidised further as this would involve breaking a strong, covalent C—C bond.

Tertiary alcohols, such as 2-methylpropan-2-ol, do not oxidise because they do not have a hydrogen atom on the carbon atom to which the –OH group is attached.

▼ **Table 3** *Oxidation of alcohols*

Type of alcohol	Product(s) of oxidation with acidified potassium dichromate(VI) solution	Final colour of reaction mixture
primary	aldehyde (with no excess oxidising agent), carboxylic acid (with excess oxidising agent)	green
secondary	ketone	green
tertiary	does not oxidise	orange

Heating under reflux
Refluxing is a safe method for heating reactions involving volatile and flammable liquids. The liquid is boiled with a vertically mounted condenser so the vapour condenses and returns back into the reaction mixture (Figure 6).

Aldehydes and ketones
Both aldehydes and ketones contain a carbonyl group, C=O. In an aldehyde, the carbonyl group is at the end of an alkane chain – the functional group is:

$$\overset{\displaystyle R}{\underset{\displaystyle H}{>}}C=O$$

Aldehydes are named using the suffix –al (Figure 7).

In a ketone, the carbonyl group is within an alkane chain – the functional group is:

$$R-\overset{\overset{\displaystyle O}{\|}}{C}-R$$

Ketones are named using the suffix –one (Figure 8).

Activity WM 1.1

In this activity you will have the opportunity to oxidise an alcohol.

Study tip

Sometimes an oxidising agent is represented as [O] in organic reactions.

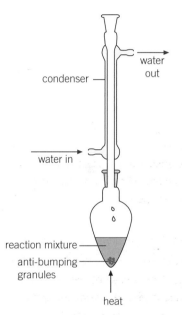

▲ **Figure 6** *Heating under reflux*

$$H-\overset{\overset{\displaystyle H}{|}}{\underset{\underset{\displaystyle H}{|}}{C}}-\overset{\overset{\displaystyle}{}}{C}\overset{\displaystyle O}{\underset{\displaystyle H}{\diagup\!\!\!\diagup}}$$

▲ **Figure 7** *ethanal*

$$H-\overset{\overset{\displaystyle H}{|}}{\underset{\underset{\displaystyle H}{|}}{C}}-\overset{\overset{\displaystyle O}{\|}}{C}-\overset{\overset{\displaystyle H}{|}}{\underset{\underset{\displaystyle H}{|}}{C}}-H$$

▲ **Figure 8** *propanone*

Elimination reaction

A reaction where a small molecule is removed from a larger molecule leaving an unsaturated molecule. In the case of alcohols the small molecule is water.

Activity WM 1.2

In this activity you can practise making a liquid haloalkane from an alcohol and purifying the product.

Dehydration of alcohols

Many alcohols can lose a molecule of water to form an alkene. The conditions needed for this type of reaction are a heated catalyst of alumina, Al_2O_3, at $300\,°C$ or reflux with concentrated sulfuric acid. Under either conditions the corresponding alkene and water are produced.

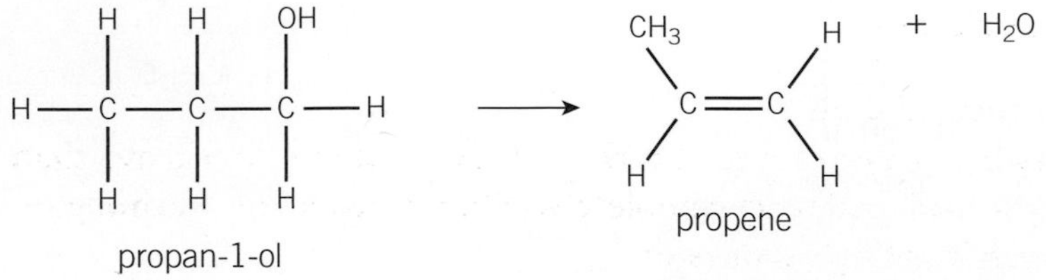

propan-1-ol → propene + H_2O

It can be easier to follow a reaction using the skeletal formulae:

The reaction is described as **dehydration** since it involves the removal of a water molecule from a molecule of the reactant. Dehydration is an example of an **elimination** reaction.

You can also think of an elimination reaction as being the reverse of an addition reaction.

Substitution reactions

Alcohols undergo nucleophilic substitution reactions with halide ions, in the presence of a strong acid, to produce haloalkanes. For example:

$$CH_3CH_2CH_2CH_2OH + H^+ + Br^- \rightarrow CH_3CH_2CH_2CH_2Br + H_2O$$

Formation of esters

There are two ways of converting alcohols into esters – esterification using an acid anhydride and esterification using a carboxylic acid.

Esterification

Esterification is the reaction of an alcohol with a carboxylic acid. For example, ethanol reacts with ethanoic acid to form the ester, ethyl ethanoate.

$$CH_3CH_2OH + CH_3COOH \rightleftharpoons CH_3COOCH_2CH_3 + H_2O$$
ethanol   ethanoic acid   ethyl ethanoate   water

You can write the same reaction using skeletal formulae:

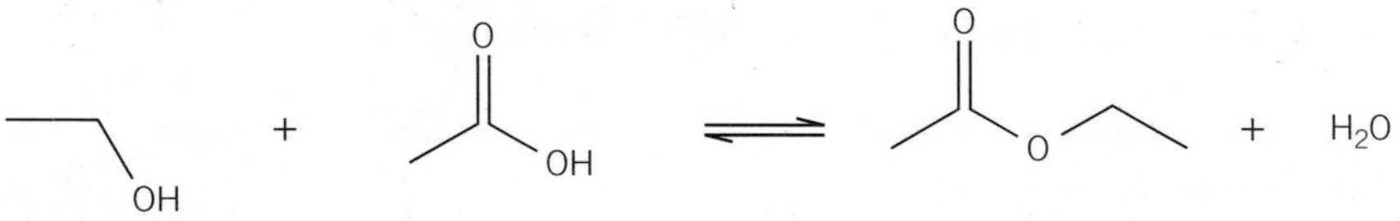

The reaction occurs extremely slowly unless a strong acid catalyst is present. A small amount of either concentrated sulfuric acid or concentrated hydrochloric acid is generally used, and the reaction mixture is heated under reflux. The ⇌ symbol in the equation means the reaction is reversible and comes to an equilibrium, where both reactants and products are present. The ester would then have to be separated from the mixture using distillation and purified.

Look at the structure of ethyl ethanoate. Esters are named from the alcohol and acid that form them – 'ethyl' from ethanol and 'ethanoate' from ethanoic acid.

Using an acid anhydride

Acid anhydrides, derivatives of carboxylic acids, are more reactive than a carboxylic acid, and react completely with an alcohol on warming to give a much higher yield of ester.

$$(CH_3CO)_2O \quad + \quad CH_3CH_2OH \quad \rightarrow \quad CH_3COOCH_2CH_3 \quad + \quad CH_3COOH$$

ethanoic anhydride   ethanol   ethyl ethanoate   ethanoic acid

Ethers

Be careful not to confuse alcohols with **ethers**. Ethers have the same molecular formula as alcohols but have a different structure – they are structural isomers. The general formula of ethers is R—O—R. They are derived from alkanes by substituting an alkoxy group (–OR) for a hydrogen atom. For example:

$$CH_3CH_2OCH_3CH_2$$

When naming an ether, the longer hydrocarbon chain is chosen as the parent alkane.

Purification of organic liquid products

Most common haloalkanes, esters, and ethers are organic liquids. During synthesis only a crude product is produced. This has to be purified before it can be used.

The main techniques are use of a separating funnel, use of drying agents, and simple distillation.

Synoptic link
You first encountered equilibrium in Chapter 3, Elements from the sea.

Synoptic link
You will learn more about naming esters in Chapter 7, Polymers and life.

Activity WM 1.3
This activity allows you to check your understanding of alcohol reactions.

Synoptic link
You can find out more about the purification of organic liquid products in Techniques and procedures.

Summary questions

1 a For each of the skeletal formula below give the type of function group present.

i (1 mark)

ii (1 mark)

iii (1 mark)

iv (1 mark)

v

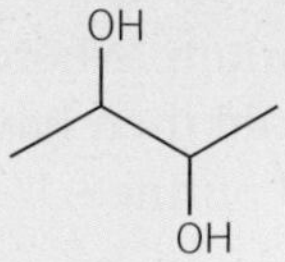

(1 mark)

vi

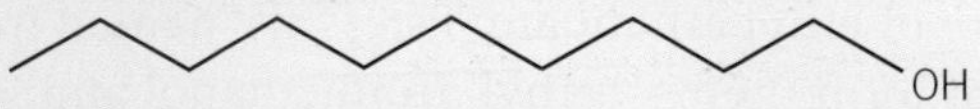

(1 mark)

vii

(1 mark)

b Give the systematic names of compounds **i** to **vi**. (7 marks)

2 Why does ethanol mix with water but hexanol does not? (2 marks)

3 Look at the following compounds:

A   B 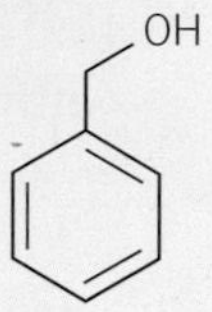   C

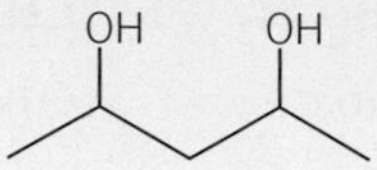

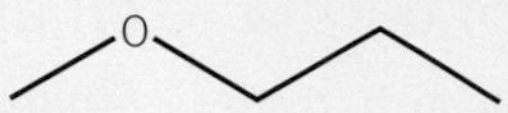

D 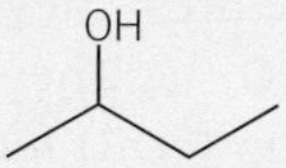   E

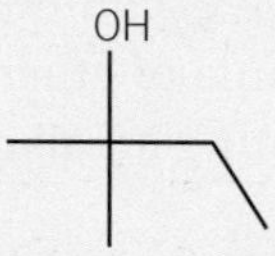

a Which compound(s) is(are) alcohols? (1 mark)
b Which compound(s) is(are) ethers? (1 mark)
c Which compound(s) is(are) diols? (1 mark)
d Which compounds are isomers? (1 mark)
e Which compound do you think will be the most volatile? (1 mark)
f Which compound would you expect to be the most soluble in water? (1 mark)
g Which compound(s) would form carboxylic acid on refluxing with excess acidified potassium dichromate(VI)? (1 mark)
h Which compound would not react on refluxing with excess acidified potassium dichromate(VI)? (1 mark)

4 Here are the boiling points and relative molecular masses (M_r) of a number of substances:

Substance	Boiling point / °C	M_r
water, H_2O	100.0	18.0
ethane, CH_3CH_3	−88.5	30.0
ethanol, CH_3CH_2OH	78.0	46.0
butan-1-ol, $CH_3CH_2CH_2CH_2OH$	117.0	74.0
ethoxyethane, $CH_3CH_2OCH_2CH_3$	35.0	74.0

Use ideas about bonds between molecules to explain why
a ethanol has a higher boiling point than ethane (2 marks)
b water has a higher boiling point than ethanol (2 marks)
c butan-1-ol has a higher boiling point than ethanol (2 marks)
d butan-1-ol has a higher boiling point than ethoxyethane. (2 marks)

5 **a** Identify a compound that would give but-1-ene on dehydration.

(1 mark)

b Write an equation for any dehydration identified. *(2 marks)*

6 For each of the compounds **i** to **v**, what would you see on:

a addition of sodium hydroxide solution *(1 mark)*

b reflux with excess potassium chromate(VII) solution. *(1 mark)*

i (phenol structure with OH)

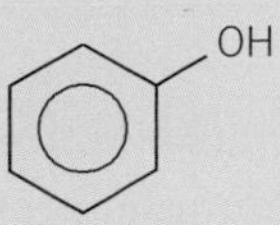

ii (ethanol structure with OH)

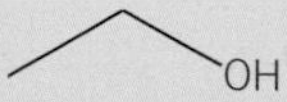

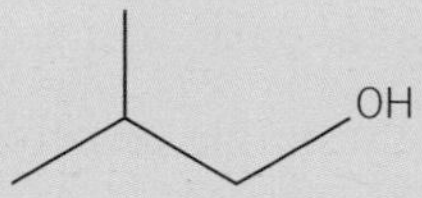

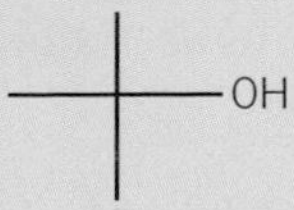

iii (2-methylpropan-1-ol structure with OH)

iv (2-methylpropan-2-ol structure with OH)

v (ethanoic acid structure with O and OH)

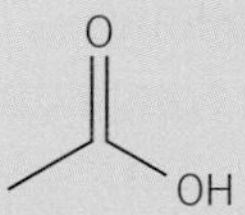

7 For each of the following reactions decide:

a what type of reaction will occur.

b what type of product will be formed.

i propan-1-ol warmed with ethanoic anhydride *(2 marks)*

ii butan-1-ol refluxed with excess potassium dichromate(VI) solution *(2 marks)*

iii butan-2-ol refluxed with concentrated sulfuric acid *(2 marks)*

iv 2-methyl-propan-2-ol refluxed with excess potassium dichromate(VI) solution *(2 marks)*

v butan-1-ol refluxed with propanoic acid with a few drops of concentrated sulfuric acid added *(2 marks)*

vi ethanol (in excess) heated with sodium dichromate(VI) solution, with products distilled out as the reaction proceeds *(2 marks)*

vii butan-2-ol heated to 300 °C over aluminium oxide, Al_2O_3 *(2 marks)*

viii 2-methyl-propan-2-ol shaken together with concentrated hydrochloric acid *(2 marks)*

ix butan-2-ol refluxed with potassium dichromate(VI) solution *(2 marks)*

WM 2 Identifying the active chemical in willow bark

Specification reference: WM(a), WM(c)

How do chemists find out the chemical structure of compounds like salicylic acid? One way is to use chemical reactions to test for the presence of particular functional groups.

Relatively simple test-tube experiments can often be used effectively in the identification of unknown substances. For example, orange/brown bromine solutions go colourless on the addition of molecules with a double bond between carbon atoms.

Three chemical tests are particularly helpful in providing clues about the structure of salicylic acid:

1. An aqueous solution of the compound is weakly acidic.

2. Salicylic acid reacts with alcohols (such as ethanol) to produce compounds called esters. Esters have strong pleasant odours, often of fruit or flowers.

3. A neutral solution of iron(III) chloride turns an intense pink colour when salicylic acid is added.

Tests 1 and 2 are characteristic of carboxylic acids (compounds containing the –COOH functional group). Test 3 indicates the presence of a phenol group (an –OH group attached to a benzene ring).

Chemical ideas: Organic chemistry: modifiers 13.4

Carboxylic acids and phenols

Carboxylic acids and their derivatives

Carboxylic acids contain the carboxyl group (Figure 1). This formula is often abbreviated to –COOH, although the two oxygen atoms are not joined together. The structure of the rest of the molecule can vary widely and this gives rise to a large number of different carboxylic acids. When the remainder of the molecule is an alkyl group, the acids can be represented by the general formula R—COOH.

A carboxyl group can also be attached to a benzene ring (Figure 4).

The –OH group in the carboxyl group can be replaced by other groups to give a whole range of carboxylic acid derivatives (Table 1).

▲ **Figure 1** *Carboxylic acid functional group*

▼ Table 1 *Some examples of carboxylic acid derivatives*

Acid derivative	Functional group	Example
ester RCOOR	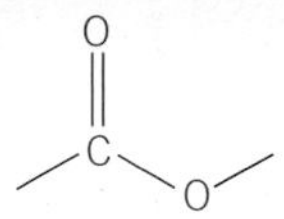	ethyl ethanoate 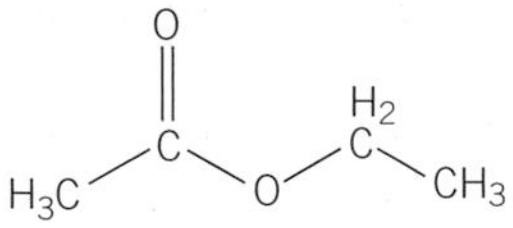
acid anhydride $(RCO)_2O$	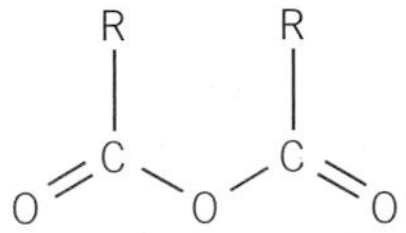	ethanoic anhydride 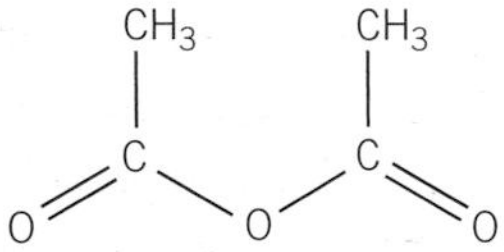

The –OH group in alcohols, phenols, and carboxylic acids

The hydroxyl group, –OH, can occur in three different environments in organic molecules.

- As part of a carboxyl group in carboxylic acids.
- Attached to an alkane chain in alcohols. There are three types of alcohols – primary, secondary, and tertiary – according to the position of the –OH group (Topic WM 1).
- Attached to a benzene ring in phenols (Figure 5).

Although phenols look similar to alcohols, they behave very differently. It is generally true that functional groups behave differently when attached to an aromatic ring from when they are attached to an alkyl group.

Acidic properties of the –OH group

The –OH group reacts with water.

$$R\text{—}OH + H_2O \rightleftharpoons R\text{—}O^- + H_3O^+$$

R stands for the group of atoms which makes up the rest of the molecule.

Water dissociates to a very small extent, so, at any one time a small number of water molecules donate H^+ ions to other water molecules – water behaves as a weak acid.

$$H\text{—}OH + H_2O \rightleftharpoons H\text{—}O^- + H_3O^+$$

A similar reaction occurs with ethanol, but to a lesser extent. The equilibrium lies further to the left, and ethanol is a weaker acid than water.

With phenol, the equilibrium lies further to the right than in water – phenol is slightly more acidic than water. Carboxylic acids are even more acidic, though still weak. The order of acid strength is

ethanol < water < phenol < carboxylic acids

Phenols and carboxylic acids are strong enough acids to react with strong bases such as sodium hydroxide, NaOH, to form salts (Figure 7).

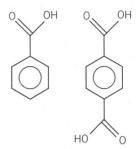

▲ **Figure 2** *Benzenecarboxylic acid or benzoic acid (left) and benzene-1,4-dicarboxylic acid (right)*

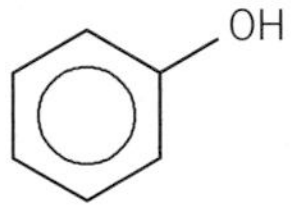

▲ **Figure 3** *Phenol*

Activity WM 2.1

This activity helps you recognise the formulae and structure of a range of different homologous series you have met so far.

Synoptic link

You first looked at the properties of acids in Chapter 1, Elements of life.

Synoptic link

You first encountered equilibrium in Chapter 3, Elements from the sea.

Synoptic link

Acid–base equilibria are explored more in Chapter 8, Oceans.

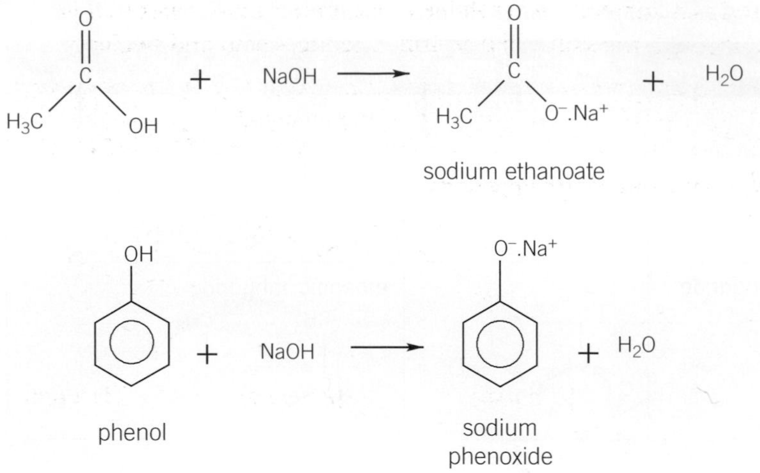

$$\text{(structure: } H_3C-C(=O)-OH\text{)} + NaOH \longrightarrow \text{(structure: } H_3C-C(=O)-O^-.Na^+\text{)} + H_2O$$

sodium ethanoate

Study tip

- Carboxylic acids and phenols react with NaOH(aq) to form sodium salts.

- Carboxylic acids, but not phenols, react with Na_2CO_3(aq) to form the sodium salt of the acid. The reaction fizzes and carbon dioxide is released.

- Alcohols do not react with either NaOH(aq) or Na_2CO_3(aq).

Synoptic link

You will find out more about complexes in the A level module on Developing metals.

Activity WM 2.2

In this activity you can investigate the reactions of phenols and carboxylic acids.

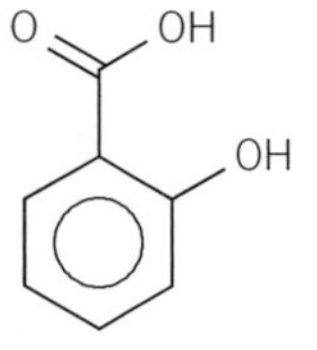

▲ **Figure 8** *2-hydroxybenzoic acid (salicylic acid)*

▲ **Figure 9** *Aspirin*

OH (structure) + NaOH ⟶ O⁻.Na⁺ (structure) + H_2O

phenol   sodium phenoxide

▲ **Figure 7** *The reaction of ethanoic acid and phenol with sodium hydroxide – a strong base*

The salts produced are ionic and remain in solution after the reaction. Only carboxylic acids have high enough concentrations of H^+(aq) ions to produce carbon dioxide on reaction with carbonates.

$$CO_3{}^{2-}(aq) + 2H^+(aq) \rightarrow CO_2(g) + H_2O(l)$$

As such carboxylic acids make carbonates fizz, but alcohols and phenols do not. (The H_3O^+ ion is often written as H^+(aq).)

The iron(III) chloride test for phenol and its derivatives

Some groupings of atoms can become attached to metal ions and form **complexes**. The $-C=C-OH$ group (called the enol group) can form a purple complex with Fe^{3+} ions in neutral solution. Only phenol and its derivatives have such an arrangement of atoms and they are the only ones to give a colour with neutral iron(III) chloride solution. This is used as a test for phenol and its derivatives. Similar complexes are used to make the colours of some inks.

Ester formation

The formation of esters using alcohols was described in Topic WM 1. Esters can also be made using phenols and acid anhydrides in alkaline conditions.

$$C_6H_5OH + (CH_3CO)_2O \rightarrow CH_3COOC_6H_5 + CH_3COOH$$
phenol   ethanoic anhydride   phenyl ethanoate   ethanoic acid

Phenols *do not* react with carboxylic acids to produce esters.

Esters from salicylic acid

Figure 8 shows the structure of 2-hydroxybenzoic acid (salicylic acid). There are two ways of esterifying it. The phenol –OH group can react with an acid anhydride, or the –COOH group can react with an alcohol.

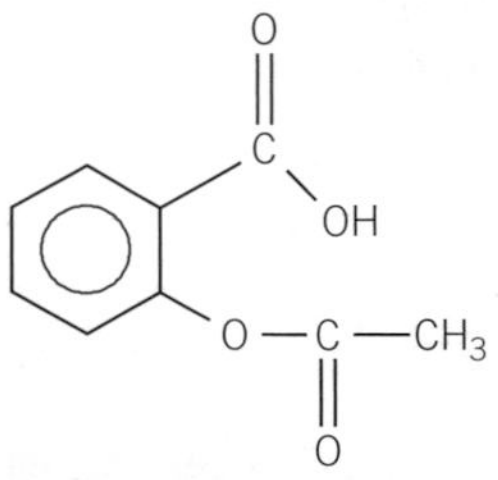

Aspirin is the product of esterifying the phenol –OH group to form 2-ethanoyloxybenzoic acid (Figure 9). It is quite soluble in water, so it can be absorbed into the bloodstream through the stomach wall.

The product of reacting the –COOH group with methanol is called methyl 2-hydroxybenzoate (Figure 10). This is better known as oil of wintergreen and is used as a linament. It is soluble in fats rather than water so it is absorbed through the skin – like aspirin, it reduces pain and swelling.

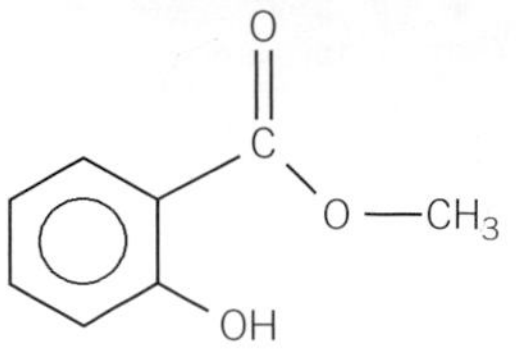

▲ **Figure 10** *methyl 2-hydroxybenzoate*

Summary questions

1 a Name the following compounds.

i

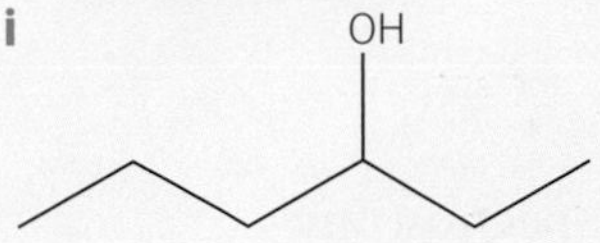

(*2 marks*)

ii

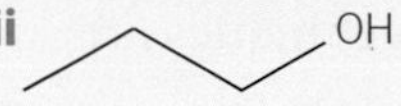

(*2 marks*)

iii

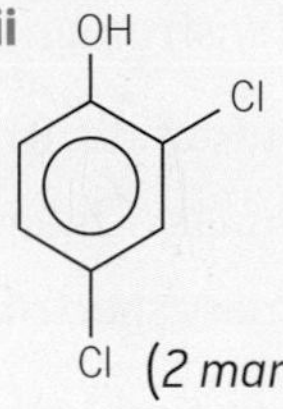

(*2 marks*)

iv

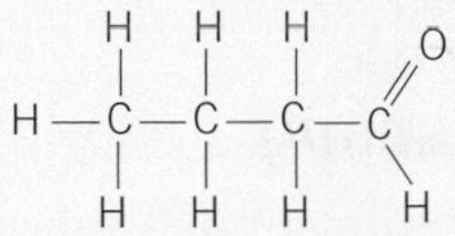

(*2 marks*)

v

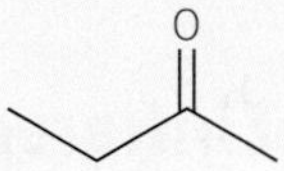

(*2 marks*)

vi $CH_3CH(CH_3)(OH)CH_2CH_3$　　　　　　(*2 marks*)

vii HCOOH　　　　　　　　　　　　　　(*2 marks*)

b From compounds **i** to **vii** identify:

a a secondary alcohol　(*1 mark*)　　**c** a phenol　　　　(*1 mark*)

b an aldehyde　　　　(*1 mark*)　　**d** a ketone　　　　(*1 mark*)

e an aliphatic alcohol that is not easily oxidised on heating with acidified potassium dichromate(VI)　(*1 mark*)

f produces a purple colour with neutral aqueous iron(III) chloride　(*1 mark*)

g gives carbon dioxide with sodium carbonate　(*1 mark*)

h produces a carboxylic acid on refluxing with excess acidified potassium dichromate(VI)　(*1 mark*)

i can be produced by the oxidation of methanol.　(*1 mark*)

Study tip

You may need to look back at Topic WM 1 to answer question 2.

2 Draw the structure of the organic product when each of the following reacts with excess NaOH.

a

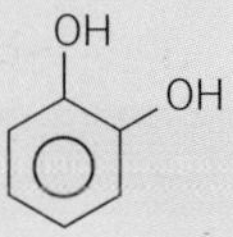

(*2 marks*)

b

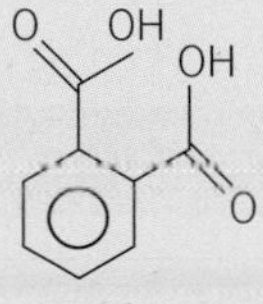

(*2 marks*)

3 Phenols and carboxylic acids are weak acids and show typical acid properties. Write a balanced equation for each of the following reactions.

a phenol and sodium hydroxide　(*2 marks*)

b propanoic acid and potassium hydroxide　(*2 marks*)

c butanoic acid and sodium carbonate　(*2 marks*)

4 Draw the structure of the ester produced in the following reactions.

a

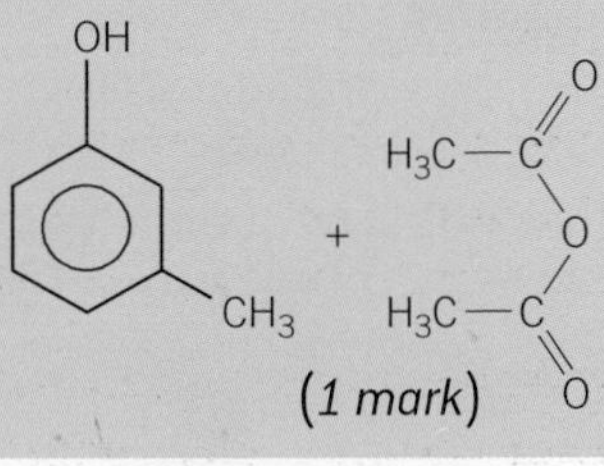

(*1 mark*)

b

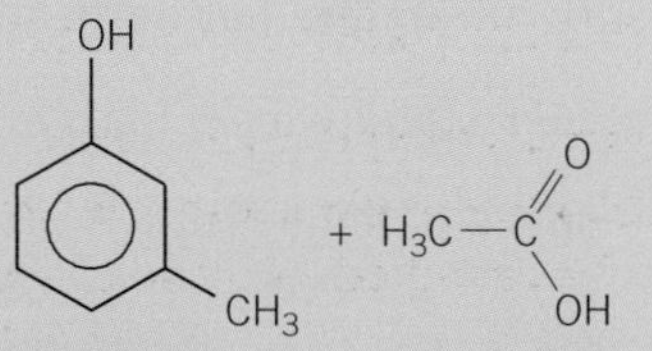

(*1 mark*)

WM 3 Infrared spectroscopy

Specification reference: WM(j)

Although chemical tests provide evidence for the presence of carboxylic acid and phenol groups in salicylic acid, instrumental techniques are today's most efficient research tools. Three frequently used instrumental techniques are:

- infrared (IR) spectroscopy
- nuclear magnetic resonance (nmr) spectroscopy
- mass spectroscopy.

This section focuses on infrared spectroscopy.

Making use of infrared spectroscopy

One of the very first things done with any unidentified substance is to record an infrared (IR) spectrum. Figure 1 shows the IR spectrum of salicylic acid. Different function groups give different peaks on an IR spectrum.

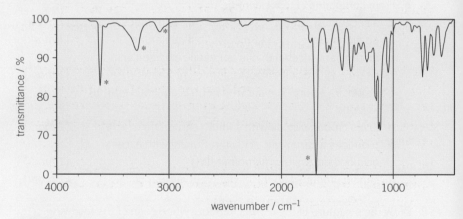

▲ **Figure 1** *Infrared spectrum of salicylic acid*

Chemical ideas: Radiation and matter 6.4

Infrared spectroscopy

The energy possessed by molecules is **quantised** – molecules must take a small number of definite energy values rather than any energy value.

In **infrared spectroscopy** substances are exposed to radiation in the frequency range 10^{14}–10^{13} Hz, that is, wavelengths 2.5–15 μm. This makes vibrational energy changes occur in the molecules, which absorb infrared radiation of specific frequencies.

Frequency and wavelength are related.

speed of light c (m s^{-1}) = wavelength λ (m) × frequency v (s^{-1})

Synoptic link

You first encountered the equation $c = \lambda v$ in Chapter 1, Elements of life.

The speed of light c is a constant – $3.00 \times 10^8\,\mathrm{m\,s^{-1}}$. From this equation, a direct measure of frequency is $\frac{1}{\lambda}$. This is the wavenumber of the radiation and usually measured in $\mathrm{cm^{-1}}$. This is the unit recorded on an **infrared spectrum**. Figure 2 shows the relationship between wavenumber, wavelength, and frequency. Note the directions of the arrows.

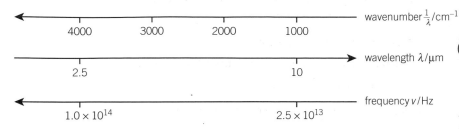

▲ **Figure 2** *The relationship between wavenumber, wavelength, and frequency*

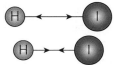

▲ **Figure 3** *Stretching in the HI molecule*

Bond deformation

Simple diatomic molecules such as HCl, HBr, and HI can only vibrate in one way – stretching – where the atoms pull apart and then push together again (Figure 3). For these molecules there is only one vibrational infrared absorption. This corresponds to the molecules changing from their lowest vibrational energy state to the next higher level, in which the vibration is more vigorous.

The frequencies of the absorptions are different for each molecule. This is because the energy needed to excite a vibration depends on the strength of the bond holding the atoms together – weaker bonds require less energy. Table 1 shows how the bond strength is related to infrared absorption.

In more complex molecules, more bond deformations are possible. Most of these involve more than two atoms. For example, carbon dioxide can vibrate as shown in Figure 4.

More complex molecules can have many vibrational modes, descriptions such as rocking, scissoring, twisting, and wagging. Add these to the fact that the molecules also contain bonds with different bond enthalpies and you may have a very complicated spectrum.

▼ **Table 1** *Bond enthalpies and infrared absorptions for the hydrogen halides*

Compound	Bond enthalpy / $\mathrm{kJ\,mol^{-1}}$	Infrared absorption / $\mathrm{cm^{-1}}$
HCl	+432	2886
HBr	+366	2559
HI	+298	2230

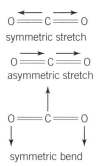

▲ **Figure 4** *Vibrations in the CO_2 molecule*

The important point to remember about infrared spectroscopy is that you do not try to explain the whole spectrum – you look for one or two signals that are characteristic of particular bonds.

Figure 5 shows the infrared spectrum of ethanol and some of the vibrations that give rise to the signals.

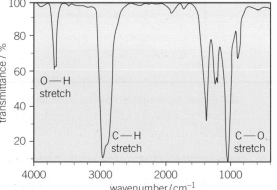
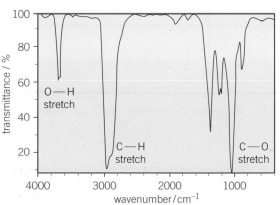

▲ **Figure 5** *The infrared spectrum of ethanol in the gas phase (left) and the ethanol molecule, showing the vibrations which give rise to some characteristic absorptions*

Interpreting the spectra

An infrared spectrometer detects absorptions and produces an infrared spectrum. The infrared spectrum of but-1-ene (Figure 7) shows most of the characteristic absorptions of an unsaturated hydrocarbon containing a C=C double bond. These have been marked on the spectrum to show the bonds which are responsible.

Activity WM 3.1

In this activity you will be able to use your knowledge to interpret infrared spectra

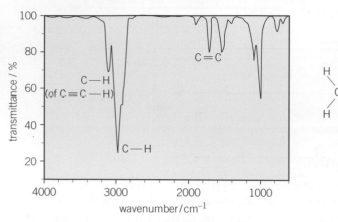

▲ **Figure 7** *Infrared spectrum of but-1-ene*

In general, you can match a particular bond to a particular absorption region. Table 2 gives some examples. The precise position of an absorption depends on the environment of the bond in the molecule, so the wavenumber is only quoted for regions where the absorptions are expected to arise.

Figure 8 shows the infrared spectrum of propanone. The absorption at around $1740 \, cm^{-1}$ is characteristic of the carbonyl, C=O, group. The absorption is very intense compared with the C=C absorption in a similar region for but-1-ene (Figure 7).

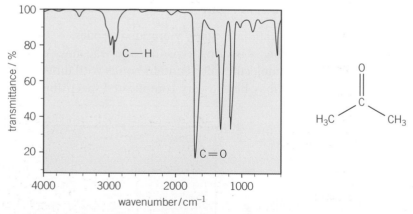

▲ **Figure 8** *Infrared spectrum of propanone*

Why are some absorptions intense while others are weaker? The strongest infrared absorptions arise when there is a large change in bond polarity associated with the vibration. Therefore O—H, C—O, and C=O bonds, which are very polar, give more intense absorptions than the non-polar C—H, C—C, and C=C bonds.

Hydrogen bonding affects the absorption due to the O—H stretching vibration. Figure 5 showed the infrared spectrum of ethanol in the gas phase. There is little hydrogen bonding and the O—H absorption is a sharp peak at 3670 cm^{-1}.

Now look at Figure 9, which shows the infrared spectrum of a liquid film of ethanol. Hydrogen bonding between the hydroxyl groups changes the O—H vibration and the absorption becomes much broader. It is also shifted to a lower wavenumber and shows maximum absorption at 3340 cm^{-1}.

Synoptic link

You found out about bond polarity in Chapter 4, The ozone story.

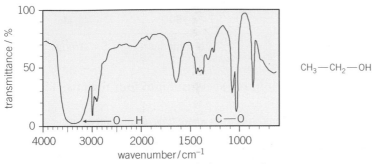

CH_3—CH_2—OH

▲ **Figure 9** *Infrared spectrum of ethanol (liquid film)*

Study tip

The spectrum in Figure 9 has a non-linear scale for the horizontal axis. This is common for many instruments, so always check the scale carefully when reading off values of wavenumbers for absorptions.

Using the combination of wavenumber and intensity it is possible to interpret simple infrared spectra. Reference tables, like Table 2, are used when interpreting simple infrared spectra.

▼ **Table 2** *Characteristic infrared absorptions in organic molecules*

Bond	Location	Wavenumber/cm^{-1}	Intensity
C—H	alkanes	2850–2950	M—S
	alkenes, arenes,	3000–3100	M—S
	alkynes	*ca* 3300	S
C=C	alkenes	1620–1680	M
	arenes	several peaks in range 1450–1650	variable
C≡C	alkynes	2100–2260	M
C=O	aldehydes	1720–1740	S
	ketones	1705–1725	S
	carboxylic acids	1700–1725	S
	esters	1735–1750	S
	amides	1630–1700	M
C—O	alcohols, ethers, esters	1050–1300	S
C≡N	nitriles	2200–2260	M
C—F	fluoroalkanes	1000–1400	S
C—Cl	chloroalkanes	600–800	S
C—Br	bromoalkanes	500–600	S

Study tip

Remember the infrared spectrum helps you decide the type of bonds present. It cannot tell you where they are in the molecule or how many of each type there are.

▼ Table 2 *(continued)*

Bond	Location	Wavenumber / cm^{-1}	Intensity
O—H	alcohols, phenols	3600–3640	S
	*alcohols, phenols	3200–3600	S (broad)
	*carboxylic acids	2500–3200	M (broad)
N—H	primary amines	3300–3500	M—S
	amides	*ca* 3500	M

* hydrogen-bonded   M = medium   S = strong

It is helpful to divide the infrared spectrum into four regions (Table 3).

▼ Table 3 *Regions in the infrared spectrum where typical absorptions occur*

Absorption range / cm^{-1}	Bonds responsible	Examples
4000–2500	single bonds to hydrogen	O—H, C—H, N—H
2500–2000	triple bonds	C≡C, C≡N
2000–1500	double bonds	C=C, C=O
below 1500	various	C—O, C—X (halogen)

Below 1500 cm^{-1} the spectrum can be quite complex and it is more difficult to assign absorptions to particular bonds. This region is characteristic of the particular molecule and is often called the **fingerprint region**. It is useful for identification purposes, for example, if you need to compare two spectra to find out if they are spectra of the same compound. It is only rarely used to identify functional groups. Figure 10 summarises the information in Table 3.

Aromatic compounds often exhibit complex absorption patterns in the fingerprint region (see Figures 13 and 14). Such compounds can often be identified by comparing their infrared spectra with reference spectra.

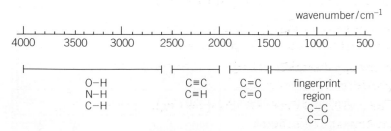

▲ Figure 10 *Typical regions of absorption in the infrared spectrum*

Worked example: Using infrared spectra – butane, $CH_3CH_2CH_2CH_3$

What are the key features shown in the infrared spectrum in Figure 11 that indicates the molecule is butane?

1   A strong absorption at 2970 cm^{-1} characteristic of C—H stretching in aliphatic compounds.

2   No indication of any functional groups.

3   An alkane fits with these features.

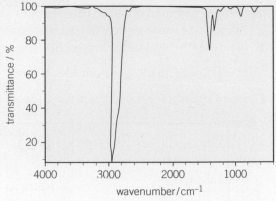

▲ Figure 11 Infrared spectrum of butane

 Worked example: Using infrared spectra – methylbenzene,

What are the key features shown in the infrared spectrum in Figure 12 that indicates this molecule is methylbenzene?

1   C—H absorptions just above 3000 cm^{-1} for the C—H on the benzene ring.

2   C—H absorption just below 3000 cm^{-1} for C—H on the methyl group.

3   An absorption pattern around 700 cm^{-1} is typical of a benzene ring with one substituted group.

4   No indication of any functional groups.

5   An alkylbenzene (arene) matches these features.

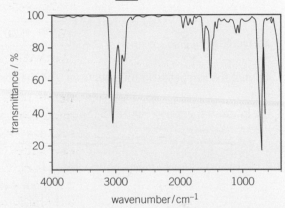

▲ Figure 12 Infrared spectrum of methylbenzene

 Worked example: Using infrared spectra – benzoic acid,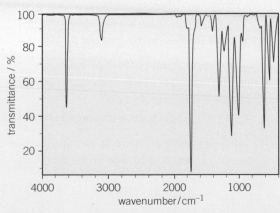

What are the key features shown in the infrared spectrum in Figure 13 that indicate this is benzoic acid?

1   A sharp absorption at 3580 cm^{-1} characteristic of an O—H bond (not hydrogen-bonded).

2   A strong absorption at 1760 cm^{-1} shows the presence of a C=O group.

3   The position of the C—H absorption suggests it is an aromatic compound.

4   An aromatic carboxylic acid fits these features. The identity of the sample could be confirmed by comparing its spectrum with that of an authentic sample of benzoic acid, checking the fingerprint region carefully.

▲ Figure 13 Infrared spectrum of benzoic acid

Summary questions

1 The infrared spectrum of carbon dioxide shows a strong absorption at $2360\,cm^{-1}$.

   a Calculate the wavelength of the radiation absorbed. Give your answer in μm.   *(1 mark)*

   b Use $c = \lambda v$ to calculate the frequency of the radiation absorbed.   *(1 mark)*

   $(c = 3.00 \times 10^8\,m\,s^{-1}$ $1\,\mu m = 1 \times 10^{-6}\,m)$

2 Figure 14 shows the infrared spectrum of phenol.

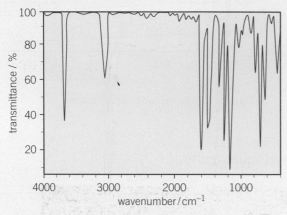

▲ **Figure 14** *Infrared spectrum of phenol*

Identify the key peaks in the spectrum, and the bond to which each corresponds.   *(3 marks)*

3 The infrared spectra in Figure 15 (Spectrum A and Spectrum B) represent butan-2-ol and butanone.

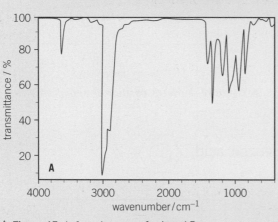 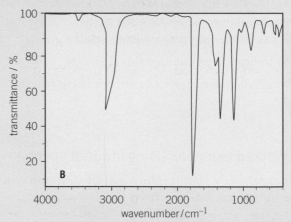

▲ **Figure 15** *Infrared spectra for A and B*

   a Draw structures for butan-2-ol and butan-2-one.   *(2 marks)*

   b Identify the key peaks in each spectrum, and the bond to which each corresponds. Give your answer in the form of a table.   *(4 marks)*

   c Decide which spectrum represents butan-2-ol and which represents butan-2-one.   *(2 marks)*

4 The infrared spectra in Figure 16 represents three compounds – C, D, and E. The compounds are an ester, a carboxylic acid, and an alcohol, though not necessarily in that order.

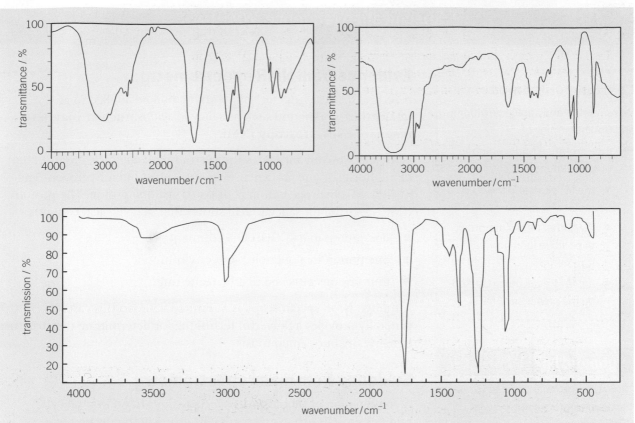

▲ Figure 16 *Infrared spectra for compounds C, D, and E*

a Identify the key peaks in each spectrum, and the bond to which each corresponds.
Give your answer in the form of a table. *(6 marks)*

b Decide which spectrum represents which type of compound. *(3 marks)*

5 Oil of wintergreen has mild pain-killing properties. Its structure is shown below.

Draw up a table to show the key peaks you would expect to see in the infrared spectrum of oil of wintergreen, and the bond that each absorption corresponds to. *(8 marks)*

WM 4 Mass spectrometry

Specification reference: WM(i)

Learning outcomes

Demonstrate and apply knowledge and understanding of:

→ interpretation and prediction of mass spectra:

- the M⁺ peak and the molecular mass
- that other peaks are due to positive ions from fragments
- the M+1 peak being caused by the presence of ^{13}C.

Synoptic link

You will learn more about nmr spectroscopy in Chapter 7, Polymers and life.

Evidence from NMR spectrometry

A second instrumental technique that could be applied to an unidentified compound such as salicylic acid is **nuclear magnetic resonance spectroscopy** (NMR).

NMR spectroscopy measures the surrounding chemical environment of the nuclei of one particular element – most often hydrogen. The nucleus of a hydrogen atom consists of just one proton. The proton NMR spectrum for salicylic acid shows that salicylic acid contains:

- one proton in a –COOH environment
- one proton in a phenolic –OH environment
- four protons attached to a benzene ring.

Although NMR spectroscopy is of limited value in this case, it generally provides a powerful technique for determining the structure of many organic compounds.

Evidence from mass spectrometry

A combination of IR and NMR spectroscopy shows that salicylic acid has an –OH group and a –COOH group both attached to a benzene ring. In other words, a better name for salicylic acid is hydroxybenzoic acid.

However, there are three possible isomeric hydroxybenzoic acids – 2-hydroxybenzoic acid, 3-hydroxybenzoic acid, and 4-hydroxybenzoic acid. A decision about which isomer salicylic acid is can be made by analysis of the mass spectrum of the compound.

Synoptic link

In Chapter 1, Elements of life, you saw how information about the relative abundance of isotopes of an element can be obtained from a mass spectrum.

Chemical ideas: Radiation and matter 6.5b

Mass spectrometry with compounds

Mass spectrometry can be used to find out the atomic mass of elements and the relative abundances of isotopes in an element. In the case of more complex molecules, these can be fragmented, ionised, and detected in a mass spectrometer.

Using mass spectra to investigate the structure of molecules

A typical mass spectrum from an organic compound, such as the one for 2-ethoxybutane (Figure 1), can be quite complex. The spectrum is dislayed as percentage intensity against a value of *m/z*. You can assume this is equivalent to the molecular mass of the ion causing the peak.

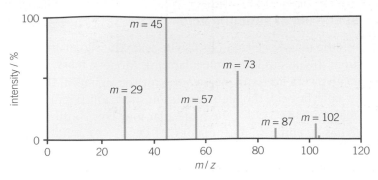

▲ **Figure 1** *Simplified mass spectrum for 2-ethoxybutane, showing the six largest peaks*

The heaviest ion (m = 102) is the one corresponding to the ethoxybutane molecule with just one electron removed – this is called the **molecular ion** M^+. The m/z value for the M^+ ion helps chemists determine the molecular mass of the compound being analysed.

However, this is not the only ion reaching the detector. The M^+ ion fragments into smaller ions and the mass spectrometer detects and analyses these too. The way in which a parent ion breaks down is characteristic of that compound – the **fragmentation pattern**. This is why there are many peaks with a m/z value below the molecular mass of the unfragmented compound.

The most abundant ion gives the strongest detector signal. This is set to 100% in the spectrum and is referred to as the **base peak**. The intensities of all the other peaks are expressed as a percentage of its value. It is not uncommon for the molecular ion peak to be so weak as to be unnoticeable – the spectrum obtained then consists entirely of fragments.

Often a peak appears at a value of M+1. This is because a small proportion of carbon (approximately 1.1%) exist as ^{13}C. If a molecule has incorporated one of these carbon-13 atoms, the molecular mass of the compound will be M+1 (where M is the molecular mass of carbon-12). Because there are proportionally a small number of carbon-13 atoms compared to carbon-12 atoms, the M+1 peak will be small when compared to the peak caused by the M^+ ion.

Interpreting the mass spectrum of 2-ethoxybutane (Figure 1)

The peaks at 29, 45, 57, 73, and 87 are caused by fragmentation of the ion of the parent molecule. This produces a number of different fragment ions of different masses.

The molecular mass of the sample can be easily seen to be 102 and is a result of ionisation of the parent molecule (M^+), in this case $[CH_3CH_2(CH_3)OCH_2CH_3]^+$.

The small peak at 103 is caused by the M^+ ion with a carbon-13 atom present in each ion being detected.

> **Synoptic link**
>
> You will learn how to interpret fragmentation patterns in Chapter 7, Polymers and life.

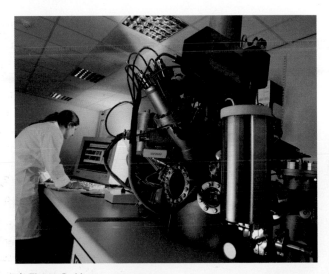

▲ **Figure 2** *Mass spectrometer*

The mass spectrum of salicylic acid

Figure 3 shows the mass spectrum of salicylic acid. The signal at mass 138 is from the parent ion (molecular ion). Modern mass spectrometers give very accurate mass values for the signals in the spectrum. For salicylic acid, the high-resolution spectrum (Figure 3, top) gives a mass of 138.0317 for the molecular ion. This confirms that the substance has an empirical formula of $C_7H_6O_3$.

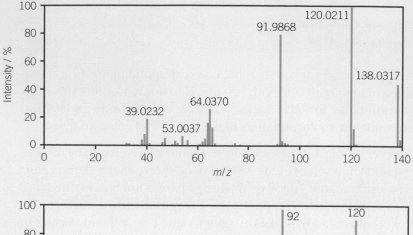

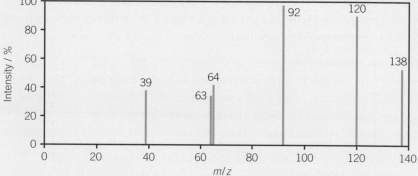

▲ **Figure 3** *Mass spectra of salicylic acid at high-resolution (top) and low-resolution (bottom). The relative heights are not the same because different experimental conditions have been used*

Considering the way that the 3-hydroxybenzoic acid and the 4-hydroxybenzoic acid isomers would break down suggests that that these isomers could not form some of the fragments observed in the mass spectrum of salicylic acid. For example, the signal at mass 120 could not be formed from either of these isomers. Therefore salicylic acid must be 2-hydroxybenzoic acid. Comparison of the fragmentation pattern with a database of known mass spectra identifies salicylic acid as 2-hydroxybenzoic acid.

Drawing the evidence about the structure of salicylic acid together

Chemical tests showed the presence of a phenolic, –OH, group and a carboxylic acid, –COOH, group. Infrared spectroscopy showed that the –OH and –C═O groups were present. Nuclear magnetic resonance spectroscopy confirmed that there were hydrogen atoms in three types of environment: –COOH, –OH attached directly to a benzene ring and

–H attached directly to a benzene ring. Mass spectroscopy showed that salicylic acid was the same compound as that stored in the database as 2-hydroxybenzoic acid (Figure 4).

The mass spectrum fragmentation pattern showed that the structure could not be 3-hydroxybenzoic acid or 4-hydroxybenzoic acid (Figure 5).

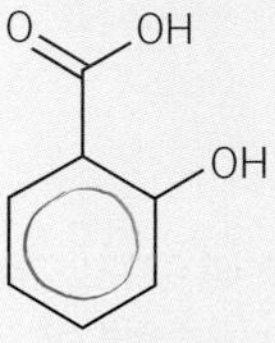

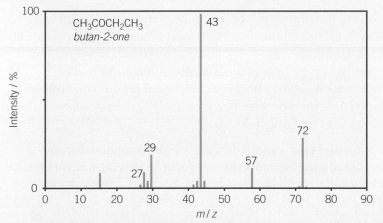

▲ **Figure 4** *The structure of 2-hydroxybenzoic acid*

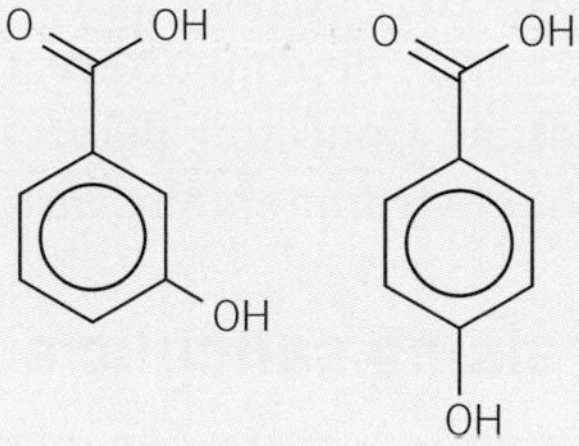

▲ **Figure 5** *The structure of 3-hydroxybenzoic acid (left) and 4-hydroxybenzoic acid (right)*

Summary questions

1 Look at the mass spectrum of butan-2-one.

$CH_3COCH_2CH_3$
butan-2-one

(mass spectrum showing peaks at m/z values: 27, 29, 43, 57, 72; Intensity / % on y-axis, m/z on x-axis)

a Identify the molecular ion peak and write the formula for it. (*1 mark*)
b What causes the peak at $m/z = 73$? (*1 mark*)
c Why are there peaks at m/z values of 27, 29, 43, 57, and 72? (*2 marks*)

2 Butan-2-ol has a number of different structural isomers. How would the mass spectra of these isomers:
a be the same as the mass spectrum for butan-2-ol? (*2 marks*)
b differ from the mass spectrum for butan-2-ol? (*2 marks*)

Activity WM 4.1

In this activity you can practise the use of both mass spectra and IR spectra to determine the structure of organic compounds.

WM 5 The synthesis of salicylic acid and aspirin

Specification reference: WM(e), WM(g)

Learning outcomes

Demonstrate and apply knowledge and understanding of:

→ the principles of green chemistry in industrial processes

→ techniques and procedures for making a solid organic product and for purifying it using filtration under reduced pressure and recrystallisation (including choice of solvent and how impurities are removed); techniques and procedures for melting point determination and thin layer chromatography.

Medicines that are natural products – those that come directly from plants – may be difficult to obtain when needed. Supply may be seasonal, depend on weather conditions, and be liable to contamination. Collecting plants from their natural habitat is also usually not environmentally sustainable.

Synthesising salicylic acid

Chemists do not want to rely on willow trees as their source of salicylic acid (2-hydroxybenzoic acid). Once the chemical structure of the active compound in a plant is known, chemists begin to search for ways of producing it artificially.

Simple inorganic substances, such as aluminium chloride, can be synthesised directly from their elements, but larger, more complex molecules cannot be made directly in this way. Instead, a known compound with a similar structure is identified, and the structure modified.

At the end of the nineteenth century, the compound phenol was already well known in the pharmaceutical industry – it has germicidal properties. It was also readily available as a product from heating coal in gas-works. Its molecular structure differs from that of 2-hydroxybenzoic acid by only one functional group. The problem in synthesis is to introduce this extra group in the right position without disrupting the rest of the molecule.

In this particular case, carbon dioxide can be combined directly with phenol to give 2-hydroxybenzoic acid by careful control of the conditions.

$$C_6H_5OH \quad + \quad CO_2 \quad \rightarrow \quad C_6H_4(OH)COOH$$

This general method is known as the Kolbe synthesis and an industrial version of this addition reaction was developed by the German chemist Felix Hoffmann. This is an early example of 'green chemistry' with an atom economy of 100% because there are no leaving atoms or molecules.

Chemical ideas: Greener industry 15.3

Key principles of green chemistry

Although green chemistry can mean different things to different people it is simply developing chemical products and especially chemical processes that are as sustainable and as environmentally friendly as possible. There are twelve key principles of green chemistry (Table 1).

▼ **Table 1** *The 12 principles of green chemistry*

Principles	Explanation
Better atom economy	Means more of the feedstock is incorporated into the product and less waste products are produced.
Prevention of waste products	This is better than treating and disposing of waste.
Less hazardous chemical synthesis	Using less hazardous chemicals in the chemical reaction.
Design safer chemical products	Less toxic and hazardous chemical products.
Use safer solvents	Minimise the use of organic solvents.
Lower energy usage	Lower temperature and pressure processes.
Use renewable feedstocks	Instead of depleting natural resources.
Reduce reagents used and the number of steps	As these can generate waste.
Use catalysts and more selective catalysts	These generally reduce energy usage and waste products.
Design chemical products for degradation	When released into the environment should break down into innocuous products.
Employ real time process monitoring	Better monitoring of chemical processes reduces waste products.
Use safer chemical processes	Choose processes that minimise the potential for releasing gases, fires and explosions.

Synoptic link

You first encountered atom economy in Chapter 3, Elements from the sea.

It follows that in any chemical process the key factors to consider are cost, impact on the environment, and health and safety. All individual chemical processes involve a balance of all of these. It cannot be assumed that application of the twelve principles will automatically results in a less expensive (therefore more profitable) process.

Examples of use if the green chemistry principles include

- The original manufacture of Taxol (a chemotherapy drug) was made by extraction from the tree bark of an extremely slow growing Pacific yew. This needed vast quantities of organic solvent and often killed the tree. A newer method involves growing tree cells in a fermentation vat followed by recrystallization, so reducing the amount of solvent and preventing the death of the yew trees. There are developments in using the bacterium *E. coli* to increase synthetic yields.

- The cholesterol reducing drug atorvastatin is now synthesised using an enzyme that catalyses chemical reactions in water, reducing the need for potentially polluting organic solvents.

- The painkiller Ibuprofen was originally synthesised in a 12 step synthesis with a 40% atom economy. It is now synthesised in a four stage process with an atom economy of of 77%.

▲ **Figure 1** *The industrial production of medicines*

Synoptic link

You will look at more industrial applications of chemistry in Chapter 6, The chemical industry.

Each one of these changes will follow one or more of the green chemistry principles outlined above. However, there will always be other aspects to take into account. For example:

- By changing the reagents to improve atom economy, does this increase the cost of the new reactants too much?
- Has the replacement of one reagent with another raised new health as safety issues?
- Even though the number of steps in a synthesis have been reduced what is the overall yield?
- Maybe a lowerd temperature (thereby reducing energy costs) increases the time for the reaction to reach equilibrium (thereby increasing the overall cost).

Chemists are continually balancing out costs, health and safety, and the impact on the environment. It would be rare to meet a synthesis that holds to all 12 green principles.

An example of using some of these principles can be seen in the approaches taken in the manufacture of 4-aminodiphenylamine.

> **Study tip**
>
> You do not need to learn the example below, but you should be able to analyse and use given information in a range of given examples.

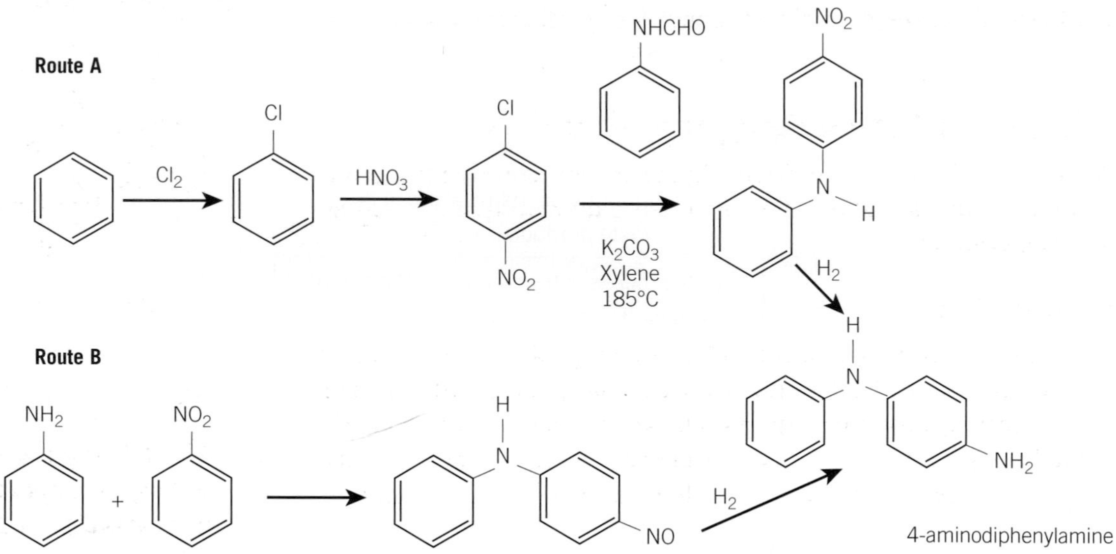

The older route (Route A) involves the use of chlorine, with its associated hazards and is a four step process. The newer route (Route B) has an improved atom economy, reduces the amount of waste and improved the risks associated with the process.

Making aspirin

Through Hoffmann's work, synthetic 2-hydroxybenzoic acid of reliable purity became available. Synthetic 2-hydroxybenzoic acid was widely used for curing fevers and suppressing pain, but reports began to accumulate of irritating effects on the mouth, gullet, and stomach. Clearly the new wonder medicine had unpleasant side-effects. Chemists had a new problem – could they modify the structure to reduce the irritating effects, whilst still retaining the beneficial ones?

Hoffmann prepared a range of compounds by making slight modifications to the structure of 2-hydroxybenzoic acid. His father was a sufferer from chronic rheumatism and Hoffmann tried out each of the new preparations on him to test its effects. This was a bit more primitive than the modern testing of medicines. It is not recorded what Hoffmann's father thought of all this, but he survived long enough for his son to prepare a derivative that was as effective as 2-hydroxybenzoic acid and had less unpleasant side effects.

The effective product was 2-ethanoxybenzoic acid (sometimes called 2-ethanoylhydroxybenzoic acid or acetylsalicylic acid). This is now known as aspirin and it is both an ester and a carboxylic acid (Figure 2).

However aspirin is not very soluble in water. It was first available as a powder in sachets – Bayer then decided to pellet the powder and aspirin became the first medicine to be sold as tablets.

Drug purity

When an organic solid product is synthesised it is in the crude form and it contains impurities, such as reactants, reagents, and solvents. The product needs to be purified and its purity checked, particularly if it is to be used as a drug. This is often more time consuming than the synthesis. There are several techniques for purifying products.

Techniques for purifying organic solids

The three main techniques for purifying and/or identifying organic solids are recrystallisation, thin layer chromatography, and melting point determination.

Recrystallisation

This technique is used to purify solid, crude organic products. Only the desired compound dissolves (to an appreciable extent) in the chosen hot solvent, leaving insoluble impurities to be filtered off. On cooling, the desired compound will **crystallise** out, leaving any soluble impurities in solution. The pure crystals can be filtered off, washed, and dried.

Thin layer chromatography

Thin layer chromatography is used to separate small quantities of organic compounds, purify or check the purity of organic substances, and follow the progress of a reaction over time.

A suitable solvent must be chosen. Different organic compounds have different affinities for a particular solvent, and so will be carried through the chromatography medium (plate) at different rates. When chromatography is carried out using a silica plate, it is known as **thin layer chromatography**.

Melting point determination

Measuring melting points is used as evidence of a solid organic compound's identity and purity. The value obtained is compared to the published value. A pure compound should melt within 0.5 °C of its true melting point.

▲ **Figure 1** *Meadowsweet (Spiraea ulmaria), from which salicylic acid was first extracted in 1835 – aspirin got its name from 'a' for acetyl (an older word for ethanoyl) and 'spirin' for spirsaüre (the German word for salicylic acid)*

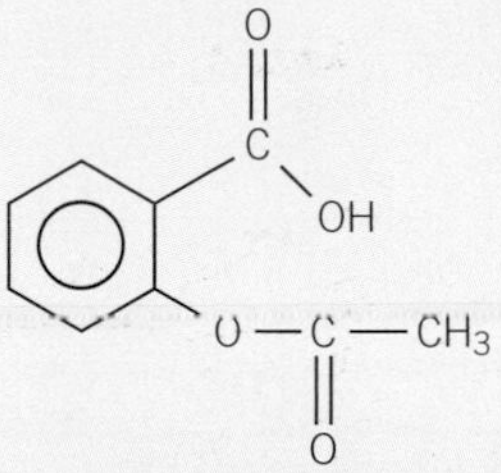

▲ **Figure 2** *The structure of aspirin*

Synoptic link

You will find out more detail on recrystallisation, thin layer chromatography, and melting point determination in Techniques and procedures.

Activity WM 5.1

This activity gives you the opportunity to synthesis aspirin in a two stage process starting from oil of wintergarden.

Activity WM 5.2

In this activity you will recrystallise aspirin and check its purity.

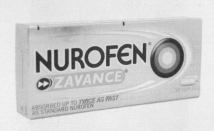

▲ **Figure 3** *Ibuprofen is available in a number of different formulations and trade names. It is relatively cheap now as the patent has expired*

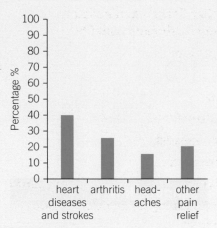

▲ **Figure 4** *Current uses of aspirin*

Developing a new medicine costs an enormous amount of money. The selling price charged by the pharmaceutical company must be sufficient not only to cover the costs of production and marketing, but also to recover the development costs. If other companies could simply copy the medicine, they would be able to sell it at a much lower price. This is where patents become important.

Protecting the discovery

When a pharmaceutical company discovers a new medicine, it takes out patents to protect the discovery. Patents apply in only one country, so several patents must be taken out to prevent companies in other countries manufacturing the medicine. Patents last only for a specific amount of time, but whilst the patent is in force no other company can manufacture the medicine in that country. By the time a patent runs out, the company that discovered the medicine will hopefully have sold enough of it to cover its development costs. Afterwards, any company can produce and sell the medicine. Ibuprofen – a drug used to treat pain, inflammation, and fever – is sold under a variety of trade names as tablets and in pain relief gels (Figure 3).

New uses of aspirin

The life cycle of modern medicines is often short because medical and pharmaceutical research offers better remedies all the time. Aspirin is now over 100 years old but new uses continue to emerge all the time. For example, the discovery of its potential in treating heart disease came from observations by doctors on their patients. By careful analysis, Dr Laurence Craven in the USA noticed that his male patients who took aspirin suffered fewer heart attacks. Similar work is being carried out to observe its effects on certain cancers, diabetes, deep vein thrombosis, and Alzheimer's disease, among others. This research is being carried out by doctors and universities using statistical analysis of medical records, and not by pharmaceutical companies, who have nothing to gain by it.

Summary questions

1 Write a step by step method for recrystallisation, using your practical work as a guide. *(3 marks)*

2 What principles could be used to reduce waste products in industrial processes for manufacturing chemicals? *(2 marks)*

3 List three different types of organic reaction and comment on their atom economy. *(3 marks)*

4 Research different techniques for determining melting points and give a brief description. What advantages has the technique chosen by your educational institution? *(3 marks)*

Practice questions

1 How many esters are there with formula $C_4H_8O_2$?

A 1

B 2

C 3

D 4   (*1 mark*)

2 Which pair of compounds would react to form CH_3COOCH_3?

A	CH_3CH_2OH	HCOOH
B	CH_3OH	$(CH_3CO)_2O$
C	CH_3COOH	CH_3CH_2OH
D	CH_3Cl	CH_3COOH

(*1 mark*)

3 What will be the main product when propan-1-ol is distilled with acidified dichromate?

A CH_3COCH_3

B CH_3CH_2COOH

C CH_3CH_2CHO

D $CH_3CH_2CH_2OH$ (*1 mark*)

4 A student is recrystallising a solid using water as solvent. The solid is dissolved in the minimum of hot water and allowed to crystallise.

Which is the correct sequence that follows?

A wash with water, filter, and dry the crystals

B remove the wet crystals and let them dry on a hotplate

C filter the crystals and dry them

D filter the crystals, wash, and dry.   (*1 mark*)

5 Which of the following sequences would work for purifying a liquid organic product?

A wash, separate, dry, distil

B distil, wash, separate, dry

C separate, dry, wash, distil

D dry, distil, wash, separate   (*1 mark*)

6 Which of the following is **not** true about a mass spectrum?

A The M+1 peak is caused by impurities.

B The M$^+$ peak indicates the M_r of the compound.

C Peaks of smaller mass are caused by fragments of the molecule.

D Only positive ions are detected.   (*1 mark*)

7 In which of the following rows does the name correctly describe the formula?

A	$(CH_3CO)_2O$	ester
B	CH_3CHO	ketone
C	CH_3OCH_3	ether
D	CH_3COCH_3	acid anhydride

(*1 mark*)

8 CH_3COOH can be made by the oxidation of which of the following:

1 CH_3CHO

2 CH_3CH_2OH

3 CH_3COCH_3

A 1, 2, and 3 correct

B 1 and 2 are correct

C 2 and 3 are correct

D Only 1 is correct   (*1 mark*)

9 A substance that fizzes when added to aqueous sodium carbonate could be:

1 a carboxylic acid

2 a phenol

3 an alcohol.

A 1, 2, and 3 correct

B 1 and 2 are correct

C 2 and 3 are correct

D Only 1 is correct   (*1 mark*)

10 Which of the following is/are correct about a tertiary alcohol?

1 It cannot be oxidised using acid dichromate.

2 It will form esters.

3 It can be dehydrated.

A 1, 2, and 3 correct

B 1 and 2 are correct

C 2 and 3 are correct

D Only 1 is correct   (*1 mark*)

11 Some students set out to make a sample of aspirin.

They react salicylic acid with ethanoic anhydride, using a phosphoric acid catalyst.

a Complete the equation for the reaction by drawing out the skeletal structures of the products. *(2 marks)*

$$C_7H_6O_3 \quad + \quad C_4H_6O_3 \quad \longrightarrow$$

b The impure product is filtered off, washed, and dried, and its melting point taken.

(i) Name the method of purification that the students would use. *(1 mark)*

(ii) How would the melting point change (if at all) after this purification?

(1 mark)

(iii) The students start with 10.0 g of salicylic acid and finish with 5.0 g of purified product. Calculate the percentage yield. *(2 marks)*

(iv) The students use thin layer chromatography to check they have made aspirin and to see if there is any unreacted salicylic acid in the sample of aspirin. Give details of their method. *(6 marks)*

c (i) Which of aspirin and salicylic acid will react with neutral iron(III) chloride? Give your reason and state the colour of a positive test.

(3 marks)

(ii) Write the equation for the reaction of salicylic acid with sodium carbonate, using structural formulae for organic substances. *(2 marks)*

d The students titrate 1.05 g of their purified aspirin with sodium hydroxide.

24.70 cm³ of 0.100 mol dm⁻³ NaOH was required.

(i) Calculate the purity of the aspirin, assuming that one mole of aspirin reacts with one mole of NaOH under these conditions. *(2 marks)*

(ii) How might NaOH react with aspirin under other conditions? *(2 marks)*

(iii) Which impurity, if present in the aspirin, would make the results of your calculation in (i) invalid?

(1 mark)

12 Butan-2-ol is an intermediate in the industrial formation of 'MEK', a widely-used solvent.

$$\underset{\text{butan-2-ol}}{CH_3CH_2CH(OH)CH_3} \quad \xrightarrow{\text{oxidation}} \quad MEK$$

a (i) Classify butan-2-ol as a primary, secondary, or tertiary alcohol, giving a reason. *(2 marks)*

b (i) Give the reagents and conditions that would be used in a laboratory to make MEK from butan-2-ol.

(2 marks)

(ii) Draw the structure of MEK and name its functional group. *(2 marks)*

c   A chemist takes a sample from the plant
    making MEK from butan-2-ol and runs
    the infrared spectrum shown below.

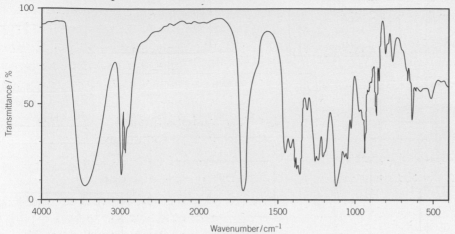

Explain the conclusions the chemist can
draw about the progress of the reaction.
                                    (*4 marks*)

d   (i)   A chemist runs a mass spectrum of
          butan-2-ol. Two of the peaks obtained
          are shown in the table below:

          Give the type of ion responsible for
          each peak

Peak	Ion
74	
75 (very small)	

                                    (*3 marks*)

    (ii) Other peaks are found at m/z values
          below 74. How do these form?
                                    (*1 mark*)

e   A compound **A** gives just butan-2-ol
    when it reacts with water.

    Give the full structural formula for
    compound **A**.   (*1 mark*)

f   Butan-2-ol can be dehydrated in the
    laboratory to a mixture of compound **A**
    and one other compound.

    (i)   Give the reagents and conditions
          for this reaction.   (*1 mark*)

    (ii) Give the skeletal formula of the other
          compound that is formed.   (*1 mark*)

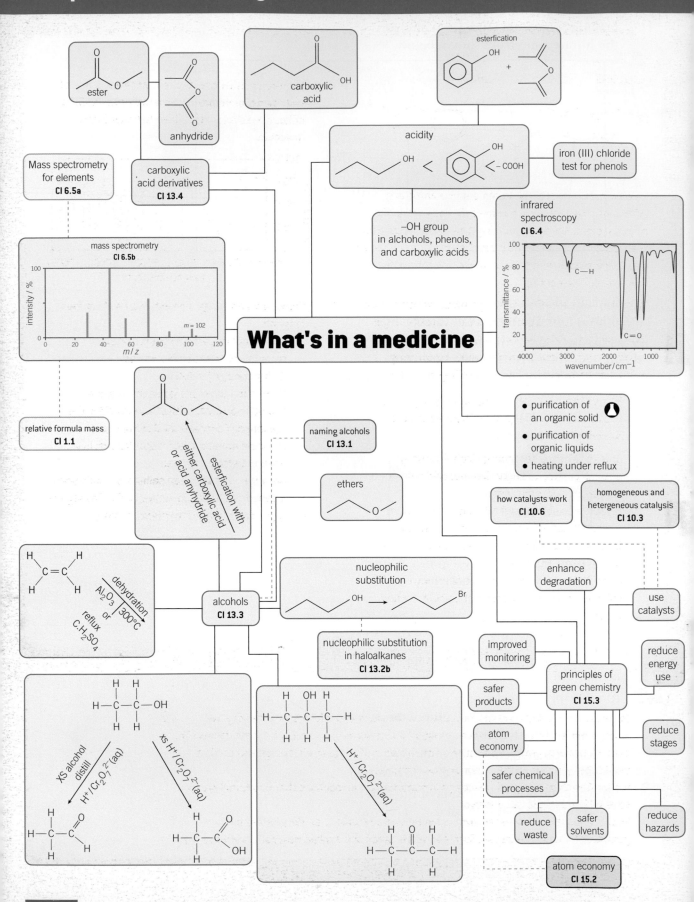

What's in a medicine

ester

anhydride

carboxylic acid

esterification

Mass spectrometry for elements
CI 6.5a

carboxylic acid derivatives
CI 13.4

acidity

OH < OH < –COOH

iron (III) chloride test for phenols

mass spectrometry
CI 6.5b

intensity / %

m = 102

m/z

–OH group in alchohols, phenols, and carboxylic acids

infrared spectroscopy
CI 6.4

transmittance / %

C—H

C=O

wavenumber/cm⁻¹

relative formula mass
CI 1.1

esterification with either carboxylic acid or acid anhydride

naming alcohols
CI 13.1

ethers

• purification of an organic solid
• purification of organic liquids
• heating under reflux

how catalysts work
CI 10.6

homogeneous and hetergeneous catalysis
CI 10.3

enhance degradation

dehydration
Al₂O₃ 300°C
or
reflux
C.H₂SO₄

alcohols
CI 13.3

nucleophilic substitution

OH → Br

use catalysts

nucleophilic substitution in haloalkanes
CI 13.2b

improved monitoring

safer products

principles of green chemistry
CI 15.3

reduce energy use

H—C—C—OH

atom economy

H—C—C—C—H

H⁺/Cr₂O₇²⁻(aq)

reduce stages

XS alcohol distill
H⁺/Cr₂O₇²⁻(aq)

xs H⁺/Cr₂O₇²⁻(aq)

safer chemical processes

H—C—C=O
H

H—C—C—OH

H—C—C—C—H

reduce waste

safer solvents

reduce hazards

atom economy
CI 15.2

Salbutamol and ibuprofen

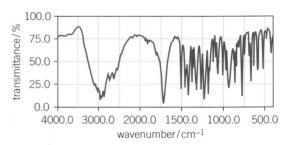

salbutamol   ibuprofen

Salbutamol is used for the treatment of asthma and other bronchial diseases. It relaxes the muscles in the air passages of the lungs, making it easier to breathe by keep the airways open. Salbutamol is often prescribed as an inhaled preparation to be breathed in by the patient. This makes it fast acting.

Ibuprofen is an over-the-counter painkiller which reduces inflammation. It can be taken in tablet, capsule, or gel form and particularly works to improve movement by reducing muscle pain. Ibuprofen works by blocking production of molecules causing swelling and pain.

1 Name the oxygen-containing functional groups in salbutamol and ibuprofen. Choose from the following functional groups.

primary alcohol, secondary alcohol, tertiary alcohol, phenol, aldehyde, ketone, acid anhydride, carboxylic acid, ether, ester

2 Explain why salbutamol will react with sodium hydroxide but not sodium carbonate, whilst ibuprofen will react with both sodium hydroxide and sodium carbonate.

3 Describe and explain the effect of adding iron(III) chloride to separate samples of salbutamol and ibuprofen.

4 Identify whether the infrared spectrum is that of salbutamol or ibuprofen. Give the wavenumbers of key peaks and relate them to bonds in the molecule.

5 Salbutamol and ibuprofen will react with various reagents.

   a State what you would observe when salbutamol reacts with acidified dichromate(VI) solution.

   b Name the **type** of molecule made when salbutamol reacts with ethanoic acid in the presence of concentrated sulfuric acid.

   c Draw the product when ibuprofen reacts with ethanol to form an ester.

   d Draw the product when salbutamol undergoes a dehydration reaction using Al_2O_3. Classify the type of reaction mechanism occurring.

Extension

1 Research how salbutamol and ibuprofen can be made in the laboratory. Identify the starting materials, and describe as many of the steps as you can. Look for the conversion of one functional group into another in each step and suggest what reagents could be used to bring about each functional group change.

2 Draw a flowchart of organic reactions to summarise the reactions that functional groups covered in this topic can undergo.

3 Find out about the processes involved in testing a possible new drug before it is permitted for commercial use. Summarise the steps and explain why each is necessary.

Techniques and procedures

Understanding and being familiar with practical techniques and procedures is an important part of being an effective chemist. This section outlines the techniques and procedures you need to know about as part of your course.

Measurement

Weighing a solid

Typically an accurate weighing will use a balance that records to two or three decimal places.

1   Zero the balance (sometimes called tare).

2   Place a weighing bottle or similar container onto the balance and add in approximately the required mass of solid.

3   Accurately weigh the mass of solid plus weighing bottle and record this information.

4   Empty the solid into the glassware where you will be using it.

5   Accurately reweigh the empty weighing bottle.

6   Subtract the recorded mass for the empty weighing bottle from the mass recorded for the solid and the weighing bottle.

Measuring volumes of liquids

Beakers or measuring cylinders can be used to give rough measurements of liquids. In order to measure volumes of liquids *accurately* two methods can be used – a pipette or a burette.

Pipette

A pipette is used for accurately dispensing a *fixed* volume of a liquid (typically $1.0\,cm^3$ to $50\,cm^3$ or $25\,cm^3$).

1   Ensure the pipette is completely clean by rinsing out with water and then a small volume of the solution to be pipetted.

2   Dip the pipette into the solution to be pipetted and, using a pipette filler, draw enough liquid into the pipette until is it exactly the right volume – when the bottom of the meniscus is level with the line on the neck of the pipette when viewed at eye level.

3   Run the liquid out of the pipette into the piece of glassware the solution is being transferred to. When all of the liquid has run out

4   Allow the liquid to run out of the pipette until it stops. Touch the end of the pipette on the side of the conical flask and remove. There will still be a drop in the pipette – this is how it should be. The precise volume you require will have been dispensed.

Burette

1   Clean the burette by rinsing out with water and then a small volume of liquid of the solution to be used.

▲ **Figure 1** *Measuring the volume of liquid in a pipette*

labels: filler, meniscus, line on stern, pipette

2   Make sure the burette tap is closed. Pour the solution into the burette using a small funnel. Fill the burette above the zero line.

3   Use a clamp to hold your burette in place and allow some of the solution to run into a beaker until there are no air bubbles in the jet of the burette. Record the burette reading to the nearest $0.05\,cm^3$.

4   Carry out the titration to the end point.

5   Record the reading on the burette to the nearest $0.05\,cm^3$. Subtract the reading taken at the beginning of the titration from this reading taken at the end. This is known as the titre.

Measuring volumes of gases

The volume of a gas produced in a chemical reaction can be measured using either a gas syringe or an inverted buratte. The latter is called 'collecting gas over water'.

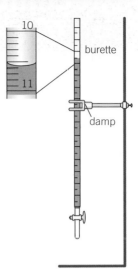

▲ **Figure 2** *Measuring the volume of liquid using a burette*

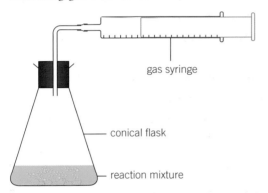

▲ **Figure 3** *Collecting gas using a gas syringe*

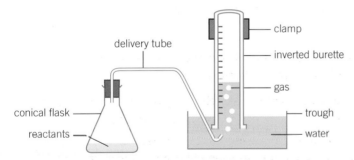

▲ **Figure 4** *Collecting a gas using a measuring cylinder or inverted burette*

In order for as much as possible of the gas to be collected the system needs to be gas tight.

The volume of gas collected in an inverted burette is the initial volume minus the final volume of gas.

Synthesis

Heating under reflux

Heating under reflux is used for reactions involving volatile liquids. It ensures that reactants and/or products do not escape whilst the reaction is in progress.

> ### Study tip
>
> If the gas is soluble in water you need to use the gas syringe method, as some of the gas would dissolve before reaching the burette in the second method.

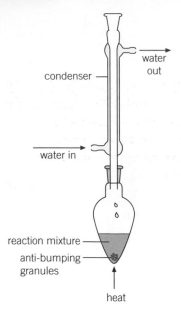

▲ **Figure 1** *Reflux apparatus*

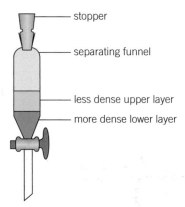

▲ **Figure 2** *Separating funnel*

Study tip

Take care to keep the correct layer. An easy way to test which layer is the aqueous layer is to add water to the separating funnel and wait for the layers to separate again. The layer that is bigger will be the aqueous layer.

1 Put the reactants into a pear-shaped or round-bottomed flask and add a few anti-bumping granules – these granules burst the bubbles in the boiling mixture and reduce the chance of boiling over.

2 Do not stopper the flask – doing this would cause pressure to build up and the glassware could crack or the stopper could fly out. In either case, a serious accident could result.

3 Attach a condenser vertically to the flask so that water flows into the condenser at the bottom and out of the condenser at the top. This ensures that the condenser is always full of cold water.

4 Heat so that the reaction mixture boils gently, using a Bunsen flame or heating mantle. When refluxing correctly, any vapours should reach no more than half way up the condenser before condensing back into a liquid. The liquid should drip back into the reaction flask steadily.

Purifying an organic liquid product

Organic liquid products have to be purified after it has been synthesised.

1 When the organic product is mixed with another immiscible liquid (often an aqueous liquid) the two layers can be separated using a separating funnel. The layers separate, with the denser liquid forming the lower layer. Allow the layers to settle and then run off and dispose of the aqueous layer. Run the organic (product) layer into a clean conical flask.

2 If acidic impurities are present, add sodium hydrogen carbonate solution and shake well to remove them. If the crude product is alkaline and needs neutralising then add a dilute acid until the mixture is neutral.

3 Dry the crude product by adding anhydrous sodium sulfate and swirling the mixture. It is possible to use other anhydrous salts, such as calcium chloride, to dry organic compounds.

4 The pure product can then be separated by distillation.

Making water-soluble inorganic salts

Soluble salts can be made by two techniques – reacting an acid with a soluble base (alkali) or by reacting an acid with an insoluble base.

Reacting an acid and a soluble base (alkali)

Acid–base titrations can be used to produce a soluble salt.

1 Carry out an acid–base titration to find out how much acid solution is needed to neutralise $25\,cm^3$ of the alkaline solution.

2 Transfer $25.0\,cm^3$ of the alkaline solution to a clean conical flask.

3 Using a burette, add the correct amount of acid to neutralise the alkali. *Do not* add any indicator.

4 Transfer the neutralised solution to a clean evaporating basin and heat over a Bunsen flame to evaporate the water. Take care not to heat too strongly, in order to avoid spitting. Once crystals of solid start to appear stop heating.

5 Leave the mixture to cool in the evaporating basin.

6 Filter the mixture.

7 Wash the solid residue with cold distilled water.

8 Transfer the residue to a watch glass and heat in an oven to dry the solid. Ensure the oven is set at a temperature below the melting point of the salt you have prepared.

9 At regular intervals remove the watch glass and solid, cool in a desiccator, and weigh. Once the solid has dried to a constant mass (the mass between two readings does not change) stop heating the solid salt and leave to cool in a desiccator.

> **Study tip**
>
> A desiccator allows materials to cool in a dry atmosphere, so preventing the reabsorption of moisture.

Reacting an acid and an insoluble base

1 In a beaker, warm excess insoluble base in dilute acid.

2 Continue to warm (but not boil) until the solution is neutral (use universal indicator paper for this step), adding more solid base if needed.

3 Leave to cool.

4 Filter off the excess base and transfer the filtrate to a clean, dry evaporating basin.

5 Heat the evaporating basin until salt crystals begin to appear on the sides of the basin.

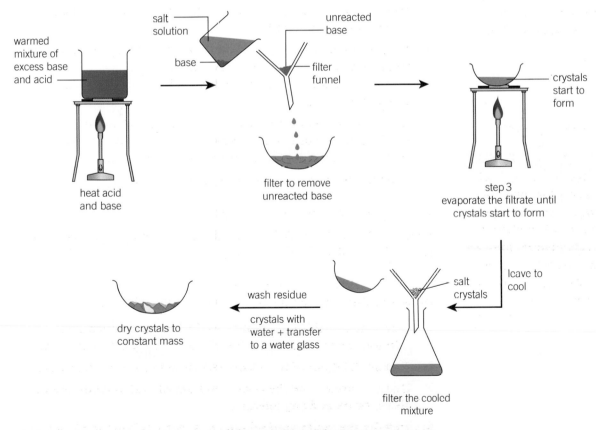

summary of the preparation of an inorganic salt by reacting an molecule base cool an acid

▲ **Figure 3** *Summary of the preparation of an inorganic salt by reaction of an insoluble base with an acid*

6   Cool the basin and contents.

7   Filter the mixture, and discard the filtrate.

8   Wash the solid residue with cold distilled water.

9   Transfer the residue to a watch glass and heat in an oven to dry the solid. Ensure the oven is set at a temperature below the melting point of the salt you have prepared.

10 At regular intervals remove the watch glass and solid, cool in a desiccator, and weigh. Once the solid has dried to a constant mass (the mass between two readings does not change) stop heating the solid salt and leave to cool in a desiccator.

Making water-insoluble inorganic salts

Insoluble salts can be prepared by the reaction of two soluble salts in solution.

1   Add equal volumes of the desired salt solutions in a beaker to form a precipitate of the insoluble salt.

2   Filter the precipitate

3   Wash the precipitate several times with cold deionized water.

4   Transfer the filtered, washed precipitate to a clean watch glass and place in a drying oven. Ensure the oven is set at a temperature below the melting point of the salt you have prepared

5   At regular intervals remove the watch glass and solid, cool in a desiccator, and weigh. Once the solid has dried to a constant mass (the mass between two readings does not change) stop heating the solid salt and leave to cool in a desiccator.

Purification

Simple distillation

Distillation can be used to separate a mixture of miscible liquids with unique boiling points. By heating the mixture, each pure component is vaporised, condensed, and collected. The components will evaporate in the order of their boiling points – the one with the lowest boiling point will evaporate first. Quickfit glassware is commonly used for distillation. It has ground glass joints that can be sealed using grease to prevent the loss of reagents. Other small scale systems are also available.

1   Put the mixture into a pear-shaped flask and add a few anti-bumping granules. Set up the distillation apparatus as shown in Figure 4. The position of the thermometer is important – it gives an accurate reading of the vapour temperature.

2   Heat the mixture until it boils gently, using a Bunsen flame or heating mantle. Heating mantles are safer to use when heating flammable liquids as they do not have a naked flame.

3   When the vapour temperature is approximately two degrees below the boiling point of the liquid you are about to collect, put the collecting beaker in place. Collecting the distilled liquid until

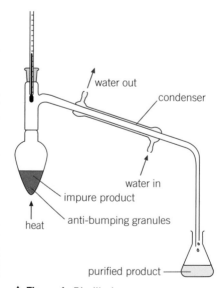

water out

condenser

water in

impure product

heat

anti-bumping granules

purified product

▲ **Figure 4** *Distillation apparatus*

the temperature of the vapour rises above the boiling point of the liquid you are collecting. Stop heating.

4   If another compound is required of a higher boiling point, repeat step **3** using a clean collecting beaker.

Thin layer chromatography

Thin layer chromatography (t.l.c.) is used to separate small quantities of organic compounds, purify organic substances, and follow the progress of a reaction over time. The method relies on the fact that different organic compounds have different affinities for a particular solvent, and so will be carried through the chromatography medium at different rates. Chromatography carried out using a silica plate is known as thin layer chromatography.

1   Spot the test mixture and reference samples on a pencil line 1 cm from the base of the thin layer chromatography plate. Pencil is used because it will not run into the solvent.

2   Suspend the plate in a beaker containing the solvent (Figure 5) and cover the beaker with a watch glass to prevent the solvent from evaporating.

3   Remove the plate when the solvent front is near the top. Mark how far the solvent has reached. Allow the plate to dry.

4   Locate any spots with iodine, ninhydrin, or under an ultraviolet lamp.

5   Match the heights reached, or R_f values, with those of known compounds musing the same chromatography solvent mix.

Recrystallisation

Recrystallisation is used to purify solid crude organic products with small amounts of impurities. A suitable hot solvent is chosen that only the desired compound dissolves in to an appreciable extent. When cooled, the pure organic compound will drop out of solution (recrystallise) – any other soluble impurities stay in solution.

1   Select a solvent in which the desired substance is very soluble at higher temperatures and insoluble, or nearly so, at lower temperatures.

2   Dissolve the mixture in the minimum quantity of hot solvent – the smaller the amount of solvent used the better the yield of the desired substance.

3   Filter to remove any insoluble impurities and retain the filtrate. It is best to preheat the filter funnel and conical flask to prevent any solid crystallising out at this stage.

4   Leave the filtrate to cool until crystals form.

5   Collect the crystals by vacuum filtration.

6   Dry the crystals in an oven or by leaving them in the open, covered by an inverted filter funnel.

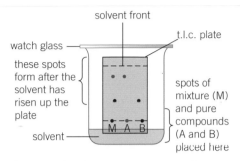

▲ **Figure 5** *Thin-layer chromatography*

> ## Study tip
>
> R_f value is the distance travelled by the spot you are interested in divided by the distance moved by the solvent. It is always a value if 1 or less.

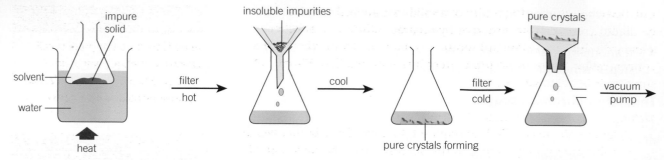

▲ **Figure 6** *Recrystallisation of an impure solid*

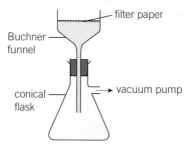

▲ **Figure 7** *Vacuum filtration*

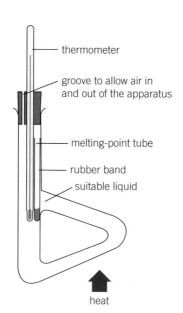

▲ **Figure 8** *Melting point apparatus*

Vacuum filtration

Vacuum filtration is used to separate a solid from a filtrate rapidly.

1   Connect a conical flask to a vacuum pump via the side arm (Figure 7). Do not switch the pump on yet.

2   Dampen a piece of filter paper and place it flat in the Buchner funnel.

3   Switch the vacuum pump on and then carefully pour in the mixture to be filtered. The pump creates a partial vacuum so that the filtrate gets 'pulled through' quickly.

4   Disconnect the flask from the vacuum pump before turning the pump off – this avoids 'suck back'.

Analysis

Determining melting points

This technique is used to determine the melting point of organic solids. The melting point can then be used as evidence of a product's identity and purity.

1   Seal the end of a glass melting point tube by heating it to melting in a Bunsen flame.

2   Tap the open end of the tube into the solid so a small amount goes into the tube. Tab the tube so that the solid falls to the bottom of the sealed end.

3   Fix the tube in the melting point apparatus and heat the surrounding liquid gently, stirring to ensure even heating throughout. The temperature should rise very slowly.

4   Note the temperature at which the solid starts and finishes melting. The difference between the highest and lowest temperatures recorded is known as the melting range.

5   Compare the experimental value to the published value for the melting point. The wider the melting range, the more impure the substance. A pure compound will melt within 0.5 °C of the true melting point.

Making a standard solution

A standard solution is one where its concentration is accurately known. It can be used to determine the concentration of another unknown solution or the purity of a solid. A standard solution can be made up either from a solid or by accurate dilution of another standard solution. A standard solution is made up fresh whenever a concentration has to be accurate. Here it is assumed that $250 \, cm^3$ of solution is being made up.

Making a standards solution from a solid

1 Calculate the mass of solute required. In a weighing bottle, weigh out this amount accurately, to the nearest $0.01 \, g$. Make a note of the mass of the weighing bottle and solute.

2 Pour $100 \, cm^3$ of deionised water into a $250 \, cm^3$ beaker. Carefully transfer the weighed solute into the water from the weighing bottle.

3 Reweigh the weighing bottle. The difference between the mass of the weighing bottle and solute and the mass of just the weighing bottle is the mass of solute transferred.

4 Stir the mixture in the beaker to ensure complete dissolving of the solute.

5 Transfer the solution to a clean $250 \, cm^3$ volumetric flask. Rinse the beaker and stirring rod well with deionised water, making sure that all the washings go into the volumetric flask.

6 Add deionised water to the solution, swirling at intervals to mix the contents, until the level is within about $1 \, cm$ of the $250 \, cm^3$ mark on the neck of the volumetric flask.

7 Using a dropping pipette, add deionised water so that the bottom of the meniscus is level with the mark on the neck of the flask when looking at it at eye level.

8 Insert the stopper in the flask and invert it, shaking thoroughly to ensure complete mixing.

Making a standard solution by dilution

If the concentration of an existing solution is too high it can be adjusted. The existing solution is called the stock solution. In this example, the concentration is diluted by a factor of 10 to produce $250 \, cm^3$ of a $0.1 \, mol \, dm^{-3}$ solution from a standard solution of $1.0 \, mol \, dm^{-3}$

1 Rinse a clean dry beaker with the stock solution, and then half fill it.

2 Use a pipette filler to rinse a clean $25.0 \, cm^3$ pipette with some of the stock solution. Fill the pipette to the $25.0 \, cm^3$ mark – with the bottom of the meniscus exactly on the mark.

3 Run the solution into a $250 \, cm^3$ volumetric flask.

4 Add deionised water to the solution, swirling at intervals to mix the contents, until the level is within about $1 \, cm$ of the mark on the neck of the volumetric flask.

5 Using a dropping pipette, add deionised water so that the bottom of the meniscus is level with the mark on the neck of the flask.

6 Insert the stopper in the flask and invert it, shaking thoroughly to ensure complete mixing. You now have your new diluted standard solution.

Study tip

Impurities in a crude product will cause the melting point to be lowered. Following recrystallization the melting point will be higher.

Study tip

The solute should be pure. Pure reagents can be supplied, and are known as Analar reagents. For standard solutions use Analar reagents.

Calculating the volumes needed

In order to calculate the volumes needed you need to use the equation:

concentration of diluted solution being made C_1 (mol dm^{-3})

× volume of diluted solution being made V_1 (dm^3)

= concentration of stock solution C_2 (mol dm^{-3})

× volume of stock solution V_1 (dm^3)

 Worked example: Standard solution by dilution

To prepare 50 m^3 of 1.0 mol dm^{-3} solution from a stock solution of concentration 2.0 mol dm^{-3}, what volume of stock solution is required?

Step 1: Use the equation $C_1V_1 = C_2V_2$ and insert known values.
$$1.0 \text{ mol dm}^{-3} \times 0.05 \text{ dm}^3 = 2.0 \text{ mol dm}^{-3} \times Z \text{ dm}^3$$

Step 2: Rearrange the equation to calculate the required volume of stock solution.
$$Z = \frac{1.0 \text{ mol dm}^{-3} \times 0.05 \text{ dm}^3}{2.0 \text{ mol dm}^{-3}} = 0.025 \text{ dm}^3$$

Step 3: Convert dm^3 reading to cm^3 reading – 25 cm^3 of stock solution needed.

Acid–base titration

Acid–base titration is used to determine the concentration of an acid or an alkali accurately. The method described below assumes that you are titrating an alkali of known concentration against an acid to calculate the concentration of the acid.

1 Rinse a burette with some of the acid solution then fill it with the acid. Run a little of the acid through the burette into a waste beaker to fill the tip. Record the initial burette reading to the nearest 0.05 cm^3 (Figure 9).

2 Fill a clean 25.0 cm^3 pipette with some of the alkaline solution.

3 Run the alkaline solution into a clean 250 cm^3 conical flask.

4 Add two or three drops of a suitable indicator and swirl to mix

5 Run the acid from the burette into the flask. Swirl the flask continually and watch for the first hint of the solution changing colour. This first titration should be used as a trial run to give a rough indication of the amount of acid required. Record the final burette reading – the volume of acid used is called the titre.

6 Refill the burette and record the initial burette reading.

7 Using the pipette, transfer 25.0 cm^3 of the alkaline solution to a clean conical flask. Add two to three drops of the indicator and swirl to mix.

8 Run in the acid solution to 1 cm^3 below the rough titre. Then add the acid dropwise, swirling after each drop, until the colour of the indicator changes.

5 Repeat steps **6**, **7**, and **8** until there are three concordant results, that is, three results within 0.10 cm^3 of each other.

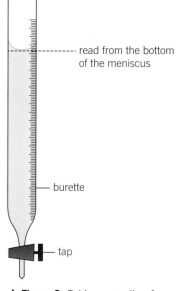

read from the bottom of the meniscus

burette

tap

▲ **Figure 9** *Taking a reading from a burette*

The volume and concentration of the standard alkaline solution required for neutralisation can be used to calculate the concentration of the hydrochloric acid.

Iodine–thiosulfate titration

Iodine-thiosulfate titrations involve redox reactions. They are used to find the concentration of a chemical that is a strong enough oxidizing agent to oxidise iodide ions to iodine. The liberated iodine is titrated with thiosulfate ions, with starch as an indicator. Once the amount of thiosulfate ions are calculated the amount of the chemical being analysed (in the example below chlorate(I) found in domestic bleach solutions) can be calculated.

Step 1

Chlorate(I) ions are strong enough oxidising agents to reduce iodide ions to iodine. An iodine-thiosulfate titration can be used to find the concentration of a solution of chlorate(I), for example, sodium chlorate(I).

$$ClO^- + 2I- + 2H^+ \rightarrow I_2 + Cl^- + H_2O \qquad \textbf{Equation 1}$$

Cl	+1		−1	reduced
I		−1	0	oxidised

1   Pour some of the chlorate(I) solution into a clean and dry beaker.

2   Rinse a $25.0\,cm^3$ volumetric pipette with water and then the chlorate(I) solution. This ensures that the pipette is clean and that the chlorate(I) solution hasn't been diluted.

3   Transfer a carefully measured $25.0\,cm^3$ **aliquot** of the chlorate(I) solution to a conical flask using a volumetric pipette and filler. This piece of equipment is very precise – it has a low precision error. Ensure that the bottom of the meniscus is exactly on the mark when filling. Before emptying the solution into the flask, dry the outside with a paper towel and after emptying, touch the tip of the pipette to the inside of the conical flask. These procedures ensure that the $25.0\,cm^3$ are measured as accurately as possible.

4   Add excess iodide ions using a measuring cylinder to transfer $15\,cm^3$ of $0.5\,mol\,dm^{-3}$ potassium iodide to the conical flask. Since the amount of I_2 liberated is determined by the amount of chlorate(I) used, the volume of potassium iodide does not need to be precise. Add excess hydrogen ions using a measuring cylinder to transfer $20\,cm^3$ of $1\,mol\,dm^{-3}$ sulfuric acid.

5   The contents of the flask will be brown due to the iodine produced. (Figure 5a).

Step 2

Looking at Equation 1, you can see that 1 mole of chlorate(I) ions make 1 mole of iodine molecules.

$$I_2 + 2S_2O_3^{2-} \rightarrow 2I^- + S_4O_6^{2-} \qquad \textbf{Equation 2}$$

I	0		−1	Reduced
S		+2	+2.5	Oxidised

You can titrate the iodine produced with sodium thiosulfate solution to work out the number of moles of iodine that are produced in Equation 1. Equation 2 is also redox. (Don't worry about the +2.5 oxidation state for the sulfur.)

1   Wash a burette is with water followed by a standard solution of $0.100\,mol\,dm^{-3}$ sodium thiosulfate solution. This ensures that the burette is clean and that the sodium thiosulfate solution hasn't been diluted. The burette readings will be precise because the precision error for it is low.

2   Fill the burette with the sodium thiosulfate solution, making sure that the jet is full. The titre will be inaccurate if you start with an empty jet.

3   Put the conical flask on a white tile to make the end point easier to see.

4   Record the initial burette reading to the nearest $0.05\,cm^3$. It will be easier to read the burette if you put a white tile behind the graduations to see where the bottom of the meniscus is.

5   Start the rough titration. This gives you a rough idea of how much sodium thiosulfate is needed. You can add $1\,cm^3$ of sodium thiosulfate at a time. Near the end point the contents of the conical flask will be a pale straw colour (Figure 5b). When you see this it is the best time to add a few drops of starch solution. Now the contents of the flask will be blue black (Figure 5c). It will be easier to tell when you have reached the end point of colourless now (Figure 5d).

6   Record the final burette reading and calculate your rough **titre** by subtraction.

7   Wash the conical flask with distilled water. You don't need to dry it. Add the contents described in Step 1 to the conical flask.

8   Now carry out some accurate titrations. You can run the sodium thiosulfate into the conical flask until it is $1\,cm^3$ below the rough titre.

9   Now add the sodium thiosulfate drop wise so that you don't overshoot the end point. Also wash the inside of the flask and the end of the burette down regularly with distilled water from a wash bottle to ensure that all the reactants are together in the bottom of the conical flask.

10 Continue with the accurate titres until they are **concordant** – within $0.1\,cm^3$ of each other.

▲ **Figure 10** *Stages in an iodine–thiosulfate titration*

Synoptic link

See Topic EL 9, How Salty? for how to calculate the concentration of the acid.

Using a colorimeter

Colorimeter is used to determine the concentration of a coloured solution. Coloured solutions can absorb certain wavelengths of light. The amount of this light that is either absorbed or transmitted can be measured. This is proportional to the concentration of the solution.

1   Select a filter with the complementary colour to the solution being tested. For example, a purple solution absorbs orange light so you would choose an orange filter. This allows only those

wavelengths absorbed most strongly by the solution to pass through to the sample.

2   Make up a range of standard solutions of the test solution. There should be solutions both above and below the concentration of the unknown solution.

3   Zero the colorimeter using a tube/cuvette of pure solvent – this will be water in most cases.

4   Measure the absorbance of each of the standard solutions and plot a calibration curve of concentration against absorbance.

5   Measure the absorbance of the unknown sample and use the calibration curve to determine the concentration of the unknown solution.

By determining the concentrations of a reactant at different time intervals in a reaction you can follow the progress of that reaction.

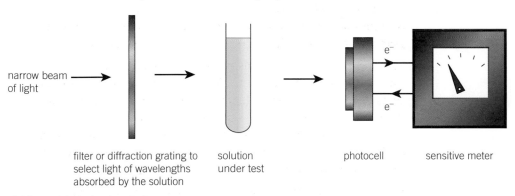

narrow beam of light

filter or diffraction grating to select light of wavelengths absorbed by the solution

solution under test

photocell   sensitive meter

▲ **Figure 11** *A simple calorimeter*

Measuring the energy transferred in a chemical reaction

Measuring the energy transfer when a fuel burns

This technique is used to determine the enthalpy change of combustion when a fuel is burnt. The values obtained experimentally can then be compared with theoretical values.

1   Using a measuring cylinder, pour a known volume of water into a copper calorimeter. Record its temperature.

2   Weigh a spirit burner – keep the cap on the burner to reduce loss of the fuel by evaporation.

3   Support the calorimeter over a spirit burner containing the fuel to be tested. Surround it with a draught excluder to help to reduce energy losses (Figure 12).

4   Remove the cap of the spirit burner and light the wick.

5   Use the thermometer to stir the water all the time it is being heated – carry on heating until the temperature has risen by 15 to 20 °C.

6   Extinguish the spirit burner and put the cap back in place. Keep stirring the water and make a note of the highest temperature reached.

7   Weigh the burner again.

The results can be used to calculate the enthalpy of combustion for the fuel under test.

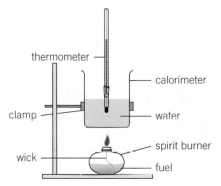

thermometer
calorimeter
clamp
water
wick
spirit burner
fuel

▲ **Figure 12** *Apparatus to determine the enthalpy of combustion*

Synoptic link

Look at Topic DF2, How much energy? For how to calculate the enthalpy of combustion.

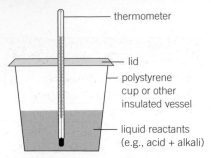

▲ **Figure 13** *Apparatus to determine the enthalpy of neutralisation*

Measuring the energy transferred when reactions occur in solution

This technique allows the enthalpy change of a reaction to be calculated through the measurement of changes in temperature when known quantities of reactants react together, for example, the enthalpy of neutralisation.

1 Using a measuring cylinder, add a known volume of a known concentration of acid to an insulated vessel and take the temperature.

2 Using a measuring cylinder, add a known volume of a known concentration of the alkali. Stir well to mix the reactants.

3 Top the vessel with a lid with a hole in (Figure 13).

4 Place the thermometer through the hole in the lid and record changes in temperature every 30 seconds until there are no further changes in temperature.

5 Calculate the maximum increase in temperature

The reactions take place in insulated vessels to minimize the transfer of thermal energy to the surroundings. Use this information to calculate the enthalpy change of neutralisation.

These reactions may involve two solutions or solids reacting with solutions

Reactions of solids and solutions

1 Using a measuring cylinder add a known volume of a known concentration of the reactant solution into an insulated vessel. Take the temperature.

2 Add a known mass of solid reactant. This should be in excess.

3 Top the vessel with a lid with a hole in.

4 Place the thermometer through the hole in the lid and record changes in temperature every 30 seconds until there are no further changes in temperature.

In order to calculate the maximum change in temperature, you will need to plot a graph of temperature against time (Figure 14)

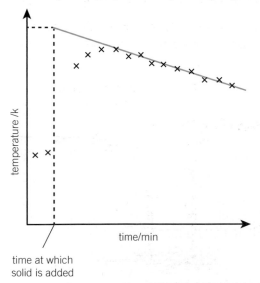

▲ **Figure 14** *Graph of temperature against time showing a line of best fit and extrapolation*

Once a line of best fit has been drawn you need to be able to estimate the temperature rise immediately following the mixing of reagents.

The values recorded experimentally are too low because the reaction does not occur instantaneously but over a period of time. Throughout the reaction, heat is released and some is lost from the reaction vessel to the wider surroundings. The theoretical maximum temperature change is never achieved experimentally and must be found by extending the line of best fit back to the time at which the reactants were mixed. Extending data points beyond those actually measured is called **extrapolation**.

Other techniques

Carrying out the electrolysis of aqueous solutions

During electrolysis of an aqueous solution, an electric current is passed through the electrolyte. For electrolysis to occur in an aqueous solution an electrical circuit needs to be set up (Figure 15)

The dc power supply may be a powerpack or batteries. The most common material for electrodes (anode and cathode) is graphite. This conducts electricity, is cheap, and is relatively inert (unreactive). Platinum may also be used but it is expensive.

If the products of electrolysis are gaseous they can be collected using the apparatus in Figure 16.

The test tubes are filled with water at the start of the reaction. The gaseous products displace the water and are collected in the test tubes. Once the powerpack is switched on the electrolysis will proceed.

If electrolysis is to be carried out for the purpose of purification of a metal, the anode will need to be made of the impure metal, the electrolyte must contain ions of that metal, and the cathode should be made of the pure metal.

Cracking a hydrocarbon vapour over a heated catalyst

Cracking is used to break longer molecules down into smaller molecules.

1   Set up the apparatus below, ensuring there is space above the catalyst (Al_2O_3) to allow gases to pass freely over over it.

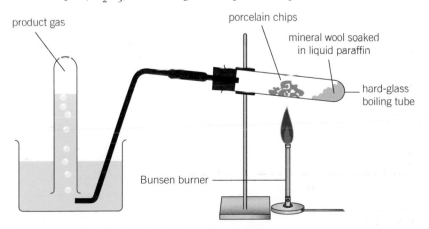

▲ **Figure 17** *Apparatus for cracking a mixture of alkanes*

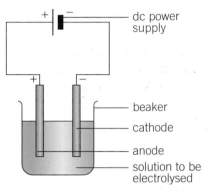

▲ **Figure 15** *Apparatus for electrolysis*

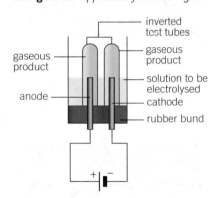

▲ **Figure 16** *Apparatus to collect gaseous products in electrolysis*

2   Place several test tubes in the water in the collection trough.

3   Heat the catalyst strongly. This ensures that when the alkane vapour passes over it the temperature is high enough for the cracking reactions to take place.

4   Heat the alkane gently, collecting any gases that pass into the collection tubes, changing and corking full tubes. Continue to heat whilst changing the collection tubes. This prevents 'suck-back'.

5   Discard the first tube of gas. This will just be displaced air, rather than product.

6   Continue heating both the catalyst and the alkane mixture to be cracked until you have collected several tubes of gas or until no more gas is produced.

7   Remove the delivery tube from the collection trough before stopping heating the catalyst and alkane mixture. This prevents suck back.

8   Leave to cool then dismantle the apparatus.

9   Test any liquid product from the middle collection tube with bromine water. The bromine water should remain yellow/brown.

10 Test the gas collected in tubes by shaking with bromine water. The bromine water should decolourise.

Testing for the presence of unsaturation in alkenes

The addition of bromine water to an alkene is a reliable test for the presence of carbon-carbon double bonds.

1   To a few drops of the unknown liquid or gas add a few drops of bromine water and shake well.

2   If the unknown is an alkene the bromine water will change from orange/yellow to colourless (it is decolourised). If the sample is not unsaturated it will not decolourise the bromine water and it will remain orange/yellow.

Periodic table

(1)	(2)	(3)	(4)	(5)	(6)	(7)						(3)	(4)	(5)	(6)	(7)	(0)
1	2	3	4	5	6	7	8	9	10	11	12	13	14	15	16	17	18
1 **H** hydrogen 1.0																	2 **He** helium 4.0
3 **Li** lithium 6.9	4 **Be** beryllium 9.0											5 **B** boron 10.8	6 **C** carbon 12.0	7 **N** nitrogen 14.0	8 **O** oxygen 16.0	9 **F** fluorine 19.0	10 **Ne** neon 20.2
11 **Na** sodium 23.0	12 **Mg** magnesium 24.3											13 **Al** aluminium 27.0	14 **Si** silicon 28.1	15 **P** phosphorus 31.0	16 **S** sulfur 32.1	17 **Cl** chlorine 35.5	18 **Ar** argon 39.9
19 **K** potassium 39.1	20 **Ca** calcium 40.1	21 **Sc** scandium 45.0	22 **Ti** titanium 47.9	23 **V** vanadium 50.9	24 **Cr** chromium 52.0	25 **Mn** manganese 54.9	26 **Fe** iron 55.8	27 **Co** cobalt 58.9	28 **Ni** nickel 58.7	29 **Cu** copper 63.5	30 **Zn** zinc 65.4	31 **Ga** gallium 69.7	32 **Ge** germanium 72.6	33 **As** arsenic 74.9	34 **Se** selenium 79.0	35 **Br** bromine 79.9	36 **Kr** krypton 83.8
37 **Rb** rubidium 85.5	38 **Sr** strontium 87.6	39 **Y** yttrium 88.9	40 **Zr** zirconium 91.2	41 **Nb** niobium 92.9	42 **Mo** molybdenum 95.9	43 **Tc** technetium	44 **Ru** ruthenium 101.1	45 **Rh** rhodium 102.9	46 **Pd** palladium 106.4	47 **Ag** silver 107.9	48 **Cd** cadmium 112.4	49 **In** indium 114.8	50 **Sn** tin 118.7	51 **Sb** antimony 121.8	52 **Te** tellurium 127.6	53 **I** iodine 126.9	54 **Xe** xenon 131.3
55 **Cs** caesium 132.9	56 **Ba** barium 137.3	57–71 lanthanoids	72 **Hf** hafnium 178.5	73 **Ta** tantalum 180.9	74 **W** tungsten 183.8	75 **Re** rhenium 186.2	76 **Os** osmium 190.2	77 **Ir** iridium 192.2	78 **Pt** platinum 195.1	79 **Au** gold 197.0	80 **Hg** mercury 200.6	81 **Tl** thallium 204.4	82 **Pb** lead 207.2	83 **Bi** bismuth 209.0	84 **Po** polonium	85 **At** astatine	86 **Rn** radon
87 **Fr** francium	88 **Ra** radium	89–103 actinoids	104 **Rf** rutherfordium	105 **Db** dubnium	106 **Sg** seaborgium	107 **Bh** bohrium	108 **Hs** hassium	109 **Mt** meitnerium	110 **Ds** darmstadtium	111 **Rg** roentgenium	112 **Cn** copernicium		114 **Fl** flerovium		116 **Lv** livermorium		

57 **La** lanthanum 138.9	58 **Ce** cerium 140.1	59 **Pr** praseodymium 140.9	60 **Nd** neodymium 144.2	61 **Pm** promethium 144.9	62 **Sm** samarium 150.4	63 **Eu** europium 152.0	64 **Gd** gadolinium 157.2	65 **Tb** terbium 158.9	66 **Dy** dysprosium 162.5	67 **Ho** holmium 164.9	68 **Er** erbium 167.3	69 **Tm** thulium 168.9	70 **Yb** ytterbium 173.0	71 **Lu** lutetium 175.0
89 **Ac** actinium	90 **Th** thorium 232.0	91 **Pa** protactinium	92 **U** uranium 238.1	93 **Np** neptunium	94 **Pu** plutonium	95 **Am** americium	96 **Cm** curium	97 **Bk** berkelium	98 **Cf** californium	99 **Es** einsteinium	100 **Fm** fermium	101 **Md** mendelevium	102 **No** nobelium	103 **Lr** lawrencium

Answers

EL 1

1

Isotope	Symbol	Atomic Number	Mass Number	Number of neutrons
carbon-12	$^{12}_{6}C$	6	12.0	6
carbon-13	$^{13}_{6}C$	6	13.0	7
oxygen-16	$^{16}_{8}O$	8	16.0	8
strontium-90	$^{90}_{38}Sr$	38	90.0	52
iodine-131	$^{131}_{53}I$	53	131.0	78
iodine-121	$^{121}_{53}I$	53	121.0	68

2
  a 35 protons, 44 neutrons, 35 electrons

  b 35 protons, 46 neutrons, 35 electrons

  c 17 protons, 18 neutrons, 17 electrons

  d 17 protons, 20 neutrons, 17 electrons

3
  a $A_r(Br) = 80.0$

  b $A_r(Ca) = 40.1$

4
  a $100 - x$

  b $193x$

  c $191(100 - x)$

  d $193x + 191(100 - x)$

  e $[193x + 191(100 - x)] \div 100$

  f 60% iridium-193, 40% iridium-191

5
  a $^{7}_{3}Li + ^{1}_{1}p \rightarrow 2\,^{4}_{2}He$

  b $^{14}_{7}N + ^{1}_{0}n \rightarrow ^{14}_{6}C + ^{1}_{1}p$

6 40% antimony-123, 60% antimony-121

EL 2

1 green

2
  a bright red   **d** brick red

  b yellow   **e** apple green

  c lilac   **f** green-blue

3 Similarities – line spectrum; lines in same place (same frequency); lines get closer up; frequency

Differences – black lines on a bright background for absorption spectrum

4
  a Arrow points down. Shorter arrow represents red line.

  b Arrow points down. Longer arrow represents green line

(Both arrows must start and finish on lines; does not matter which levels they go between.)

5 $5.5 \times 10^{14}\,s^{-1}$

EL 3

1
  a The electron is in the first electron shell.

  b The electrons are in an s type orbital.

  c There are two electrons in this orbital.

2
  a $1s^2 2s^2 2p^1$

  b $1s^2 2s^2 2p^6 3s^2 3p^3$

3 $Z = 16$. The element is sulfur.

4
  a chlorine   **c** titanium

  b potassium   **d** tin

EL 4

1
  a 2,1

  b 2,8,5

  c 2,8,8,2

2 A, C, and E (Group 2)

3

Electronic shell configuration	Group	Period
2,8,7	7	3
2,3	3	2
2,8,6	6	3
2,1	1	2
2,8,1	1	3
2,8,8,1	1	4

4
  a s-block   **c** p-block

  b p-block   **d** s-block

5
  a d-block   **c** f-block

  b p-block   **d** s-block

EL 5

1
  a

  b

  c

  d

  e

  f

2 a

H· ×C ××× C× ·H (structure with two C, four H)

b

H ×C ××× C× H

c

H ×C ×O× H (with H above C and below)

3

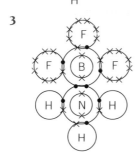

4 a   109.5°

(Si central with four H, dot-and-cross diagram)

b H ×S× H   104.5°

c H ×P× H   107°
     H

d ×O× C× O×   180°

e F ×S× F   104.5°

f Cl, B, Cl, Cl   120°

g H ×C ××× C× H   180°

5 a

H—C—C—H   is written CH₃CH₃   109.5°
  | |
  H H

H H

b

109.5° H
H—C—O—H   104.5°
    |
    H

c

H—C—N—H   109.5°   107°
  |
  H H H

d

H   H   120°
 \ /
  C=C
 / \
H   H

e

H   H   180°
 \
  C—C≡N:
 /
H   109.5° H

6 a pyramidal

b trigonal planar

c octahedral

EL 6

1 a 144

b neodymium, Nd

2 a $2Mg + O_2 \rightarrow 2MgO$

b $2H_2 + O_2 \rightarrow 2H_2O$

c $CaCO_3 + 2HCl \rightarrow CaCl_2 + CO_2 + H_2O$

d $2HCl + Ca(OH)_2 \rightarrow CaCl_2 + 2H_2O$

e $2CH_3OH + 3O_2 \rightarrow 2CO_2 + 4H_2O$

3 a $Zn(s) + H_2SO_4(aq) \rightarrow ZnSO_4(aq) + H_2(g)$

b $MgCO_3(s) \rightarrow MgO(s) + CO_2(g)$

c $BaO(s) + 2HCl(aq) \rightarrow BaCl_2(aq) + H_2O(l)$

4 a The mass of the sample is needed to be sure that iodine and oxygen are the only elements in the compound.

b The relative number of moles of iodine and oxygen.

c To change the relative number of moles into the ratio of moles of oxygen relative to 1 mole of iodine.

d In order to produce a ratio involving whole numbers

e I_2O_5, I_4O_{10}, I_6O_{15}, etc.

f The molar mass is needed.

5 a 1 **b** 0.5

c Atoms of copper are approximately twice as heavy as atoms of sulphur. Thus the same mass contains only half as many moles of copper as it does of sulphur.

6 a 2 **b** 0.02 **c** 5 **d** 1.0×10^6

EL 7

1 a $\left[Li \right]^+ \left[H× \right]^-$ **b** $\left[K \right]^+ \left[×F× \right]^-$

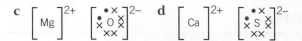

c $\left[Mg \right]^{2+}$ $\left[\overset{\bullet\times}{\underset{\times\times}{\times O \times}} \right]^{2-}$ **d** $\left[Ca \right]^{2+}$ $\left[\overset{\bullet\times}{\underset{\times\times}{\times S \times}} \right]^{2-}$

2 a macromolecular (covalent molecular)

b simple molecular (covalent molecular)

c ionic (giant lattice)

3 a $\left[Ca \right]^{2+}$ $\left[\overset{\times\times}{\underset{\times\times}{\bullet Cl \times}} \right]^{-}$ $\left[\overset{\times\times}{\underset{\times\times}{\bullet Cl \times}} \right]^{-}$

b $\left[Na \right]^{+}$ $\left[Na \right]^{+}$ $\left[\overset{\bullet\times}{\underset{\times\times}{\times S \times}} \right]^{2-}$

4 a In a normal covalent bond, each atom supplies a single electron to make up the pair of electrons involved in the bond. In a dative covalent bond one atom supplies both electrons.

b $H \overset{\bullet\bullet}{\underset{H}{\times N \times}} H$ $\left[H \overset{\bullet\bullet}{\underset{H}{\times N \times}} H \right]^{+}$

5 a Ionic (giant lattice)

b Metallic (giant lattice)

6 a $\left[Na \right]^{+}$ $\left[Na \right]^{+}$ $\left[Na \right]^{+}$ $\left[\overset{\bullet\times}{\underset{\bullet\times}{\times N \times}} \right]^{3-}$

b $\left[Al \right]^{3+}$ $\left[\overset{\bullet\times}{\underset{\times\times}{\times F \times}} \right]^{-}$ $\left[\overset{\bullet\times}{\underset{\times\times}{\times F \times}} \right]^{-}$ $\left[\overset{\bullet\times}{\underset{\times\times}{\times F \times}} \right]^{-}$

EL 8

1 charge on the ion and radius of the ion

2 a magnesium + steam →
magnesium hydroxide + hydrogen
$Mg(s) + 2H_2O(g) \rightarrow Mg(OH)_2(s) + H_2(g)$

b calcium oxide + hydrochloric acid →
calcium chloride + water
$CaO(s) + 2HCl(aq) \rightarrow CaCl_2(aq) + H_2O(l)$

c beryllium carbonate →
beryllium oxide + carbon dioxide
$BeCO_3(s) \rightarrow BeO(s) + CO_2(g)$

d barium hydroxide + sulphuric acid →
barium sulfate + water
$Ba(OH)_2(aq \text{ or } s) + H_2SO_4(aq) \rightarrow$
$BaSO_4(s) + 2H_2O(l)$

3 a 1st ionisation $Ca(g) \rightarrow Ca^+(g) + e^-$
2nd ionisation $Ca^+(g) \rightarrow Ca^{2+}(g) + e^-$
3rd ionisation $Ca^{2+}(g) \rightarrow Ca^{3+}(g) + e^-$

b Once an electron has been removed the remaining electrons are held more tightly. Hence it is more difficult to remove a second electron.

c Second ionisation enthalpy involves removal of an electron from shell 4 but third involves removal of an electron from shell 3 which is closer to the nucleus.

EL 9

1

Ion	Concentration / $g\,dm^{-3}$	Concentration / $mol\,dm^{-3}$
Cl^-	183.0	**5.15**
Mg^{2+}	36.2	**1.51**
Na^+	**31.5**	1.37
Ca^{2+}	**13.4**	0.355
K^+	6.8	**0.174**
Br^-	5.2	**0.065**
SO_4^{2-}	**0.6**	0.006 25

2 a 117 g

b 3.95 g

c 1.4 g

d 9930 g

e 0.0024 g

3 a 4.4×10^{-3} moles of NaOH(aq) in 25 cm³ gives concentration of 0.176 mol dm⁻³

b No effect on next titre because pipette delivers same amount of NaOH(aq) as before.

4 a A solution of accurately known concentration.

b Solid NaOH has the property of absorbing water (and carbon dioxide) from the air so it is not possible to accurately weigh NaOH.

DF 1

1 a Enthalpy change when one mole of the compound is burnt completely in oxygen, under standard conditions.

b Enthalpy change when one mole of a compound is formed from its elements, with both the compound and its elements being in their standard states.

2 Formation of a compound from its elements may be an exothermic reaction ($\Delta_f H$ negative) or an endothermic reaction ($\Delta_f H$ positive). However, energy is liberated whenever a substance burns, so combustion reactions are always exothermic ($\Delta_c H$ negative)

3

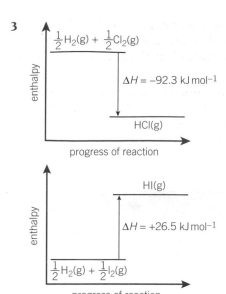

4 $\Delta_c H^{\ominus}(C)$ is enthalpy change when one mole of carbon is burnt completely under standard conditions.

$\Delta_f H^{\ominus}(CO_2)$ is enthalpy change when one mole of carbon dioxide is formed from its elements with both the carbon dioxide and its constituent elements in their standard states.

Same equation for both:
$C(s) + O_2(g) \rightarrow CO_2(g)$

5 a $H_2(g) + \frac{1}{2} O_2(g) \rightarrow H_2O(l)$

b $\Delta_f H(H_2O) = -286\,kJ\,mol^{-1}$

c $1g\ H_2 = \frac{1}{2}$ mol $= -143\,kJ$

assumptions: complete reaction occurred and no heat is lost

d $\Delta H = +286\,kJ\,mol^{-1}$

6 a $n(NaOH) = n(HCl) = 2 \times 10^{-3}\,mol$

$58\,000 \times 2 \times 10^{-3}\ (=116)\,J$

$\Delta T = \dfrac{116}{40 \times 4.18} = 0.69\,°C$

b amount (mol) doubles in same volume temp doubles $= 1.4\,°C$

c $n(NaOH) = 1 \times 10^{-3}\,mol$ (limiting)

$\Delta T = \dfrac{58}{30 \times 4.18} = 0.46\,°C$

d same reaction $H^+ + OH^- \rightarrow H_2O$

same temp rise as **a** $= 0.69\,°C$

e $n(NaOH) = 2 \times 10^{-3}$ (limiting)

volume same as **a**, so temp rise same $= 0.69\,°C$

f $n(NaOH) = n(H^+)$
$n(H_2O)$ formed $= 4 \times 10^{-3}\,mol$

$58\,000 \times 4 \times 10^{-3} = 232\,J$

$\Delta T = \dfrac{232}{60 \times 4.18} = 0.93\,°C$

DF 2

Using a bomb calorimeter to accurately measure energy changes

1 Increased number of successful collisions leading to complete combustion of the whole sample.

2 Much lower heat losses in a bomb calorimeter.

3 Energy requirements to keep the volume constant are different to the energy requirements to keep the pressure constant.

1 a thermometer, measuring cylinder, gas meter

b volume of water used, temperature rise of water, volume of gas used

c cooling losses, impurities in gas, incomplete combustion, non-standard conditions

2 a $4C(s) + 5H_2(g) \rightarrow C_4H_{10}(g)$

b

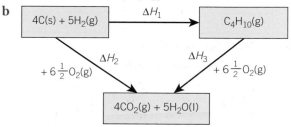

c $\Delta H_1 = \Delta H_2 - \Delta H_3$
$= 4(-393) + 5(-286) - (-2877)$
$= -125\,kJ\,mol^{-1}$

3 a The water starts dissolving products or reactants before completing the reaction.

b $n(H_2O) = 5 \times n(CuSO_4) = 5 \times 0.025$

mass H_2O = vol $H_2O = 5 \times 0.025 \times 18 = 2.25\,cm^3$

This is $(50 - 2)$ to the precision with which the volume is measured.

c first expt: $50 \times 4.18 \times \dfrac{1.28}{0.025} = +10.7\,kJ\,mol^{-1}$

second expt: $50 \times 4.18 \times \dfrac{7.88}{0.025} = -65.9\,kJ\,mol^{-1}$

d no heat losses

(specific heat capacity of water × vol of water) = (specific heat capacity of solution × mass of solution)

calorimeter absorbs negligible thermal energy

e

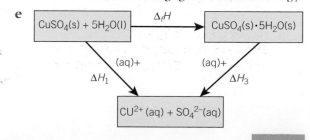

f $\Delta_f H = \Delta H_1 - \Delta H_2 = -65.9 - 10.7 = -76.6$

g $H = -77 \pm 1 \text{ kJ mol}^{-1}$

h the measuring cylinder used to measure the volume of the water

I -77.4 lies within -77 ± 1.

In this experiment systematic errors such as 'heat losses' have been reduced to below the level of apparatus uncertainty. So to improve the acuracy of the measurement, these uncertainties whould be reduced.

DF 3

1

Empirical formula	Molecular formula	M_r
C_3H_8	C_3H_8	44.0
CH_2	$C_{12}H_{22}$	168.0
CH	C_6H_6	78
$C_{10}H_{21}$	$C_{20}H_{42}$	282.0
CH_2	C_5H_{10}	70
CH	C_2H_2	26.0
C_5H_4	$C_{10}H_8$	128

2 a C_7H_{16} **b** $C_{16}H_{34}$ **c** $C_{20}H_{42}$

3 a

$$H \overset{\bullet\times}{\underset{\bullet\times}{\overset{\displaystyle H}{\underset{\displaystyle H}{C}}}} \times \overset{\times\bullet}{\underset{\bullet\times}{\overset{\displaystyle H}{\underset{\displaystyle H}{C}}}} \overset{\times}{\bullet} H$$

b

$$\overset{\displaystyle H}{\underset{\displaystyle H}{C}} \overset{\times}{\bullet} \overset{\displaystyle }{C} \overset{\displaystyle H}{\underset{\displaystyle H}{}}$$

c

$$H \overset{\bullet\times}{\underset{\bullet\times}{\overset{\displaystyle H}{\underset{\displaystyle H}{C}}}} \times \overset{\bullet\times}{\underset{\bullet\times}{\overset{\displaystyle H}{\underset{\displaystyle H}{C}}}} \times \overset{\bullet\times}{\underset{\bullet\times}{\overset{\displaystyle H}{\underset{\displaystyle H}{C}}}} \overset{\times}{\bullet} H$$

4 a CH_2 **b** C_2H_4

5 a C_2H_5

b Mathematically, the molecular formula could be any multiple of C_2H_5 but the only one that is chemically feasible is C_4H_{10}.

DF 4

1 a $C_3H_8(g) + 5O_2(g) \rightarrow 3CO_2(g) + 4H_2O(g)$

b

$$H-\overset{\displaystyle H}{\underset{\displaystyle H}{C}}-\overset{\displaystyle H}{\underset{\displaystyle H}{C}}-\overset{\displaystyle H}{\underset{\displaystyle H}{C}}-H + 5O=O \rightarrow 3O=C=O + 4H-O-H$$

c 2 C—C, 8 C—H, 5 O=O

d 6 C=O, 8 O—H **f** $-8542 \text{ kJ mol}^{-1}$

e $+6489.5 \text{ kJ mol}^{-1}$ **g** $-2052 \text{ kJ mol}^{-1}$

2 $\Delta H = -122 \text{ kJ mol}^{-1}$

3 $\Delta H = -71 \text{ kJ mol}^{-1}$

4 a $-CH_2 + 1\frac{1}{2} O_2 \rightarrow CO_2 + H_2O$

b $\Delta H = -618 \text{ kJ mol}^{-1}$

c Bond enthalpies are averages
Alcohols and water are in the liquid state (which has not been allowed for)

DF 5

1 a zeolite – heterogeneous

b platinum on aluminium oxide – heterogeneous

2 a $2CO + 2NO \rightarrow 2CO_2 + N_2$

b Both are toxic/form photochemical smog.

c chemically bound to the surface of another substance

d the reaction they catalyse is slow at low temperatures **or** do not work properly until at high temperatures

3 The mechanism should show CO and NO being adsorbed to the catalyst surface.

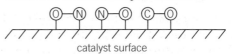

catalyst surface

The bonds in the molecules are weakened and new bonds form between the atoms.

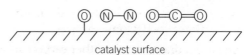

catalyst surface

Products are released from the surface.

DF 6

Name	Structure	Skeletal formula
pent-2-ene	$CH_3CHCHCH_2CH_3$	
hepta-1,4-diene	$CH_2CHCH_2CHCHCH_2$ CH_3	
4-methylpent-2-ene		
3-ethylhept-1-ene	$CH_2CHCH(CH_2CH_3)$ $CH_2CH_2CH_2CH_3$	
cyclopentene		
cyclopenta-1,3-diene		

2 a

b

3

4 a bromine, room temperature

b steam with phosphoric acid catalyst
high temperature and pressure

c (hydrogen) with platinum catalyst
room temperature and pressure

d hydrogen with nickel catalyst
150 °C, 5 atm pressure

5 shake the organic compound in a test-tube with bromine water (or 'aqueous bromine'). colour change from brown/orange/yellow to colourless

6 a polarised hydrogen bromine approaches $C=C$ bond. Curly arrow from $C=C$ bond to $H^{\delta+}$ and curly arrow from $H—Br$ bond to $Br^{\delta-}$. Curly arrow from electron pair on Br^- to C^+ of carbocation. Bromo-propene.

b bromine atom drawn on the other carbon atom of the $C=C$ bond compared to answer to **a**

7 a Polarised bromine approaches $C=C$ bond. Curly arrow from C=C bond to $Br^{\delta+}$ and another curly arrow from $Br—Br$ bond to $Br^{\delta-}$. Curly arrow from Cl^- to C^+ of carbocation. $CH_3CHBrCH_2Cl$ formed.

b Bromine and chlorine atom positions swapped. Mechanism reflects this.

c The second bromine is introduced by Br^- reacting with the carbocation.

The carbocation has the positive charge on the 'other' carbon (without Br).

DF 7

1 a

b

2

3 but-1-ene and propene

4 poly(ethene), poly(propene), poly(chloroethene)

5

DF 8

1 The particles in a gas are much further apart than in a liquid or solid. In a gas, therefore, the volume of the particles is a very small part of the total volume and does not significantly affect it. In a liquid or solid the particles are close together and their volumes must be taken into account when deciding on the total volume.

2 a $H_2(g) + \frac{1}{2}O_2(g) \rightarrow H_2O(l)$

b $5\,cm^3$

3 a 3 **b** 3 **c** 4 **d** 2 **e** C_3H_4

4 $1.19\,dm^3$ ($1190\,cm^3$)

5 a 0.25 **b** 2 **c** $48\,dm^3$ **d** $229\,dm^3$
e $30\,dm^3$

6 $17\,m^3$

DF 9

1 a C_6H_{14} C_6H_{12} not isomers

b both C_4H_9Cl isomers

c C_3H_8O C_3H_6O not isomers

d both C_7H_8O isomers

e both C_3H_9N isomers

2 a

b

c

d butane   methylbutane

3

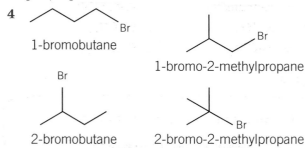

bond angle H—C—H is 109.5°

bond angle C—O—H is 104.5°

Four groups/pairs of electrons repel and get as far away as they can.

Lone pair–lone pair repulsion is greater than lone pair–bonding pair (or bonding pair–bonding pair) repulsion.

4

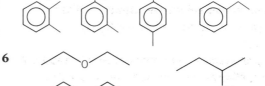

1-bromobutane

1-bromo-2-methylpropane

2-bromobutane

2-bromo-2-methylpropane

5

6

butan-1-ol

butan-2-ol

2-methylpropan-1-ol

2-methylpropan-2-ol

7

E-chloroethene

Z-chloroethene

E-chloroethene has the higher boiling point because of stronger intermolecular forces.

8 a E/Z isomers **b** 2

c citronellol is partially hydrogenated or citronellol has one fewer double bonds

d structural isomers

DF 10

1 a Primary pollutants are released directly into the atmosphere. Secondary pollutants are made by reactions of primary pollutants.

b Ozone made by Sun or nitrogen oxides and hydrocarbon/SO_3 made by the oxidation of SO_2.

c Irritating, toxic gas/contributes to photochemical smog/weakens body's immune system/greenhouse gas/oxidises rubber.

2 a so they remove pollutants just after the car starts (when gases are cool)

b something that blocks the surface of a catalyst

c give maximum surface area

d they gradually get poisoned/degraded

e CO_2 is formed and this is a greenhouse gas.

3 a catalyst

$$2CO + 2NO \rightarrow 2CO_2 + N_2$$

b excess oxygen in the exhaust so NO cannot be reduced

c lowering temperature by recycling exhaust gases through cylinder/using ammonia

$$4NO + 4NH_3 + O_2 \rightarrow 4N_2 + 6H_2O$$

4 a nitrogen and oxygen from the air react

$$N_2 + O_2 \rightarrow 2NO$$

b incomplete combustion of hydrocarbons

$$C_7H_{16} + 9O_2 \rightarrow 4CO_2 + 2CO + 8H_2O$$

c unburnt hydrocarbons are present in the exhaust gas (no equation)

d sulfur compounds in the fuel burn

$$S + O_2 \rightarrow SO_2$$

5 a For

- less CO_2 (per km)
- no (volatile) hydrocarbons
- less CO
- any other valid points

Against

- more particulates
- more NO_x
- any other valid points

b i $2CO + O_2 \rightarrow 2CO_2$

ii $C_{16}H_{34} + 24\frac{1}{2}O_2 \rightarrow 16CO_2 + 17H_2O$

DF 11

1 a $C_6H_{14} + 9\frac{1}{2}O_2 \rightarrow 6CO_2 + 7H_2O$

$C_{16}H_{34} + 24\frac{1}{2}O_2 \rightarrow 16CO_2 + 17H_2O$

$C_2H_5OH + 3O_2 \rightarrow 2CO_2 + 3H_2O$

b 0.035 dm³
0.036 dm³
0.035 dm³

c very similar, so other factors must be important

2 1.52×10^6 kJ

3 a 10.6 kg

  1.27 × 10⁵ dm³

 b i 1.06 × 10⁵ dm³

  ii 8.5 × 10⁴ dm³

ES 1

1 a $Cl_2(aq) + 2NaI(aq) \rightarrow 2NaCl(aq) + I_2(aq)$

 b $Br_2(aq) + 2KI(aq) \rightarrow 2KBr(aq) + I_2(aq)$

 c no reaction

2 $2Br_2(aq) + 2e^- \rightarrow 2Br^-(aq)$

  $2I^-(aq) \rightarrow I_2(aq) + 2e^-$ or $2I^-(aq) - 2e^- \rightarrow I_2(aq)$

3 a $Ag^+(aq) + I^-(aq) \rightarrow AgI(s)$

 b $Ag^+(aq) + Br^-(aq) \rightarrow AgBr(s)$

 c $Ag^+(aq) + Cl^-(aq) \rightarrow AgCl(s)$

4 a 0.065 mol dm⁻³ **b** 5.86 mol dm⁻³

 c 1: 90

5 a colourless solution **b** pale brown lower layer and violet upper layer

 c pale brown lower layer and violet upper layer

ES 2

1 a i $K \rightarrow K^+ + e^-$ **ii** $H_2 \rightarrow 2H^+ + 2e^-$
  iii $O + 2e^- \rightarrow O^{2-}$ **iv** $Cr^{3+} + e^- \rightarrow Cr^{2+}$

 b i oxidation **ii** oxidation
  iii reduction **iv** reduction

2 a silver = +1 **b** aluminium = +3, oxygen = −2

 c sulfur = +6, oxygen = −2 **d** phosphorus = 0

 e sulfur = +6, fluorine = −1

 f phosphorus = +5, oxygen = −2

3 a +4 **b** +7 **c** −1 **d** +7 **e** −1 **f** +1

4 a i Bromine is not very soluble in water and so a lower and higher density bromine layer is formed. This can be run off from the water that floats on top.

  ii Bromine is separated from chlorine in the distillation column because they have different boiling points.

 b $Cl_2(aq) + 2Br^-(aq) \rightarrow 2Cl^-(aq) + Br_2(l)$

 c 0.4 tonnes

 d 0.75 dm³

5 a i hydrogen is oxidised from 0 to +1
  Chlorine is reduced from 0 to −1

  ii iron is oxidised from +2 to +3. Elemental chlorine is reduced from 0 to −1. Oxidation state of chlorine in $FeCl_2$ remains −1.

  iii Oxygen is oxidised from −2 to 0. Fluorine is reduced from 0 to −1. Hydrogen remains +1.

 b i Cl_2, Cl_2, F_2

  ii H_2, Fe^{2+}, O^{2-}

6 a $2Br^- + 2H^+ + H_2SO_4 \rightarrow Br_2 + SO_2 + 2H_2O$

 b $8I^- + 8H^+ + H_2SO_4 \rightarrow 4I_2 + H_2S + 4H_2O$

7 a tin(IV) oxide **b** iron(II) chloride

 c nitrate(V) **d** lead(IV) chloride

 e manganese(II) hydroxide **f** chromate(VI)

 g vandate(V) **h** sulfate(IV)

8 a $KClO_2$ **c** $Fe(OH)_3$

 b $NaClO_3$ **d** $Cu(NO_3)_2$

ES 3

Extracting iodine from seaweed

1 Breakdown the cells of the seaweed.

2 Chlorine and bromine are stronger oxidising agents than iodine so less readily oxidised.

3 $I^-(aq) + H_2O_2(aq) \rightarrow I_2(aq) + H_2O(l)$

4 violet

5 Cyclohexane is more volatile than iodine as it will evaporate at room temperature whilst iodine will not.

1 a lead at cathode and bromine (not bromide) at anode

 b sodium at cathode and chlorine (not chloride) at anode

 c zinc at cathode and iodine (not iodide) at anode

2 a 25 000 mol

 b 0.5 mol

 c 887 500 g

3 a hydrogen at the cathode and bromine at the anode

 b hydrogen at the cathode and oxygen at the anode

 c zinc at the cathode and bromine at the anode

4 a cathode: $Zn^{2+} + 2e^- \rightarrow Zn$ reduction
  anode: $2Br^- \rightarrow Br_2 + 2e^-$ oxidation

 b cathode: $2H^+ + 2e^- \rightarrow H_2$ reduction
  anode: $2Br^- \rightarrow Br_2 + 2e^-$ oxidation

 c cathode: $2H^+ + 2e^- \rightarrow H_2$ reduction
  anode: $4OH^- \rightarrow O_2 + 2H_2O + 2e^-$ oxidation

d cathode: $2H^+ + 2e^- \rightarrow H_2$ reduction

anode: $2H_2O \rightarrow O_2 + 4H^+ + 4e^-$ oxidation

e cathode: $Cu^{2+} + 2e^- \rightarrow Cu$ reduction

anode: $Cu \rightarrow Cu^{2+} + 2e^-$ oxidation

ES 4

1 a $K_c = \dfrac{[NO_2]^2}{[NO]^2[O_2]}$ **b** $K_c = \dfrac{[C_2H_4][H_2]}{[C_2H_6]}$

c $K_c = \dfrac{[H_2][I_2]}{[HI]^2}$ **d** $K_c = \dfrac{[HCO_3^-][H^+]}{[CO_2][H_2O]}$

e $K_c = \dfrac{[CH_3COOC_3H_7][H_2O]}{[CH_3COOH][C_3H_7OH]}$

2 $2SO_2(g) + O_2(g) \rightleftharpoons 2SO_3(g)$

3 a $K_c = \dfrac{[NH_3]^2}{[N_2][H_2]^3}$

b 2.09

4 a $K_c = \dfrac{[PCl_3][Cl_2]}{[PCl_5]}$

b 0.196

5 Equilibrium constant is much greater than 1 so products favoured. Chloromethane concentration will be low.

6 (If acid was added) $[H^+(aq)]$ would increase. Some HCO_3^- and H^+ react making more CO_2 and H_2O. The equilibrium position moves to the left

ES 5

1 a 0.02 moles

b 2.46×10^{-4} moles

2 $0.56 \, mol \, dm^{-3}$

3 $Cl_2(g) + H_2O(l) \rightleftharpoons HCl(aq) + HClO(aq)$

4 $0.098 \, mol \, dm^{-3}$

ES 6

1 a 60.0

b 42.0

c 70%

2 a $C_4H_9Br + H_2O \rightarrow C_4H_9OH + HBr$

b 47.8%

c $C_4H_9Br + NaOH \rightarrow C_4H_9OH + NaBr$

d Decreases the atom economy (41.8%)

3 a $NH_3(g) + HI(g) \rightarrow NH_4I(s)$

b $8HI(aq) + H_2SO_4(aq) \rightarrow$
$$H_2S(g) + 4H_2O(l) + 4I_2(s)$$

4 a 31.7%

b 0.400 tonnes

5 Sodium chloride reacts with concentrated acid to make pure hydrogen chloride gas.

$NaCl(s) + H_2SO_4(aq) \rightarrow NaHSO_4(aq) + HCl(g)$

Sodium bromide first of all reacts with concentrated sulfuric acid to make hydrogen bromide.

$NaBr(s) + H_2SO_4(aq) \rightarrow NaHSO_4(aq) + HBr(g)$

However, the bromide ions produced are strong enough reducing agents to reduce the sulfuric acid which is present to sulphur dioxide.

$2H^+(aq) + 2Br^-(aq) + H_2SO_4(aq) \rightarrow$
$$SO_2(g) + 2H_2O(l) + Br_2(l)$$

This means that adding concentrated sulfuric acid to sodium bromide would not be a good way to make hydrogen bromide gas because it won't be pure.

The gas made will be a mixture of:

- Hydrogen bromide

- Sulfur dioxide

- Bromine vapour (since the reaction is exothermic).

ES 7

1 Chlorine is oxidised and oxygen is reduced

2 B

3 a to the right

b to the left

c no effect

d to the right

e to the right

4 a $2H_2(g) + O_2(g) \rightleftharpoons 2H_2O(g)$

b $K_c = \dfrac{[H_2O]^2}{[H_2]^2[O_2]}$

c i Forward reaction is exothermic. The equilibrium position moves to the left.

ii The equilibrium position moves to the right as the products side has fewer molecules.

5 a The equilibrium position moves to the right to use up the extra oxygen.

b There are fewer molecules on the left hand side of the equation than on the right. The equilibrium position moves to the right.

c The forward reaction is exothermic. The equilibrium position moves to the right.

6 Adding acid removes NaOH so NaOH concentration decreases. The equilibrium position moves to the left producing more NaOH and Cl_2. Chlorine is a toxic gas.

OZ 1

The atmosphere past, present, and future

1

	^{52}Cr	^{53}Cr
proton	24	24
neutron	28	29
electron	24	24

2 0.0063%

3 Seasons where plants are experiencing lots of growth will cause lower concentrations of carbon dioxide as they will be absorbing more.
Seasons where plants are not growing (such as in winter), concentrations of carbon dioxide will be higher as plants are not absorbing as much.

1 nitrogen, oxygen, argon

2 combustion of hydrocarbons, deforestation, cattle farming, landfill, changes in land use

3 a 10 000 ppm
b 0.001 82%

4 a 1.8×10^{-6} dm^3
b 0.000 18%

OZ 2

1 5.5×10^{13} Hz

2 1.15×10^{-9} m

3 a Specific frequencies corresponding to transitions between vibrational energy levels, making the bonds vibrate more.
b Molecules which have absorbed radiation have more kinetic energy. This energy is subsequently transferred to other molecules in the air by collisions.

4 a 5.43×10^{-20} J
b 8.20×10^{-13} Hz, infrared
c 3.66×10^{-6} m

5 a 1.88×10^4 J
b 1.62×10^{-24} J
c 0.978 J
d 19 200 moles

OZ 3

Other radical reactions

1 Radicals react with any species so hard to control where the monomer joins the polymer.
2 Under the right conditions (initiation), oxygen produces radicals allowing propagation reactions with the fuel to take place. Once all the fuel has been used up termination occurs.

1 a yes **c** no **e** yes
b no **d** yes **f** yes

2 a homolytic
b A is initiation, B and C are propagation.
c i $2O_3 \rightleftharpoons 3O_2$
ii catalyst

3 a oxidation of nitrogen in internal combustion engines
b i $O_3 + O \rightarrow 2O_2$
ii catalyst
iii $(-100) + (-192) = -292 \text{ kJ mol}^{-1}$

4 a initiation: reaction F
propagation: reactions G, H, and I
termination: reactions J and K
b i endothermic: reaction F
exothermic: reaction K
ii F is endothermic because C—C bond broken. K is exothermic because C—C bond formed
c $CH_3^\bullet$ methyl radical, $C_2H_5^\bullet$ ethyl radical, $H^\bullet$ hydrogen radical.

5 a $Cl_2 \rightarrow 2Cl^\bullet$ initiation
$CH_4 + Cl^\bullet \rightarrow CH_3^\bullet + HCl$ propagation
$CH_3^\bullet + Cl_2 \rightarrow CH_3Cl + Cl^\bullet$ propagation
$Cl^\bullet + Cl^\bullet \rightarrow Cl_2$ termination
$CH_3^\bullet + CH_3^\bullet \rightarrow C_2H_6$ termination
b The chloromethane can react with $Cl^\bullet$ radicals.
$CH_3Cl + Cl^\bullet \rightarrow CH_2Cl^\bullet + HCl$
$CH_2Cl^\bullet + Cl_2 \rightarrow CH_2Cl_2 + Cl^\bullet$
Further substitution produces $CHCl_3$ and CCl_4.

OZ 4

1 a Greater surface area, so more frequent collisions.
b Insufficient energy to overcome the activation energy.

c Particles are in fixed positions and number of collisions is low.

d Very high surface area and a spark will easily overcome the activation energy.

2 a A b A c B d Mainly B with A to a minor extent

3 a A, B, C, D

b A, B

c A, B, C

4 a The reaction has a high activation enthalpy that prevents it occurring at a significant rate at room temperature. However, the reaction is exothermic and once the spark has provided the energy needed to get it started, the reaction produces enough energy to sustain itself regardless of how much is present.

b The platinum catalyst offers an alternative pathway with a lower activation enthalpy close to the thermal energy of molecules at room temperature.

5 a Reaction profile shows reactants at higher enthalpy than products, and two lines showing activation enthalpy for uncatalysed reaction higher than activation enthalpy for catalysed reaction.

X-axis: progress of reaction, y-axis: enthalpy. E_a for catalysed reaction marked as $+36.4\,kJ\,mol^{-1}$, and for uncatalysed marked as $+49.0\,kJ\,mol^{-1}$.

b The rate will be faster for the catalysed reaction as the activation enthalpy is significantly lower, so more molecules have sufficient energy to react.

OZ 5

1 a B, C b A, D c D d B e B f D

2 a Increases in concentration mean more collisions so that there are more collisions with the minimum energy to react.

b As the activation energy is small, an increase in temperature will mean that many more collisions will have the minimum energy to react and so the rate of reaction increases rapidly.

3 a The shaded area in underneath the T_1 curve and to the right of E_a.

b The area shaded a different colour is underneath the T_2 curve and to the right of E_a, encompassing the first coloured area. The T_2 curve has a lower and broader maximum than the T_1 curve and the maximum value is shifted to the right. It tails off above the T_1 curve.

4 a Homogeneous means same physical state. Here the catalyst and reactants are gases.

b i It is a two step reaction with an activation energy for each step.

ii peaks – enough energy has been put in to start the reaction
troughs – intermediate

OZ 6

1 a trichloromethane

b 2-chloropropane

c 1,1,1-trichloro-2,2,2-trifluoroethane

d 2-chloro-1,1,1-trifluoropropane

e 2,2-dibromo-3-chlorobutane

2 In the solid or liquid state, noble gas atoms are held together by weak instantaneous dipole – induced dipole bonds. It takes very little energy to break these bonds and this results in very low melting and boiling points.

3

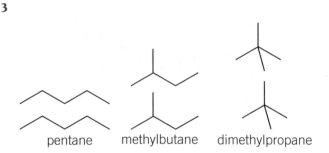

pentane   methylbutane   dimethylpropane

a Pentane is straight-chain and its molecules can approach each other closely so there are strong intermolecular forces.

Methylbutane and dimethylpropane have respectively more branching and cannot approach as closely, so the instantaneous dipole–induced dipole forces are weaker.

The stronger the intermolecular bonds, the more energy is required to break them.

b Pentane has the strongest intermolecular bonds and hence the highest boiling point. Dimethylpropane has the weakest intermolecular bonds and hence the lowest boiling point.

4 a CO_2 no dipole   d CH_3OH dipole

b $CHCl_3$ dipole   e $(CH_3)_2CO$ dipole

c C_6H_{12} (cyclohexane) no dipole

f benzene no dipole

5 a i 18 in SiH_4, 18 in H_2S.

ii The attractions will be similar because the number of electrons is the same.

iii H_2S has a permanent dipole – it is a bent molecule with two lone pairs. SiH_4 does not have an overall permanent dipole as it is a symmetrical molecule.

b Both compounds have similar instantaneous dipole–induced dipole bonds. However H_2S also has permanent dipole–permanent dipole bonds. So its boiling point is higher than that of SiH_4.

OZ 7

1 a Oxygen is more electronegative than hydrogen.

b **a** $C^{\delta+}-F^{\delta-}$ polar   **e** $H^{\delta+}-N^{\delta-}$ polar

b C–H non polar   **f** S–Br non polar

c C–S non polar   **g** $C^{\delta+}-O^{\delta-}$ polar

d $H^{\delta+}-Cl^{\delta-}$ polar

2 a

b

3 Water has two O—H bonds and two lone pairs on the oxygen so more hydrogen bonding is possible.

OZ 8

1 a $CH_3CH_2CH_2Cl(l) + NaOH(aq) \rightarrow$
$CH_3CH_2CH_2OH(aq) + NaCl(aq)$

b The chlorine atom in 1-chloropropane has been replaced by a hydroxyl group, –OH.

c Curly arrow from lone pair of electrons on OH^- to carbon in C—Cl bond.
Curly arrow from C—Cl bond to chlorine. Propanol and Cl^- formed.
$\delta+$ on carbon of C—Cl and $\delta-$ on chlorine of C—Cl bond.

2 a $C_2H_5I + OH^- \rightarrow C_2H_5OH + I^-$

b $C_2H_5Br + CN^- \rightarrow C_2H_5CN + Br^-$

c $C_5H_9Cl + OH^- \rightarrow C_5H_9OH + Cl^-$

d $CH_3C(CH_3)ClCH_3 + H_2O \rightarrow$
$CH_3C(CH_3)(OH)CH_3 + HCl$

e $CH_2BrCH_2Br + 2OH^- \rightarrow$
$CH_2OHCH_2OH + 2Br^-$

f $CH_3Br + C_2H_5O^- \rightarrow CH_3OC_2H_5 + Br^-$

g $CH_3CHClCH_3 + CH_3COO^- \rightarrow$
$CH_3CH(OOCCH_3)CH_3 + Cl^-$

3 a $CH_3CH_2Br + NH_3 \rightarrow CH_3CH_2NH_2 + H^+ + Br^-$

b

c Ammonia is the nucleophile because of its lone pair. It attacks the partial positive charge on the carbon atom and substitutes the bromine atom. There is a partial positive charge because bromine is more electronegative than carbon. After the NH_3 attacks the 1-bromoethane, a hydrogen ion is lost from the molecule to give the product ethylamine.

d Nucleophile: a species with a lone pair that can form a covalent bond.
Electronegativity: the power to attract electrons in a covalent bond.
Substitution: where one atom or group replaces another atom or group.
Curly arrow: device to show the movement of a pair of electrons.

WM 1

1 a i primary alcohol

ii secondary alcohol

iii secondary alcohol

iv tertiary alcohol

v diol

vi primary alcohol

vii ether

b i pentan-1-ol

ii heptan-3-ol

iii cyclohexanol

iv 2-methylbutan-2-ol

v butane-2,3-diol

vi decan-1-ol

2 Hydrogen bonding between ethanol and water molecules. As the hydrocarbon chain gets longer, the importance of the –OH group relative to that of the alkyl group becomes less and hexanol is unable to mix with water.

3 a A, B, D, E   **e** C

   b C   **f** A

   c A   **g** D

   d C, D   **h** E

4 a Ethanol has hydrogen bonds between molecules, ethane does not. Hydrogen bonds are the strongest intermolecular bond and so more energy is needed to break them to form gas. Therefore the boiling point is higher.

   b Water forms more hydrogen bonds than ethanol so more energy is needed to break all those in water. Therefore the boiling point is higher.

   c Both have an OH group and so form hydrogen bonds. Boiling point increases down a homologous series as M_r increases. Hence butan-1-ol has a higher boiling point than ethanol.

   d Butan-1-ol forms hydrogen bonds, ethoxyethane does not. Hence more energy is needed to break the intermolecular bonds in butan-1-ol and so the boiling point is higher.

5 a butan-1-ol

   b

   $$CH_3CH_2CH_2CH_2OH \longrightarrow CH_3CH_2CH = CH_2 + H_2O$$

6 a i fizzes

   ii no reaction

   iii no reaction

   iv no reaction

   v fizzes

   b i stays orange

   ii orange to green

   iii orange to green

   iv stays orange

   v stays orange

7 a i esterification

   ii oxidation

   iii dehydration

   iv no reaction

   v esterification

   vi oxidation

   vii dehydration

viii nucleophilic substitution

   ix oxidation

b i ester

   ii carboxylic acid

   iii alkene

   iv no reaction

   v ester

   vi aldehyde

   vii alkene

   viii haloalkane

   ix ketone

WM 2

1 a i

   b iv

   c iii

   d v

   e vi

   f iii

   g vii

   h ii and iv

   i vii

2 a

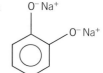

   b

3 a $C_6H_5OH + NaOH \rightarrow C_6H_5O^-Na^+ + H_2O$

   b $CH_3CH_2COOH + KOH \rightarrow$
   $$CH_3CH_2COO^-K^+ + H_2O$$

   c $2CH_3CH_2CH_2COOH + Na_2CO_3 \rightarrow$
   $$2CH_3CH_2CH_2COO^-Na^+ + CO_2 + H_2O$$

4 a

b

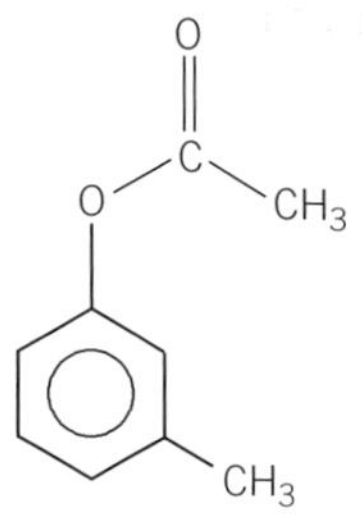

WM 3

1 a 4.24

b 7.08×10^{13} Hz

2 O—H   3660 m^{-1}
C—H(arene)   3060 cm^{-1}

3 a $CH_3CH_2CH(OH)CH_3$
$CH_3CH_2COCH_3$

b A O—H   3660   B C—H 2990
    C—H   2970   C=O 1730

c A butan-2-ol   B butanone

4 a C O—H   3580   D O—H 3670 E C—H 2990
    C—H   2990   C—H 2950   C=O 1770
    C=O   1775   (C—O 1050-1300)
                   (C—O 1050—1300)

b C carboxylic acid
   D alcohol
   E ester

5 O—H 3600-3640 phenol
C—H 2850-2950 aliphatic
C=O 1735-1750 ester
C—H 3000-3100 aromatic

WM 4

1 a $C_4H_8O^+$

b isotope peak for ^{13}C

c the first four peaks are fragments from the molecular ion peak

2 a same molecular ion peak at 72 *m/z*

b the fragmentation pattern would differ

WM 5

1 Dissolve the solute in the minimum quantity of hot solvent.
Allow to cool and crystallise.
Filter off the crystals and wash with a small quantity of cold solvent.
Dry the crystals.

2 Better atom economy
Prevention of waste products
Reduce reagents and steps
Use catalysts and more selective catalysts

3 addition   atom economy = 100%
substitution   atom economy <100%
elimination   atom economy <100%

4

Heated oil bath	cheap but messy, heated oil smells
Heated metal block	reasonably cheap, robust, melting points easily repeated
Electrically heated	expensive, robust, melting points very easily repeated

Index